Nonlinear Oscillations

Ivana Kovacic

Nonlinear Oscillations

Exact Solutions and their Approximations

Springer

Ivana Kovacic
Faculty of Technical Sciences
Centre for Vibro-Acoustic Systems
and Signal Processing
University of Novi Sad
Novi Sad, Serbia

ISBN 978-3-030-53174-4 ISBN 978-3-030-53172-0 (eBook)
https://doi.org/10.1007/978-3-030-53172-0

This Springer imprint is published by the registered company Springer Nature Switzerland AG
The registered company address is: Gewerbestrasse 11, 6330 Cham, Switzerland

Preface

This book has several representative nonlinear oscillators in its core and the rest of the material is 'knitted' around them. The mechanical models of these oscillators are collected in Chap. 1, starting with a linear oscillator, which is given as a reference point. Their corresponding mathematical models (equations of motion) are given, for which the reader needs to be familiar with the laws of mechanics and analytical dynamics. Note that for all of them, the nonlinearity stems from the restoring force. Their solutions for motion are shown in the following chapter, and what is common for all of them, is that they represent *exact solutions* for motion, and these exact solutions are the main focus and characteristic of this book—they are used through the rest of the book as a basis for other solutions and the design of systems that can exhibit them, free damped oscillators with one degree of freedom, forced oscillators with one degree of freedom, chains of oscillators, continuous systems as well as for nonlinear isochronous oscillators. It should be pointed out that although the exact solutions are given in forms of certain special functions, the prior knowledge of these functions is not needed, as the book contains several Appendices in which all necessary facts about them are collected, while the references offered at the end of each chapter can direct the reader to dig deeper into the literature on the special functions of interest. The exact solutions are either approximated or related to the known approximations, to emphasize the fact that they represent their generalizations or comprise them in a certain way, and this is why the book is entitled as it is.

This book is the outcome of my more than 15 years of research on nonlinear oscillators, and the use and development of quantitative methods for obtaining their analytical responses in approximate or exact forms. This period has been filled with good days, but it was sometimes a bumpy road indeed when I was struggling to accomplish what I was aiming for while travelling between the East and the West, living in a country with a very turbulent recent history, modest research facilities and sparse institutional support. However, I have learnt with time that it is not the place, but the people around us that matter the most. I have been lucky to be accompanied or surrounded by several colleagues from all over the world whose support, advice, suggestions or just presence and positive energy enhanced my knowledge and spirit, as they have been providing tailwind to my research boat. I will list their names in alphabetical order, as I cannot distinguish their influence on my professional development in any other way: Michael J. Brennan (Southampton, UK and Ilha Solteira, Brazil), Matthew P. Cartmell (Glasgow, UK), Ronald E. Mickens (Atlanta, Georgia, USA), Richard H. Rand (Cornell, Ithaka, USA) and Giuseppe Rega (Rome, Italy). I started as a follower of their work, learning from their own publications, but as time was passing, we became coauthors in joint publications some of which are listed in this book. It gives me an immense pleasure and pride to express my sincere gratitude and respect for this quintet of remarkable persons, gentlemen and researchers—I dedicate this book to them.

Novi Sad, Serbia Ivana Kovacic

Contents

Chapter 1
Oscillators and Oscillatory Responses in Practical and Theoretical Systems

1.1 Oscillators and Oscillatory Responses in Practical Systems

What do a swing, a music string/wire and a slider of a crank-slider mechanism (Fig. 1.1a–c) have in common? One can observe or visualize these systems to realize that their motion has a repetitive character around an equilibrium position—they all perform oscillatory motion—they oscillate!

There are numerous examples of beneficial aspects of oscillatory motion that are used in technical applications. One of them encompasses a group of pendulum-like systems. Its archetypical representative is a pendulum clock (Fig. 1.2a), which was invented by a Dutchman Christiaan Huygens in the seventeenth century, allowing for timekeeping to be more accurate than it had ever been before [1]. Another example is a Foucault pendulum (Fig. 1.2b), named after the French physicist Jean-Bernard-Léon Foucault [2]. He assembled the first pendulum of this type in Paris in 1851: it consisted of a 28 kg iron ball suspended from inside the dome of the Panthéon by a steel wire 67 m long and set in motion by drawing the ball and carefully initializing its in-plane oscillations. The rotation of this plane was the first experimental demonstration of the Earth's spin. Another example is associated with pendulum rides in amusement parks (Fig. 1.2c), where the arm of a fixed pendulum performs oscillatory motion and supports a passenger-carrying gondola, which can temporarily rotate around a certain axis in a repetitive manner. A pendulum has also had a very important role in structural engineering acting as a tuned mass-damper [4]: a device mounted in/on structures (buildings, bridges, power transmission systems, etc.) to reduce the amplitude of their mechanical vibrations, preventing their damage or failure. This concept has been used for centuries all around the globe. For instance, the wooden pagoda at the Horyu-Ji Temple (Fig. 1.3a) in Japan was built in 607AD, but stayed upright through the shaking caused by almost 50 earthquakes of magnitude 7 or greater so far [5]. It contains a 'shinbashira'—a large column built from a large pine trunk, strung from the underside of the roof that hangs down a shaft in the centre of the structure (Fig. 1.3b). The shinbashira acts as a massive tuned mass damper, helping to mitigate

I. Kovacic, *Nonlinear Oscillations*,
https://doi.org/10.1007/978-3-030-53172-0_1

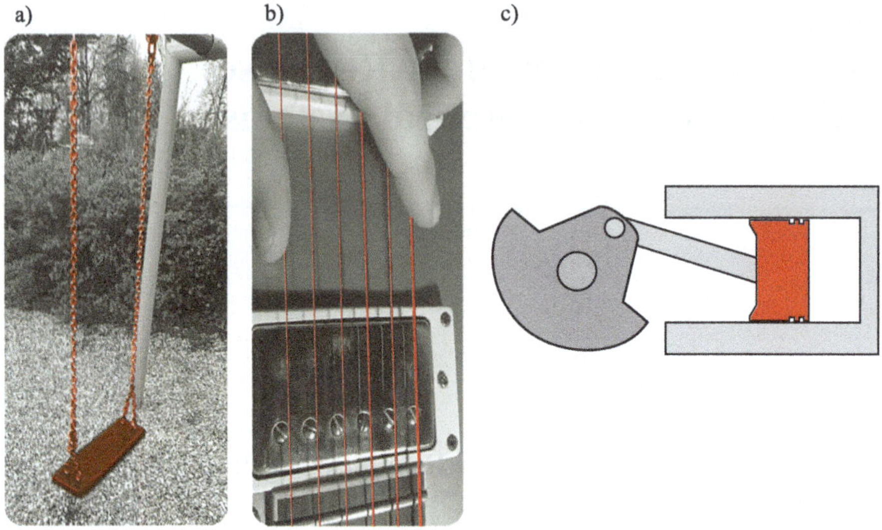

Fig. 1.1 **a** Swing; **b** Music string/wire; **c** Slider in a crank-slider mechanism

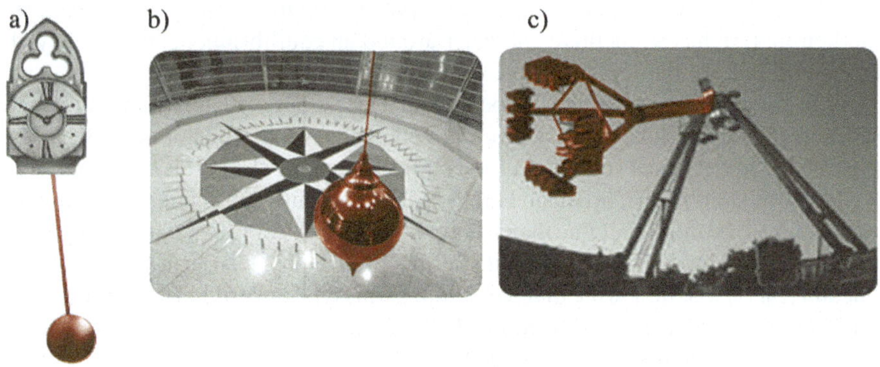

Fig. 1.2 **a** Pendulum clock; **b** Foucault pendulum (the photo from [3] is used, but it the pendulum is coloured in red here); **c** Pendulum rides in amusement parks

the earthquake's vibrations. This concept is being used nowadays as well. Taipei 101 (Fig. 1.4a) is an iconic skyscraper of 101 floors in the city of Taipei in Taiwan, which was the tallest building in the world at the time of its construction in 2004. Being located in a specific area of the Asia-Pacific, Taipei 101 was prone to earthquakes and fierce winds and certain protection measures were needed during its design. It is, thus, enhanced with a giant pendulum of a ~700-tonne steel sphere suspended from its 92nd to the 87th floor (Fig. 1.4b) to offset motions of the building [6].

There are numerous machines whose working principle is vibration-based to good effect: oscillating sanders and vibratory tumblers use vibrations to remove layers

Fig. 1.3 **a** Pagoda at the Horyu-Ji Temple in Japan; **b** Its cross-section with the 'shinbashira' in the middle

of material and finish surfaces, vibratory feeders (Fig. 1.5a, b) use vibrations to move/transfer materials, vibratory rollers help compress asphalt in paving, etc.

There are also certain positive aspects to unwanted machine vibrations. When/if measured and analysed properly, vibrations can be used in a preventive maintenance programme as an indicator of machine conditions, and help guide the maintenance professionals to undertake remedial actions before damage or failure happens. Unfortunately, vibrations and the related phenomena frequently appear as unwanted and/or potentially detrimental. In certain machines, the appearance of vibrations can indicate problems, accelerating rates of wear or deterioration in the equipment. Most common causes of such machine vibrations are [8]: (i) imbalance, when an unbalanced weight rotates around the machine's axis, creating a centrifugal force and (ii) misalignment, when machine shafts are out of line. Vibrating machinery can create noise, cause safety problems and lead to degradation in plant working conditions. Vibrations can cause machinery or their elements to consume excessive power and may damage product quality. For example, chatter vibrations, which occur between the tool and the workpiece in cutting operations, result in a poor surface finish, high-pitch noise and accelerated tool wear, which in turn reduces machine tool life, reliability and safety of the machining operation [9].

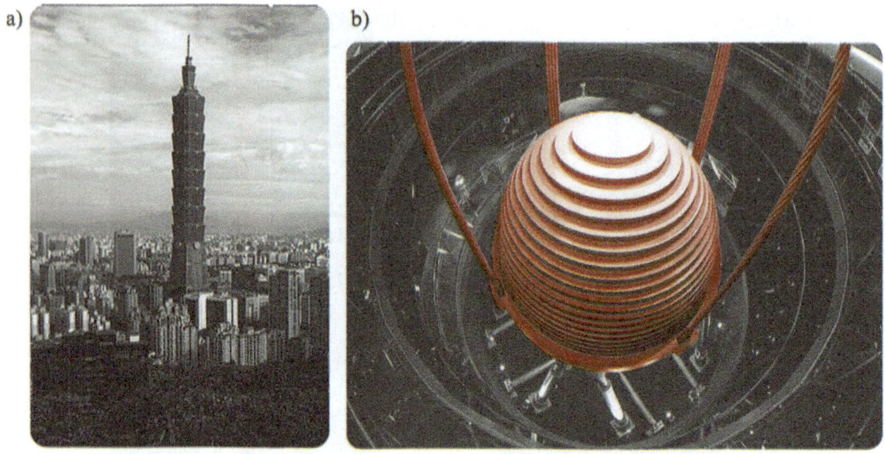

Fig. 1.4 **a** Taipei 101; **b** Giant pendulum in Taipei 101 (the photo from [7] is used, but it the pendulum is coloured in red here)

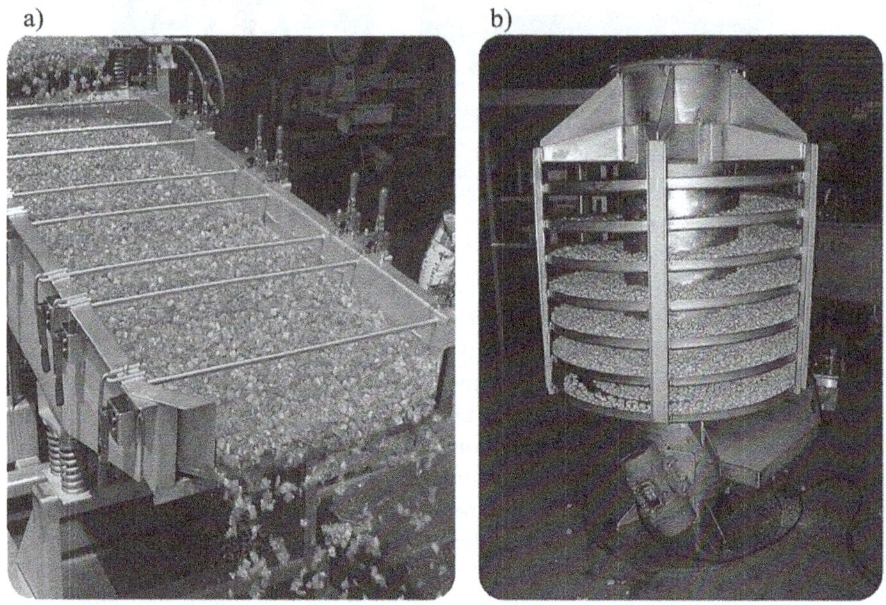

Fig. 1.5 **a** Vibrating feeder; **b** Vibratory bowl feeder

In transportation engineering, the existence of shock, vibrations and vibration-induced noise can negatively affect the driver, passengers and the vehicle's subunits. Vibrations reduce the efficiency of transport processes and deteriorate the transport safety and comfort. Traffic-induced noise and vibrations are a common source of environmental nuisance and represent a major issue in the field of environmental protection.

One of the worst aspects of vibrations is related to earthquakes—the shaking of the surface of the Earth, which can be either natural or caused by humans. A graph showing the ground motion at a measuring station as a function of time is recorded by a seismograph (Fig. 1.6a). Negative aspects can also be of aerodynamical nature, such as flutter of aircraft wings (Fig. 1.6b) or conductor galloping of overhead power lines in transmission system operators (Fig. 1.6c). The adjective 'galloping' was also associated with one of the most famous engineering disasters of all time—the collapse of the Tacoma Narrows Bridge in the USA state of Washington (Fig. 1.7). Construction on the bridge began in September 1938. From the time when the deck was built, it began to move vertically in windy conditions, which was the reason for its construction workers to give the bridge the nickname 'Galloping Gertie'. Several measures aimed at stopping the motion were ineffective and the bridge's main span finally collapsed in 64 km/h winds during the morning of 7 November 1940. The cause of the collapse has been a controversial subject [11]. The ultimate failure of the bridge was engendered by aerodynamically induced self-excitation

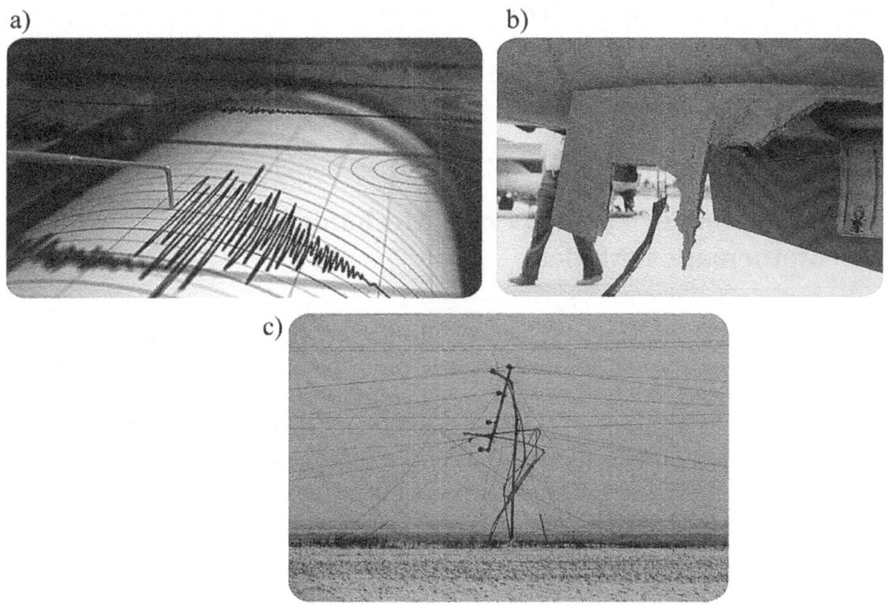

Fig. 1.6 **a** Seismogram; **b** Flutter of aircraft wings; **c** Galloping of overhead power lines (the photo is taken from [10])

Fig. 1.7 Tacoma Narrows Bridge, the photo is taken from [13]

in a torsional mode. Approximately, 700 cycles of torsional oscillations occurred during the hour prior to the collapse [12]. This event had a lasting effect on science and engineering, boosting theoretical structural aerodynamics research, and causing engineers to take extra caution to incorporate aerodynamics into structural design and to prevent unwanted structural vibrations from happening.

The examples described imply that there is a vast variety and richness of oscillatory systems, oscillatory responses and the related phenomena. They appear to be of complicated and diverse nature, but they can be mechanically modelled in a simplified way that reflects their essential and common characteristics and phenomena. The following section is focused on some of these typical mechanical models and the corresponding equations of motion.

1.2 Oscillators in Theory: From Mechanical to Mathematical Models

This section contains several representative mechanical models of oscillatory systems with one degree of freedom and their corresponding mathematical models (equations of motion), for which the reader needs to be familiar with laws of dynamics and analytical mechanics. The overview of these models starts with a linear oscillator. Then, the mechanical and mathematical models of several typical nonlinear oscillators are given, with the nonlinearity stemming from the restoring force. Their solutions for motion are presented, respectively, in Chap. 2 and all of them are characterized by the existence of the exact solution for their free response.

1.2.1 Linear (Simple Harmonic) Oscillators

Let us start from the simplest mechanical model of oscillatory systems: a linear oscillator, which is also known as a Simple Harmonic Oscillator (SHO). The reason for the latter name will be provided in Chap. 2.

A simple pendulum consists of a particle of mass m suspended from an undeformable rope or rod of negligible mass and length L (Fig. 1.8a). Its equilibrium position corresponds to the position when the rope/rod is vertical. Let us choose the generalized coordinate to be the angle x between the rope/rod and the vertical line.

Its kinetic energy is

$$E_k = \frac{1}{2} m L^2 \dot{x}^2. \tag{1.1}$$

The inertial coefficient is obviously constant and the kinetic energy is a positive definite function of the square of the generalized velocity $\dot{x}$ (the dot denotes the derivative with respect to time t, i.e. $\dot{x} = dx/dt$). This fact holds always for linear oscillators and the kinetic energy does not need to be obtained in a general position but can also be determined while the system passes through the equilibrium position, as these two forms will be equal (see [14] for more details).

The potential energy stems from the gravity force and is equal to

$$E_p = -mgL \cos x. \tag{1.2}$$

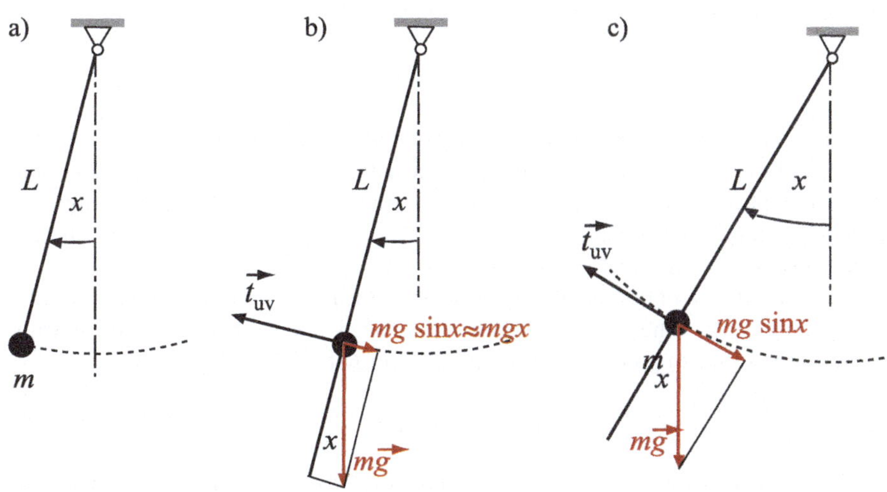

Fig. 1.8 **a** Simple pendulum; **b** Restoring force due to gravity when the oscillations are small; **c** Restoring force due to gravity when the oscillations are not small

When scleronomic systems perform small oscillations around the stable equilibrium position, the potential energy should be truncated to the quadratic terms in the generalized coordinate, as this will result in the existence of their linear form in the equation of motion [14]. Thus, the potential energy is approximated as follows:

$$E_p \approx -mgL\left(1 - \frac{x^2}{2}\right). \tag{1.3}$$

General facts that hold for this type of systems are obvious: the potential energy is a quadratic function of the generalized coordinate with a constant coefficient (note that the constant additive term can be neglected [14]). Note also that the equilibrium position is stable as $\partial^2 E_p/\partial x^2 > 0$.

By using the Lagrange equations of the second kind for systems performing small oscillations around the equilibrium position $x = 0$:

$$\frac{d}{dt}\frac{\partial E_k}{\partial \dot{x}} + \frac{\partial E_p}{\partial x} = 0, \tag{1.4}$$

one can derive

$$\ddot{x} + c_1 x = 0, \tag{1.5}$$

where $c_1 = g/L$. To emphasize the fact that this constant has a positive value, it is usually expressed as a square function of a quantity ω, which has one of the most important features of oscillatory systems—it represents its natural (angular) frequency. The equation of motion is then expressed as

$$\ddot{x} + \omega^2 x = 0. \tag{1.6}$$

The equation of motion of conservative linear oscillators Eqs. (1.5), (1.6) corresponds to a second-order linear differential equation with constant coefficients. Note that the coefficient in front of the generalized coordinate can be made equal to unity by introducing non-dimensional time $\bar{t} = \sqrt{c_1}t$ and $\bar{t} = \omega t$, but it is intentionally left here in this form for the sake of further considerations.

Regarding the essential oscillatory properties of the motion, they are assured by the restoring force. When the pendulum is displaced sideways from its equilibrium position, it is subject to a restoring force due to gravity (Fig. 1.8b):

$$F_r = -mg\sin x. \tag{1.7}$$

No matter whether the pendulum has been displaced to the right or to the left, this restoring force moves it back towards the equilibrium position. The minus sign is the consequence of the projection of the gravity force to the tangent unit vector $\vec{t}_{uv}$ and it is essential for assuring oscillatory character of motion that the restoring force has. Analogous to the approximation of the potential energy for these type of systems, the expression for the restoring force should be truncated as well, but to the linear term only:

$$F_r \approx -mgx. \tag{1.8}$$

The restoring force, Eq. (1.8), is a linear function of the generalized coordinate, which is another general characteristics of this type of systems.

Finding the scalar product of the second Newton's law with the tangent unit vector (see Fig. 1.8b) leads to

$$mL\ddot{x} = -mgx, \tag{1.9}$$

and one can end up with the same equation of motion Eqs. (1.5) and (1.6).

Another typical example of the conservative linear oscillator regards a spring–mass system: a block of mass m sliding along a smooth horizontal guide while being attached to a linear spring of stiffness k (Fig. 1.9a). The generalized coordinate x measures the displacement from the equilibrium position when the spring is unstretched/undeformed.

The kinetic energy is

$$E_k = \frac{1}{2}m\dot{x}^2. \tag{1.10}$$

Knowing that the potential energy of the linear spring is a square function of its total deflection Δl

$$E_p = \frac{1}{2}k\Delta l^2, \tag{1.11}$$

where Δl is the difference between the length of the spring in an arbitrary position l and the length of the undeformed spring l_0 ($\Delta l = l - l_0$). Recognizing that here $\Delta l \equiv x$, one gets

$$E_p = \frac{1}{2}kx^2. \tag{1.12}$$

It is seen that the forms of the kinetic and potential energies have all the characteristics mentioned above in the example of a simple pendulum. By using Eq. (1.4), one can derive Eq. (1.6), where $\omega^2 = c_1 = k/m$.

The restoring force corresponds to the spring force (Fig. 1.9b):

$$F_r \equiv F_c = -kx, \tag{1.13}$$

which has a negative sign for $x > 0$ and it is, thus, directed towards the equilibrium position; for $x < 0$, it has a positive sign and it is again directed towards the equilibrium position. So, it always 'attracts' the mass towards the equilibrium position.

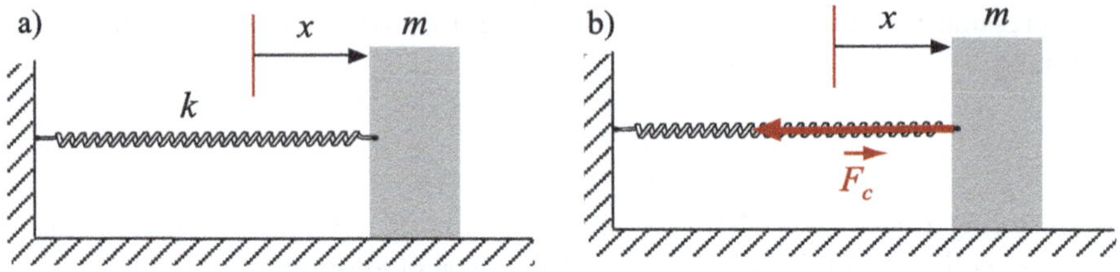

Fig. 1.9 **a** Spring–mass system; **b** Restoring force corresponding to the spring force

Note that the description of a 'linear' spring assumes that its material obeys a linear relationship between the stress and strain and, thus, between the spring force and its deformation.

More examples of the systems having the same equation of motion can be found in [14].

1.2.2 *Duffing-Type Oscillators*

Let us go back to the simple pendulum considered above, but assume that its oscillations are not small (Fig. 1.8c), which means that its restoring force will not be linearized, but one more term in the series expansion of $\sin x$ will be kept:

$$F_r \approx -mg\left(x - \frac{x^3}{6}\right). \tag{1.14}$$

The equation of motion reads as

$$mL\ddot{x} = -mg\left(x - \frac{x^3}{6}\right), \tag{1.15}$$

which can be represented in the form

$$\ddot{x} + c_1 x - c_3 x^3 = 0, \tag{1.16}$$

where, in this example, $c_1 = g/L$ and $c_3 = g/6L$. So, unlike the equation of motion of linear oscillators, this equation contains the additional cubic geometric term. Such equations correspond to a wide class of Duffing-type equations [15, 16], named after Georg Duffing, who was the first to derive them and solve them. This particular equation with a negative coefficient in front of the cubic term is of the so-called 'softening' type, and the oscillator is known as the Softening Duffing Oscillator (SDO).

It should be noted that in the case of the pendulum considered, the cubic nonlinearity is of geometric nature. This same type of nonlinearity can also stem for material elastic properties. This happens, for example, if the material of the spring in a mass–spring system obeys a nonlinear relationship between the stress and strain and, consequently, between the spring force and its deformation. One common type of the corresponding magnitude of the restoring force (which is also referred to in this book as the restoring term(s)) is

$$|F_r| \equiv F_c = k_1 x \pm k_3 x^3, \tag{1.17}$$

with k_1 and k_3 being constants. This is plotted in Fig. 1.10a together with the one for the linear case $|F_r| = k_1 x$. The minus sign in Eq. (1.17) yields a smaller force for higher x than in the linear case, so it 'softens'; the plus sign in Eq. (1.17) results in a higher force for higher x, so it 'hardens'. The corresponding equation of motion,

$$m\ddot{x} + k_1 x \pm k_3 x^3 = 0, \tag{1.18}$$

can be written as

$$\ddot{x} + c_1 x \pm c_3 x^3 = 0. \tag{1.19}$$

The system governed by the equation of motion with the positive sign in front of the cubic term is the so-called Hardening Duffing Oscillator (HDO) and the negative

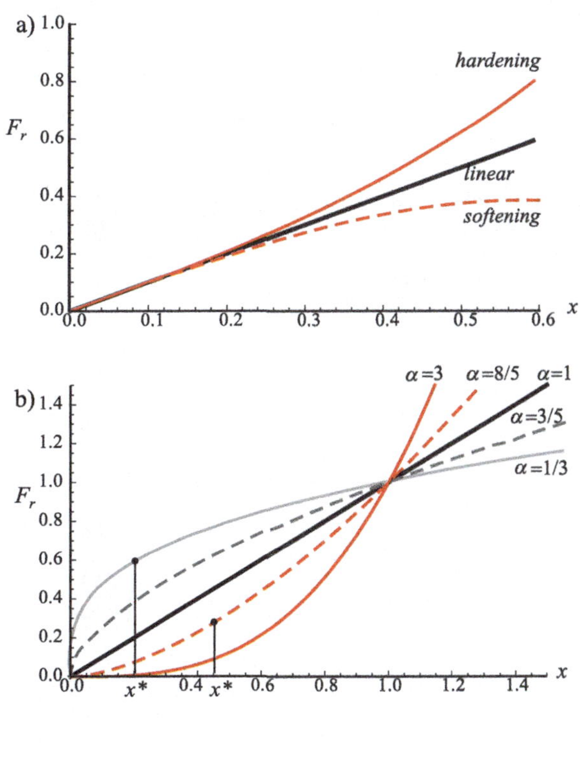

Fig. 1.10 a Force–displacement diagrams for hardening, softening and linear case; **b** Force–displacement diagrams for purely nonlinear and linear case

sign in front of the cubic term corresponds to the Softening Duffing Oscillator (SDO) mentioned earlier.

Besides hardening and softening Duffing oscillators, one can distinguish a bistable Duffing oscillator and pure cubic oscillator as well. They can also be realized in a certain pendulum-type system and a certain mass–spring system. The former is shown in Fig. 1.11a, where the pendulum system is rotated for 180° with respect to the one from Fig. 1.10a and equipped with a linear torsional spring of stiffness k. Based on Fig. 1.11b, one can create the exact equation of motion:

$$mL^2\ddot{x} = mgL\sin x - kx, \tag{1.20}$$

and then approximate the sin$x$ with two terms of its Taylor series expansion, transforming the equation of motion into

$$\ddot{x} + \left(-\frac{g}{L} + \frac{k}{mL^2}\right)x + \frac{g}{6L}x^3 = 0. \tag{1.21}$$

It is seen that the coefficient in front of the linear term can have different signs and values.

If $k > mgL$, this equation of motion corresponds to the HDO:

$$\ddot{x} + c_1x + c_3x^3 = 0, \tag{1.22}$$

and for $k < mgL$, to the Bistable Duffing Oscillator (BDO):

$$\ddot{x} - c_1x + c_3x^3 = 0. \tag{1.23}$$

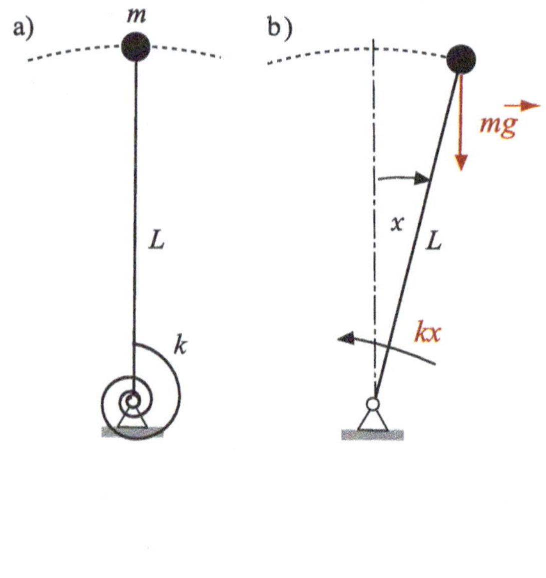

Fig. 1.11 **a** Pendulum-type system with a torsional spring; **b** System in the position displaced from the equilibrium

When $k = mgL$, this equation of motion corresponds to the Pure Cubic Oscillator (PCO):

$$\ddot{x} + c_3 x^3 = 0. \tag{1.24}$$

The BDO and PCO can be designed using a mass and a spring in the configuration presented in Fig. 1.12. The potential energy of the spring is

$$E_p = \frac{1}{2}k\Delta l^2 = \frac{1}{2}k\left(\sqrt{b^2 + x^2} - l_0\right)^2, \tag{1.25}$$

which can be approximated to

$$E_p \approx \text{const.} + \frac{1}{2}k\left(1 - \frac{l_0}{b}\right)x^2 + \frac{kl_0}{8b^3}x^4. \tag{1.26}$$

The corresponding equation of motion has the form:

$$m\ddot{x} - k\frac{l_0 - b}{b}x + \frac{kl_0}{2b^3}x^3 = 0. \tag{1.27}$$

Choosing the length of the undeformed position to be equal to the distance b ($l_0 = b$), the equation of motion will get the form of the PCO, while for $l_0 > b$ (if the spring is compressed when $x = 0$), it will correspond to the BDO.

Some other examples of Duffing-type systems, including beams, electric circuits, cables, isolators, etc., can be found in [16].

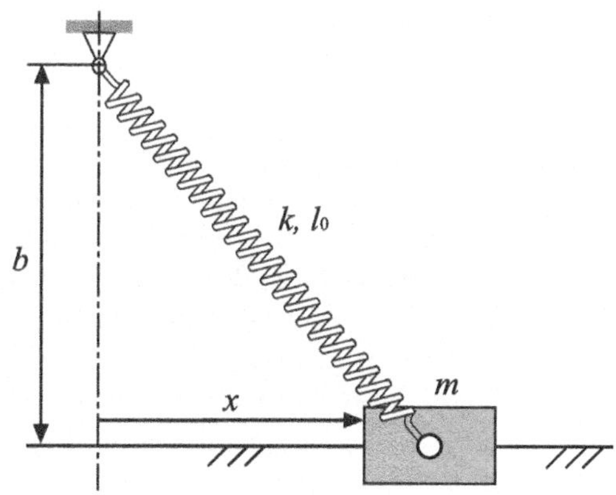

Fig. 1.12 Slider moving along a smooth guide while attached to a linear spring

1.2.3 Purely Nonlinear Oscillators

Nonlinearities due to restoring forces shown previously are modelled as polynomials, i.e. as expressions involving positive integer powers of the generalized coordinate. However, there are systems whose force–displacement relationship can include exponents that are non-integer or of a fractional order. For example, the elastic properties of some aircraft materials are of the Ramberg–Osgood type, where this exponent is a fraction higher than unity [17]. Further, the experimentally obtained exponent in the suspension system in a vehicle was found to be 3/2, approximately [18]. Also, for the elastic force due to a piano hammer, which comprises a wooden beam coated with several layers of compressed wool felt, this exponent ranges from 2.2 to 3.5 for used hammers, and from 1.5 to 2.8 for new ones [19]. PCOs can also be included in this group and they comprise the originally multi-term restoring force tuned to have a quasi-zero stiffness characteristic [20–24]. Some other examples can be found in [25]. Note that purely nonlinear oscillators are also known as non-polynomial/fractional-order/truly nonlinear oscillators [26].

Purely nonlinear oscillators are characterized by the fact that the magnitude of their restoring force $|F_r|$ is a power function of the displacement x:

$$|F_r| = k \operatorname{sgn}(x) |x|^{\alpha}, \tag{1.28}$$

where α is any positive real number. The sign and absolute value functions are used to assure that this expression represents an odd function of the displacement for all the values of α defined. It should be noted that the sign function in Eq. (1.28) is defined by

$$\operatorname{sgn}(x) = \begin{cases} 1, & x > 0 \\ 0, & x = 0 \\ -1, & x < 0 \end{cases}, \tag{1.29}$$

or, alternatively, $\operatorname{sgn}(x) = d|x|/dx$.

A force–displacement relationship which corresponds to this restoring term is shown in Fig. 1.10b for different values of the parameter α, including a linear case $\alpha = 1$ (note that the term 'force' here refers to its magnitude, i.e. to the restoring term). Stiffness of the system corresponds to the slope of a tangent to the force–displacement characteristic and is given by $k^* = k\alpha |x|^{\alpha-1}$. For the linear system, it is constant $k^*_\alpha(\alpha = 1) \equiv k$. Stiffness of the nonlinear system and that of the linear one become equal for $x^* = \alpha^{\frac{1}{1-\alpha}}$. This characteristic value of the displacement corresponding to α=1/3 and α=8/5 is depicted in Fig. 1.10b. It concludes that for α<1 (under-linear case), the stiffness is higher than in the linear system until x^*, when it becomes smaller. For α>1 (over-linear case), the opposite is true. Further, it is seen that when α<1 and the displacement increases, the slope of a tangent to the force–displacement curve becomes smaller as the displacement is increased. Thus, the displacement increases more rapidly than the restoring force, i.e. the softening effect is brought into bear. However, when $\alpha > 1$, the slope becomes higher, so that the restoring force increases more quickly than the displacement, which implies hardening behaviour.

The corresponding equation of motion can be written down in the following form:

$$\ddot{x} + c_\alpha \text{sgn}\,(x)\,|x|^\alpha = 0. \tag{1.30}$$

To assure that the restoring force is an odd function, one can use an alternative expression, which includes the use of the absolute value:

$$\ddot{x} + c_\alpha\, x\,|x|^{\alpha-1} = 0. \tag{1.31}$$

1.2.4 Oscillators with a Constant Restoring Force

Oscillators with a constant restoring force are also known as antisymmetric oscillators with a constant restoring force [26, 27] or piecewise constant restoring force [28, 29]. It was Lipscomb and Mickens who raised an interest in studying these oscillators, noting that the corresponding governing equation models, for example, the motion of an infinitesimal ball moving along a V-shaped route in a constant gravitational field [30]. This type of oscillators is of interest also in two other classes of systems: those with a constant force spring element [29, 31] or those related to vibro-impact systems [32].

Consider a particle of mass m moving in a constant gravitational field along a smooth V-shaped guide as shown in Fig. 1.13 (an infinitesimally small radius is assumed to exist at the point O; thus, it can be further assumed that there is no impact at this point and the mass smoothly passes from the right part of the guide to the left one, and vice versa). Take the y-direction to be vertical and the x-direction to be horizontal. It follows that

$$y = \tan\theta\, x, \tag{1.32}$$

where θ is a constant inclination angle.

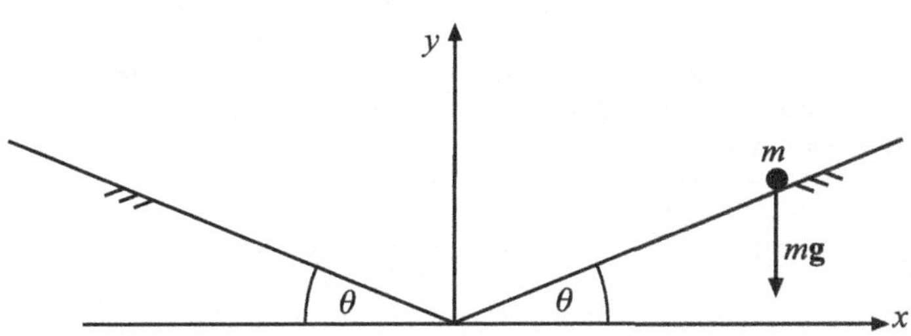

Fig. 1.13 Particle moving in a constant gravitational field along a smooth V-shaped guide

The potential energy E_p is

$$E_p = mg\, y = mg\, \tan\theta\, |x|, \tag{1.33}$$

while the kinetic energy is

$$E_k = \frac{1}{2}\left(\dot{x}^2 + \dot{y}^2\right) = \frac{1}{2}\frac{m}{\cos^2\theta}\dot{x}^2. \tag{1.34}$$

The Lagrange equation of the second kind (1.4) gives

$$m\frac{\ddot{x}}{\cos^2\theta} = -mg\, \tan\theta\, \frac{d\,|x|}{dx}. \tag{1.35}$$

Knowing $d\,|x|\,/dx$ =sgn(x), leads to the following equation of motion:

$$\ddot{x} + c\,\mathrm{sgn}\,(x) = 0, \tag{1.36}$$

where c is constant.

References

1. Huygens C (Hugenii) (1673) Horologium Oscillatorium. Apud F. Muguet, Parisiis (English translation: The pendulum clock. Iowa State University Press, Ames (1986))
2. https://www.britannica.com/biography/Leon-Foucault. Accessed 15 April 2020
3. https://commons.wikimedia.org/wiki/File:Foucault_pendulum_at_CAS_4.JPG. Accesses 16 April 2020
4. Den Hartog JP (1985) Mechanical vibrations. Dover Publications, New York (first published December 1956)
5. https://gizmodo.com/how-japans-oldest-wooden-building-survives-giant-earthq-5846501. Accessed 15 April 2020
6. https://www.youtube.com/watch?v=ohKqE_mwMmo. Accessed 15 April 2020
7. https://en.wikipedia.org/wiki/File:Taipei_101_Tuned_Mass_Damper_2010.jpg. Accessed 15 April 2020
8. https://www.reliableplant.com/Read/24117/introduction-machinery-vibration. Accessed 15 April 2020
9. Siddhpura M, Paurobally R (2012) A review of chatter vibration research in turning. Int J Mach Tools Manuf 61:27–47
10. https://www.hydro.mb.ca/outages/img/tower.jpg. Accessed 15 April 2020
11. Billah KY, Scanlan RH (1991) Resonance, Tacoma Narrows bridge failure, and undergraduate physics textbooks. Am J Phys 59:118–124
12. Plaut RH (2008) Snap loads and torsional oscillations of the original Tacoma Narrows bridge. J Sound Vib 309:613–636
13. https://structurae.net/en/photos/198837-vertical-oscillations-34-view-7-november-1940-from-16mm-film-shot-by-professor-f-b-farquharson-university-of-washington-laboratory-studies-on-the-tacoma-narrows-bridge-at-university-o. Accessed 18 April 2020
14. Kovačić I, Radomirović D (2017) Mechanical vibrations. Wiley, Hoboken
15. Duffing G (1918) Erzwungene Schwingungen bei veranderlicher Eigenfrequenz und ihre technische Bedeutung. Vieweg & Sohn, Braunschweig (In German)

16. Kovacic I, Brennan MJ (eds) (2011) The Duffing equation: nonlinear oscillators and their behavior. Wiley, Chichester
17. Prathap G, Varadan TK (1976) The inelastic large deformation of beams. J Appl Mech 43:689–690
18. Zhu Q, Ishitobi M (2004) Chaos and bifurcations in a nonlinear vehicle model. J Sound Vib 275:1136–1146
19. Russell D, Rossing T (1998) Testing the nonlinearity of piano hammers using residual shock spectra. Acta Acust United Acust 84:967–975
20. Kovacic I, Brennan MJ, Waters TP (2008) A study of a non-linear vibration isolator with quasi-zero stiffness characteristic. J Sound Vib 315:700–711
21. Gatti G, Kovacic I, Brennan MJ (2010) On the response of a harmonically excited two degree-of-freedom system consisting of linear and non-linear quasi-zero stiffness oscillators. J Sound Vib 329:823–1835
22. Kovacic I (2011) The method of multiple scales for forced oscillators with some real-power nonlinearities in the stiffness and damping force. Chaos Solitons Fractals 44:891–901
23. Rakaric Z, Kovacic I (2013) An elliptic averaging method for harmonically excited oscillators with a purely nonlinear non-negative real-power restoring force. Commun Nonlinear Sci Numer Simul 18:1888–1901
24. Alabudzev P, Gritchin A, Kim L, Migirenko G, Chon V, Stepanov P (1989) Vibration protecting and measuring systems with quasi-zero Stiffness. Hemisphere Publishing, New York
25. Rakaric Z, Kovacic I (2011) Approximations for motion of the oscillators with a non-negative real-power restoring force. J Sound Vib 30:321–336
26. Mickens RE (2010) Truly nonlinear oscillations: harmonic balance, parametric Expansions, Iteration, and Averaging Methods. World Scientific, Singapore
27. Lipscomb T, Mickens RE (1994) Exact solution to the antisymmetric, constant force oscillator equation. J Sound Vib 169:138–140
28. Pilipchuk VN (1999) An explicit form general solution for oscillators with a non-smooth restoring force, $\ddot{x}+\text{sign}(x)\, f\,(x) = 0$. J Sound Vib 226:795–798
29. Araki Y, Asai T, Masui T (2009) Vertical vibration isolator having piecewise-constant restoring force. Earthq Eng Struct Dyn 38:1505–1523
30. Personal communication between Ronald E. Mickens and Ivana Kovacic, March 2015
31. http://www.assocspring.co.uk/pi_confor.asp. Accessed 13 Aug 2014
32. Pilipchuk VN (2010) Nonlinear dynamics: between linear and impact limits. Springer, Berlin

Chapter 2
Free Conservative Oscillators: From Linear to Nonlinear Systems

2.1 Introduction

This chapter provides an overview of the archetypical nonlinear conservative oscillators whose nonlinearity stems from the restoring terms. Their equations of motion are special cases of the following general non-dimensional equation:

$$\ddot{x} + c_1\, x + c_2\, x^2 + c_3\, x^3 + c_\alpha \mathrm{sgn}\,(x)\,|x|^\alpha + c_0 \mathrm{sgn}\,(x) = 0, \tag{2.1}$$

where the dots denote the derivative with respect to time t, x is a generalized coordinate, while $c_1, c_2\;, c_3\;, c_\alpha$ and c_0 are constants. The overview starts with linear oscillators ($c_2 = c_3 = c_\alpha = c_0 = 0$), which are presented as a reference point for the proceeding nonlinear oscillators as well as for their mutual comparisons. The nonlinear oscillators overviewed include the following oscillators: Duffing-type oscillators ($c_2 = c_\alpha = c_0 = 0$), quadratic oscillators ($c_3 = c_\alpha = c_0 = 0$), purely nonlinear oscillators ($c_1 = c_2 = c_3 = c_0 = 0$) and those with a constant restoring force ($c_1 = c_2 = c_3 = c_\alpha = 0$). Note that physical examples of these systems are provided in Chap. 1, Sect. 1.2. What is common for the majority of these nonlinear oscillators is that they have the exact solution for motion known in terms of certain special functions. To enable the reader to comprehend the subject with special functions easily and to empower the readability, this chapter is associated with several appendices, so that the matter can be followed on a stand-alone basis, without a need for any direct use of other references. The references are, of course, given for further or additional reading and details. Besides the exact solutions for motion, the considerations of each group of oscillators comprise their potential energy, conservation laws, phase trajectories, the solutions for motion and their characteristics, and also a sectorial velocity in some cases. The main characteristics of each group of nonlinear oscillators are clearly pointed out and compared with the ones for linear oscillators. Many of the solutions for motion are used in subsequent chapters for further analysis of these oscillators in different arrangements or adjustments.

I. Kovacic, *Nonlinear Oscillations*,
https://doi.org/10.1007/978-3-030-53172-0_2

The initial conditions dealt with are divided into three groups. They can be general:

$$x(0) = A > 0, \quad \dot{x}(0) = v_0 > 0, \tag{2.2}$$

with A standing for the initial displacement and v_0 standing for the initial velocity. Besides these general initial conditions, two sets of special initial conditions and related oscillators are investigated separately: a Zero-Initial Displacement (ZID) condition and a Zero-Initial Velocity (ZIV) condition. The ZIV conditions are set as

$$x_{\mathrm{ZIV}}(0) = A, \quad \dot{x}_{\mathrm{ZIV}}(0) = 0, \tag{2.3}$$

where A will define the initial amplitude and the amplitude of the resulting motion, while the ZID conditions are

$$x_{\mathrm{ZID}}(0) = 0, \quad \dot{x}_{\mathrm{ZID}}(0) = v_0. \tag{2.4}$$

These subscripts will be used subsequently in certain solutions to indicate the initial conditions that they correspond to.

2.2 Linear (Simple Harmonic) Oscillators (SHOs)

Let us start with a linear oscillator governed by

$$\ddot{x} + c_1 x = 0, \tag{2.5}$$

where $c_1 > 0$, i.e. $c_1 = \omega^2$, with ω^2 representing the square of the natural frequency. Although the coefficient c_1 in front of the linear term (the square of the natural frequency) in Eq. (2.5) can be made equal to unity by an appropriate time-normalization, the equation is left in this form for the sake of clarity of further generalizations.

The diagram of the corresponding potential energy (single-well potential)

$$E_p = \frac{1}{2} c_1 x^2 \tag{2.6}$$

is shown in Fig. 2.1 The equilibrium $x = 0$ (centre), associated with a minimum of the potential energy, is stable and it is surrounded by closed phase trajectories, which imply that the corresponding motion is periodic. Each of these phase trajectories is defined by the level of the initial energy E_0, and some of them are labelled in Fig. 2.1

The equation of motion corresponds to a second-order linear ordinary differential equation and, according to the related theory [1, 2], its general solution can be represented as a sum of two harmonic functions—sine and cosine:

$$x = C_1 \cos\left(\sqrt{c_1}\, t\right) + C_2 \sin\left(\sqrt{c_1}\, t\right), \tag{2.7}$$

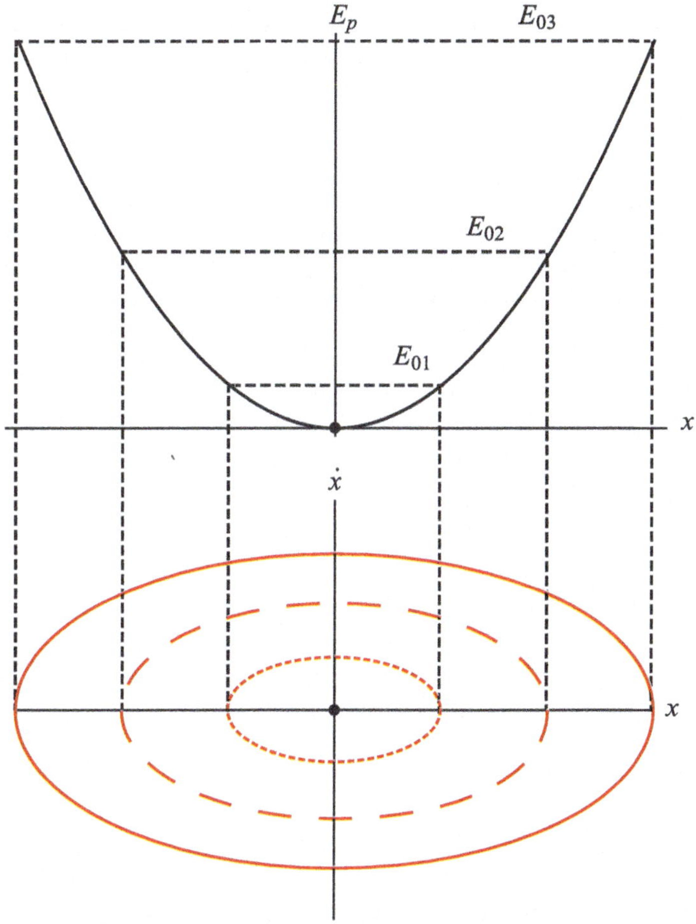

Fig. 2.1 General form of a diagram of the potential energy of the SHO and the phase trajectories corresponding to different levels of the initial energy E_0

where the values of the constants C_1 and C_2 can be calculated from the initial conditions. The fact that this solution is expressed in terms of harmonic functions is the reason why the linear oscillator is also known as a Simple Harmonic Oscillator (SHO).

If the initial conditions involve a non-zero displacement A and a zero-valued initial velocity as defined by Eq. (2.3), the solution for motion has the form

$$x_{\mathrm{SHO,\,ZIV}} = A\cos\left(\sqrt{c_1}\,t\right), \tag{2.8}$$

where the initial displacement corresponds to the amplitude of motion (see Fig. 2.2a).

If the initial conditions involve a zero-valued initial displacement and a non-zero-initial velocity, as defined by Eq. (2.4), the solution for motion is given by

$$x_{\mathrm{SHO,\,ZID}} = \frac{v_0}{\sqrt{c_1}} \sin\left(\sqrt{c_1}\, t\right), \tag{2.9}$$

where the amplitude of motion $v_0/\sqrt{c_1}$ now depends on the initial velocity v_0 and the natural frequency $\omega = \sqrt{c_1}$ (see Fig. 2.2b).

In a general case when both the initial displacement and the initial velocity are different from zero, the superposition principle holds, and the solution for motion is the sum of separate solutions given by Eqs. (2.8) and (2.9):

$$x_{\mathrm{SHO}} = A\cos\left(\sqrt{c_1}\, t\right) + \frac{v_0}{\sqrt{c_1}} \sin\left(\sqrt{c_1}\, t\right), \tag{2.10}$$

which can also be presented as

$$x = a\sin\left(\sqrt{c_1}\, t + \varphi\right), \tag{2.11}$$

with the amplitude a being

$$a = \sqrt{A^2 + \left(\frac{v_0}{\sqrt{c_1}}\right)^2}, \tag{2.12}$$

while the phase φ is defined by

$$\sin\varphi = \frac{A}{a}, \quad \cos\varphi = \frac{v_0}{a\sqrt{c_1}}. \tag{2.13}$$

Note that depending on the sign of A and v_0, one needs to calculate φ by using both expressions for $\cos\varphi$ and $\sin\varphi$ to determine which quadrant this angle belongs to.

The angular frequency existing in the solution is equal to the natural frequency. The period of vibrations (time needed for the system to exhibit one oscillation) is defined by the period of harmonic functions:

$$T = \frac{2\pi}{\sqrt{c_1}} = \frac{2\pi}{\omega}. \tag{2.14}$$

This period is labelled in Fig. 2.2.

The fact that the angular frequency of vibrations is amplitude-independent can be presented graphically in the $\omega - A$ plane as the so-called 'backbone curve', which is done in Fig. 2.3. It is seen that the backbone curve of the SHO is vertical, i.e. it represents a straight-line curve. This backbone curve will be compared with the one for some nonlinear oscillators in the subsequent sections.

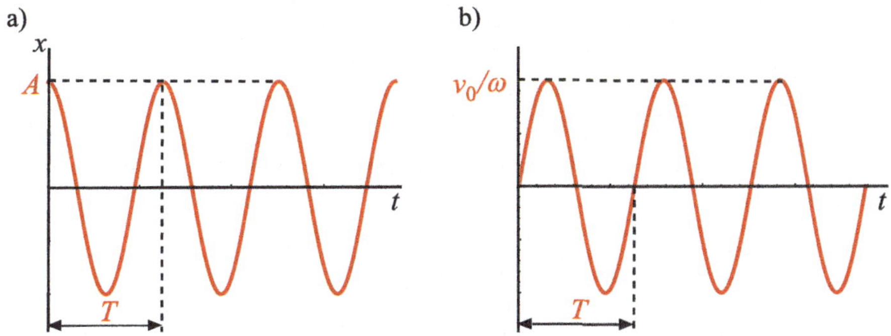

Fig. 2.2 Time history diagrams for the motion of the SHO for: **a** the ZIV and **b** the ZID conditions. The period of vibrations is also indicated

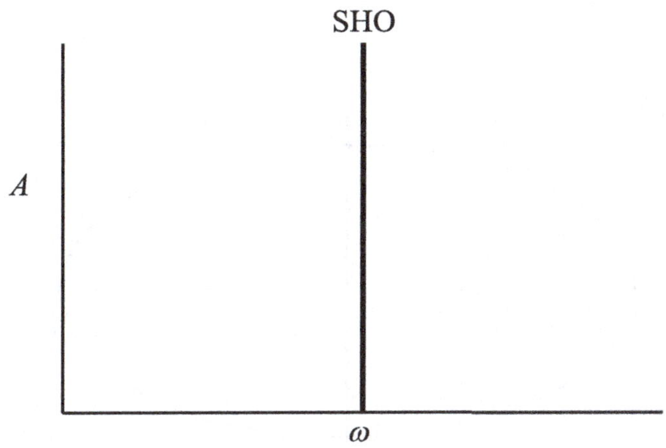

Fig. 2.3 Backbone curve of the SHO

Based on these facts, four main characteristics of the response of the free conservative SHO can be noted:

SHO1. There is an exact solution for its response and it can be expressed in terms of sine and/or cosine functions.

SHO2. The frequency and the period of vibrations are constant and do not depend on the initial conditions, i.e. they are amplitude-independent.

SHO3. The backbone curve is straight.

SHO4. There is only one frequency that exists in the response and it is related to the constant natural frequency that appears in the equation of motion.

Let us go back to the energy considerations. The SHO is conservative and its energy conservation law reads as follows:

$$\frac{1}{2}\dot{x}^2 + \frac{1}{2}c_1\, x^2 = E_0. \tag{2.15}$$

Based on Eq. (2.6), one can obtain that

$$E_0 = \frac{1}{2}c_1 a^2, \tag{2.16}$$

which gives $a^2 = 2E_0/c_1$. In this way, the solutions for motion and velocity can be written as

$$x = \sqrt{\frac{2E_0}{c_1}} \sin\left(\sqrt{c_1}\, t + \varphi\right), \tag{2.17}$$

$$\dot{x} = \sqrt{2E_0} \cos\left(\sqrt{c_1}\, t + \varphi\right). \tag{2.18}$$

Squaring them and summing them up, one obtains

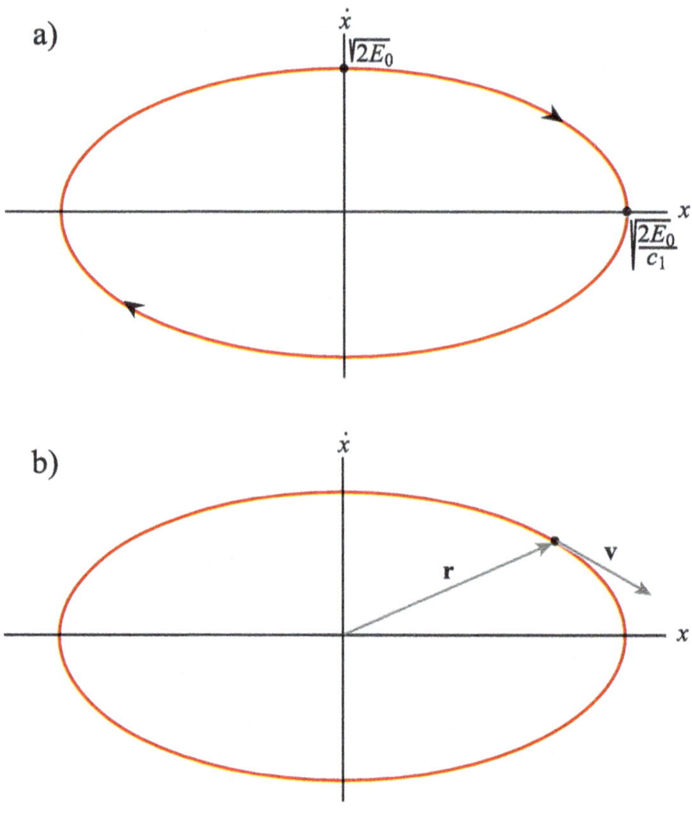

Fig. 2.4 **a** The phase trajectory of the SHO (ellipses) and its half-axes; **b** Sectorial velocity

$$\frac{x^2}{\frac{2E_0}{c_1}} + \frac{\dot{x}^2}{2E_0} = 1. \tag{2.19}$$

Thus, the phase trajectories plotted in the phase plane $x - \dot{x}$ are, in general, ellipses, whose half-axes are $\sqrt{\frac{2E_0}{c_1}} = a$ and $\sqrt{2E_0} = a\sqrt{c_1}$ (for the equation of motion non-dimensionalized to exclude c_1, the phase trajectories will be circles). The half-axes of the ellipse are shown in Fig. 2.4a. Knowing that the velocity of a point on a phase trajectory in the Cartesian coordinate system $x - \dot{x}$ is $\mathbf{v}=\dot{x}\mathbf{i}+\ddot{x}\mathbf{j}=\dot{x}\mathbf{i}-xc_1\mathbf{j}$, one can conclude that for $x > 0$, the projection of the velocity on the vertical axis is negative, i.e. the direction of motion along the phase trajectory is clockwise as indicated by the arrows in Fig. 2.4a.

The additional characteristic regards sectorial/areal velocity (the rate at which area is swept out by a point as it moves along it), shown in Fig. 2.4b:

$$\mathbf{v}_s = \frac{1}{2}(\mathbf{r} \times \mathbf{v}) = \frac{1}{2}(x\mathbf{i} + \dot{x}\mathbf{j}) \times (\dot{x}\mathbf{i} + \ddot{x}\mathbf{j}) = \frac{1}{2}\left(x\ddot{x} - \dot{x}^2\right)(\mathbf{i} \times \mathbf{j}), \tag{2.20}$$

which for the SHO becomes

$$\mathbf{v}_s = -E_0\,(\mathbf{i} \times \mathbf{j}), \tag{2.21}$$

implying that its magnitude is constant.

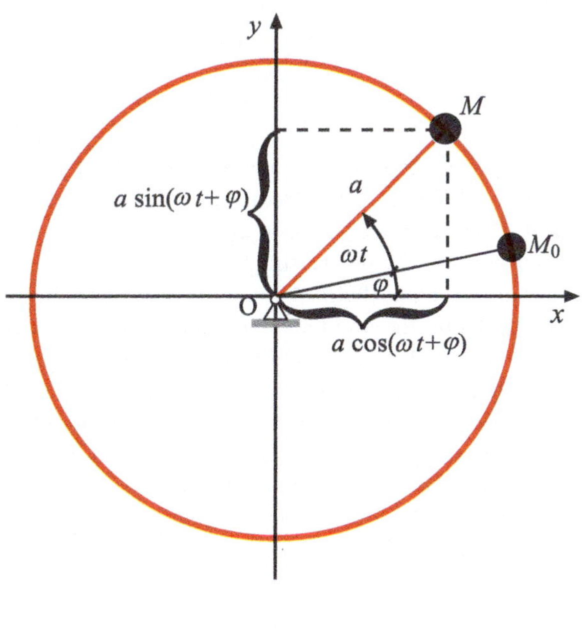

Fig. 2.5 Uniform circular motion and its projections as the analogy to the motion of the SHO

Simple harmonic motion can be considered as a one-dimensional projection of uniform circular motion (Fig. 2.5): a point M moves around a circle of radius a centred at the origin of the xy-plane, where its initial position M_0 is defined by the angle φ, while the angular velocity of the rod OM around the axis through O is constant and corresponds to ω. The motion along each coordinate is simple harmonic motion with the amplitude a and angular frequency ω. Note that its projection to the $y-$axis corresponds exactly to Eq. (2.11).

2.3 Duffing-Type Oscillators

Let us consider now a class of free conservative Duffing-type oscillators (DTOs) governed by

$$\ddot{x} + c_1 x + c_3 x^3 = 0. \tag{2.22}$$

Several oscillators can be distinguished regarding the sign of the coefficients of the linear and cubic term:
(i) Hardening Duffing Oscillators (HDOs), when $c_1 > 0$, $c_3 > 0$;
(ii) Pure Cubic Oscillators (PCOs), when $c_1 = 0$, $c_3 > 0$;
(iii) Softening Duffing Oscillators (SDOs), when $c_1 > 0$, $c_3 < 0$;
(iv) Bistable Duffing Oscillators (BDOs), when $c_1 < 0$, $c_3 > 0$.

What is of importance for this study is the fact that these DTOs have their solutions for motion in the known exact form [3–7]. These solutions can be expressed in terms of the Jacobi elliptic function (labelled by ep here) as

$$x = \text{ep}\left(\omega_{\text{ep}} t \,|m\right) \equiv \text{ep}\left(\omega_{\text{ep}} t, k\right), \tag{2.23}$$

which is a two-parameter function, having the first argument varying in time and being proportional to the frequency ω_{ep}, while the second argument m is the so-called *elliptic parameter*. Note that in some publications and software packages, the *elliptic modulus* k, whose square corresponds to the elliptic parameter $k^2 = m$, is used; in this book, both of them are used and the reader should pay attention to this fact while studying the related material. Thus, two forms of the representations exist: one is with a vertical line and the other one is with a comma (see Eq. (2.23)), although the latter can also be with a comma. The Jacobi elliptic functions that are the solutions of motion for free conservative DTOs actually correspond to one of three basic elliptic functions: cn, sn and dn. It is worth mentioning here that the cn function can be seen as a generalization of the cosine function and the sn function as a generalization of the sine function, which will be pointed out subsequently.

2.3.1 Briefly About Jacobi Elliptic Functions

Linear vibration theory is well established on the problem of the SHO. As presented in Sect. 2.1 and Fig. 2.5, its motion can be interpreted as a projection of uniform circular motion on the Cartesian coordinate system. This analogy will be presented here in a generalized version, looking for a type of nonlinear conservative oscillators (such as, for example, nonlinear springs or configurations of one or several springs attached to the mass) whose motion can be interpreted as a projection of motion along an ellipse (Fig. 2.6).

To that end, a unit circle is considered (Fig. 2.6b). The position of the point M located on the circle is defined by the angle Φ. Let this circle be extended in a horizontal direction so that it becomes an ellipse whose horizontal axis is labelled by a, while its vertical axis stays the same, i.e. $b = 1$ (Fig. 2.6d). Its equation is

$$\frac{x^2}{a^2} + y^2 = 1, \tag{2.24}$$

and it also holds

$$x^2 + y^2 = r^2. \tag{2.25}$$

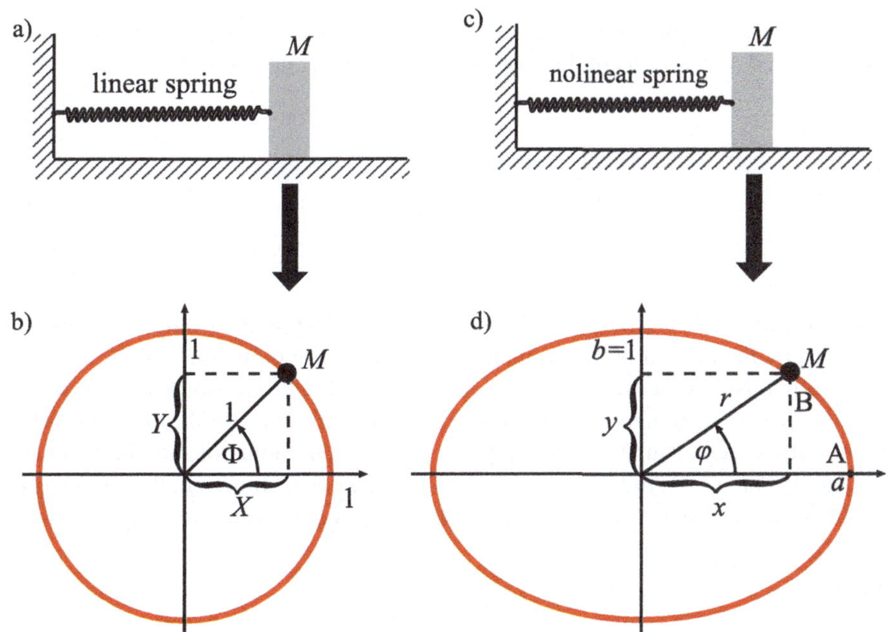

Fig. 2.6 **a** Simple harmonic oscillator; **b** Circle featured in the derivation; **c** Nonlinear oscillator; **d** Ellipse featured in derivation

The point on the ellipse is labelled again by M, and its position is defined by the angle φ. Its Cartesian coordinates can be expressed in terms of the basic trigonometric functions (cosine and sine) and are given by $x = r\cos\varphi$ and $y = r\sin\varphi$. However, unlike in the case of trigonometric functions, the angle φ is not enough to define Jacobi elliptic functions. Actually, this is one of the basic differences between them: trigonometric functions depend on one variable (argument) only, while Jacobi elliptic functions have *two* arguments. The first one is related to the angle φ and is given by

$$u = \int_A^B r\mathrm{d}\varphi = \int_A^B \frac{a}{\sqrt{1-(1-a^2)\sin^2\varphi}}\mathrm{d}\varphi, \tag{2.26}$$

where r is calculated from Eqs. (2.24) and (2.25). Note that u is neither the arc length, nor the angle.

The second argument of the Jacobi elliptic functions is the so-called *elliptic modulus* k [4, 6], which corresponds to the eccentricity of this ellipse. Using $b^2/a^2 = 1 - k^2$ and recalling that $b = 1$, one follows

$$k = \sqrt{1 - \frac{1}{a^2}}. \tag{2.27}$$

If $a = 1$, this modulus is zero and the ellipse turns into the circle. In addition, the argument u becomes the arc length, then.

With these two arguments u and k described, three basic Jacobi elliptic functions can be defined as ratios, as in the case of trigonometry:

$$\mathrm{cn}\,(u,k) = \frac{x}{a}, \tag{2.28}$$

$$\mathrm{sn}\,(u,k) = y, \tag{2.29}$$

$$\mathrm{dn}\,(u,k) = \frac{r}{a}. \tag{2.30}$$

Based on Eqs. (2.24)–(2.29), Fig. 2.7 shows how the cn and sn elliptic waves are generated by the motion of the point M along the ellipse plotted in Fig. 2.6d.

It is obvious that the first two functions (the Jacobi elliptic sine (sn) and cosine (cn) functions) generalize the sine and cosine functions, while the third one (the Jacobi elliptic delta (dn) function) appears because the radius of the ellipse is not constant. When $a = 1$, the cn function turns into the cosine function, the sn function turns into the sine function, while the dn function becomes equal to unity:

$$\mathrm{cn}\,(u,0) = \cos u, \tag{2.31}$$

$$\mathrm{sn}\,(u,0) = \sin u, \tag{2.32}$$

$$\mathrm{dn}\,(u,0) = 1. \tag{2.33}$$

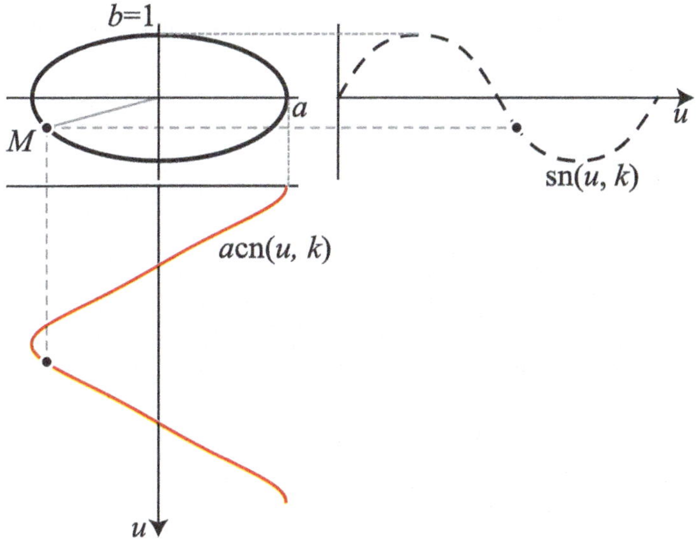

Fig. 2.7 Motion of the point M along the ellipse generating the cn and sn elliptic waves as its projections to two axes

Further, Eqs. (2.28)–(2.30) can be used together with Figs. 2.6d and 2.7 to identify that $\mathrm{cn}(0, k) = 1$, $\mathrm{sn}(0, k) = 0$, $\mathrm{dn}(0, k) = 1$. These values repeat periodically and their periods are discussed in Sect. 2.3.1.2.

Note that some facts about these functions and their characteristics are collected in Appendix B.

2.3.1.1 On the Elliptic Modulus and Associated Variables

Equation (2.27) defines the elliptic modulus. As already mentioned, there are other notations for this second argument of the Jacobi elliptic functions. One approach is to use the *elliptic parameter* $m = k^2$, which is the square of the eccentricity/elliptic modulus. The way in which elliptic functions are written down then is slightly different and is given by

$$\mathrm{cn}\,(u, k) \equiv \mathrm{cn}\,(t\,|m)\,, \tag{2.34}$$

as explained earlier related to Eq. (2.23). The analogous representations hold for other Jacobi elliptic functions, too [5].

Note that the elliptic modulus k can be complex or real, that is, the elliptic parameter m can have negative and positive real values. If the ellipse comes out from a compressed circle, i.e. $a < 1$, one has $m < 0$. In this case, which will be referred to as

Case I, the elliptic modulus is complex. To deal with real values, the *complementary modulus* k' is introduced

$$k' = \sqrt{1 - k^2}. \tag{2.35}$$

For the ellipse plotted in Fig. 2.6d, one holds $a > 1$, and, consequently, k is real, while the elliptic parameter belongs to the interval $0 < m < 1$. This will be referred to as Case II.

2.3.1.2 Elliptic Functions as Multi-frequency Trigonometric Functions: Fourier Series Representations

To provide further interpretation of the Jacobi elliptic functions, the Fourier series expansion of the cn function is considered [5]. It encompasses odd harmonics, with the frequency and their amplitudes being dependent on the elliptic parameter:

$$\mathrm{cn}\,(t\,|m\,) = \sum_{N=1}^{\infty} C_N \cos\left((2N-1)\frac{\pi}{2K}t\right), \tag{2.36}$$

$$C_N = \frac{2\pi}{K\sqrt{m}} \frac{q^{N-1/2}}{1+q^{2N-1}}, \tag{2.37}$$

where q is the Nome, which is a special function defined by

$$q = \exp\left(-\frac{\pi K'}{K}\right), \tag{2.38}$$

with K being the complete elliptic integral of the first kind and K' being its associated complete elliptic integral of the first kind. By the definition, K depends on the elliptic parameter as follows:

$$K \equiv K\,(m) = \int_0^{\pi/2} \frac{\mathrm{d}\psi}{\sqrt{1 - m\sin^2\psi}}, \tag{2.39}$$

while $K' = K\,(1-m)$. The following expansion for $K\,(m)$ is useful for further considerations [4]:

$$K\,(m) \approx \frac{\pi}{2}\left[1 + \frac{1}{4}m + \frac{9}{64}m^2 + \frac{25}{256}m^3 + O\left(m^4\right)\right]. \tag{2.40}$$

The corresponding period can be calculated from the period of the cosine function of the lowest harmonic in Eq. (2.36):

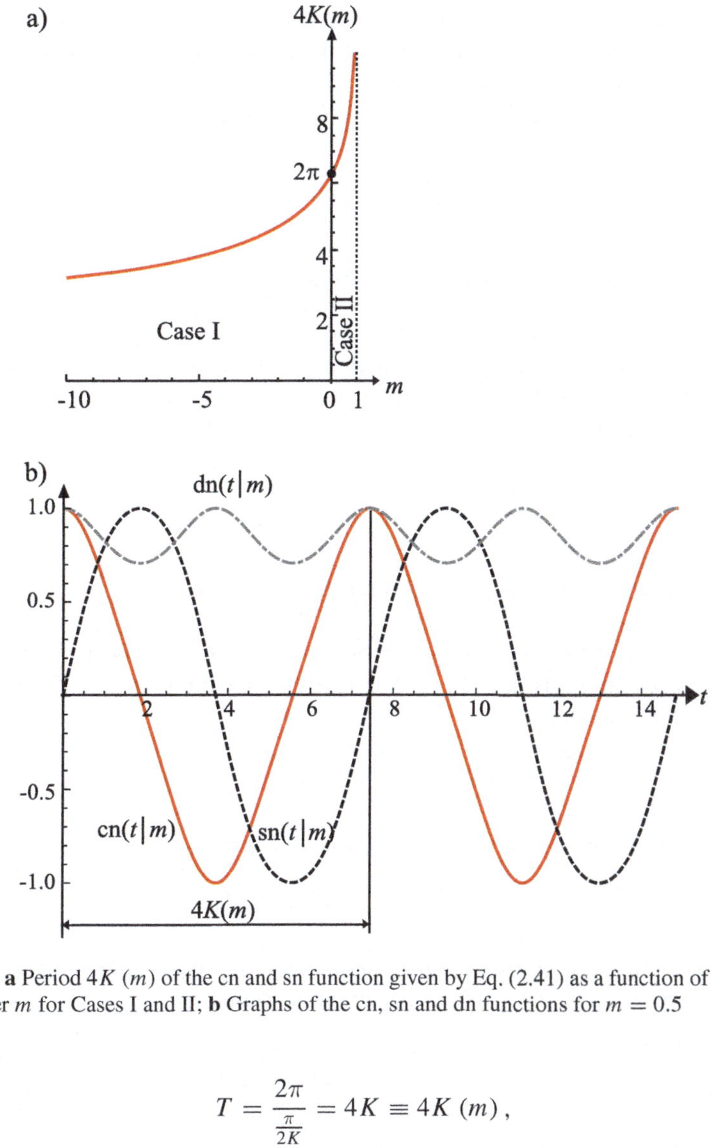

Fig. 2.8 **a** Period $4K\ (m)$ of the cn and sn function given by Eq. (2.41) as a function of the elliptic parameter m for Cases I and II; **b** Graphs of the cn, sn and dn functions for $m = 0.5$

$$T = \frac{2\pi}{\frac{\pi}{2K}} = 4K \equiv 4K\ (m)\,, \tag{2.41}$$

which is a real period of the Jacobi cn function [5] (note that its other period is complex, and due to the physics of the problems under consideration, this period is not of interest here). It is also possible to define the complete elliptic integral of the first kind as dependent on the elliptic modulus, i.e. $K \equiv K\ (k)$, but herein the notation $K\ (m)$ is used if not specified differently.

Based on Eq. (2.41), the period $4K\ (m)$ is plotted in Fig. 2.8a as a function of the elliptic parameter m. As m tends to minus infinity, the period decreases continuously and when m tends to unity, the period tends to infinity. When $m = 0$, one has $K\ (0) =$

$\pi/2$, i.e. the period is 2π, which also follows from Eq. (2.31) as the period of the cosine function. So, in Case I, the period of the cn and sn function is smaller than the one for the cosine and sine functions, while in Case II, the opposite holds.

Similarly, the Fourier series expansions of the sn and dn function are [5]

$$\mathrm{sn}\,(t\,|m\,) = \sum_{N=1}^{\infty} S_N \sin\left((2N-1)\frac{\pi}{2K}t\right), \quad S_N = \frac{2\pi}{K\sqrt{m}}\frac{q^{N-1/2}}{1-q^{2N-1}} \tag{2.42}$$

$$\mathrm{dn}\,(t\,|m\,) = D_0 + \sum_{N=1}^{\infty} D_N \cos\left(N\frac{\pi}{K}t\right), \quad D_0 = \frac{\pi}{2K}, \quad D_N = \frac{2\pi}{K}\frac{q^N}{1+q^{2N}}. \tag{2.43}$$

As it can be seen, the period of the sn function is also given by Eq. (2.41), while the one for the dn function is twice as small as this value, i.e. it is equal to $2K\ (m)$. All three functions are shown in Fig. 2.8b for $m = 0.5$ with the period $4K\ (m)$ labelled.

Note that the Fourier series for the cn, sn and dn, alongside some others, are collected in Appendix C as well.

2.3.1.3 Basic Relationships and Differential Equations

Basic relationships between the three Jacobi functions can easily be derived from the geometric presentation given in Fig. 2.6d. Thus, by substituting Eqs. (2.28), (2.29) into Eq. (2.24), one obtains

$$\mathrm{cn}^2\,(u\,|m\,) + \mathrm{sn}^2\,(u\,|m\,) = 1, \tag{2.44}$$

which generalizes $\cos^2 u + \sin^2 u = 1$. Equation (2.25) leads to

$$\mathrm{dn}^2\,(u\,|m\,) + m\mathrm{sn}^2\,(u\,|m\,) = 1. \tag{2.45}$$

It is also of interest to derive the differential relations. They can be obtained from Eq. (2.26) with $\varphi = \tan^{-1}(y/x)$ and Eq. (2.24) as follows:

$$\mathrm{d}u = r\mathrm{d}\varphi = \frac{1}{r}\,(x\,\mathrm{d}y - y\,\mathrm{d}x)\,, \tag{2.46}$$

$$\mathrm{d}x = -\frac{a^2\,y}{x}\mathrm{d}y, \tag{2.47}$$

yielding easily the expressions for $\mathrm{d}x/\mathrm{d}u$ and $\mathrm{d}y/\mathrm{d}u$. Then, by using Eqs. (2.28)–(2.30), one can derive

$$\frac{\mathrm{d}}{\mathrm{d}u}\mathrm{cn}\,(u\,|m\,) = -\mathrm{sn}\,(u\,|m\,)\,\mathrm{dn}\,(u\,|m\,)\,, \tag{2.48}$$

$$\frac{\mathrm{d}}{\mathrm{d}u}\mathrm{sn}\,(u\,|m\,) = \mathrm{cn}\,(u\,|m\,)\,\mathrm{dn}\,(u\,|m\,)\,, \tag{2.49}$$

$$\frac{\mathrm{d}}{\mathrm{d}u}\mathrm{dn}\,(u\,|m\,) = -m\mathrm{sn}\,(u\,|m\,)\,\mathrm{cn}\,(u\,|m\,)\,. \tag{2.50}$$

The first two expressions obviously relate to the trigonometric counterparts, while the third one is new and reduces to an identity when $m = 0$.

Let us now show which differential equations are satisfied by these three basic Jacobi elliptic functions. To that end, let us go back to Eq. (2.48). Substituting $x=\mathrm{cn}(u\,|m\,)$ in it, with $u = t$ squaring it and using Eqs. (2.44), (2.45), one can derive

$$\left(\frac{\mathrm{d}x}{\mathrm{d}t}\right)^2 = \left(\left[1 - m + mx^2\right]\right)\left(1 - x^2\right). \tag{2.51}$$

Taking the derivative of this expressions and simplifying it lead to

$$\frac{\mathrm{d}^2x}{\mathrm{d}t^2} + (1 - 2m)\,x + 2mx^3 = 0, \tag{2.52}$$

which is of the same form as Eq. (2.22).

Substituting $x=\mathrm{sn}(u\,|m\,)$ in Eq. (2.49) with $u = t$, squaring it and using Eqs. (2.44), (2.45), one can obtain

$$\left(\frac{\mathrm{d}x}{\mathrm{d}t}\right)^2 = \left(1 - mx^2\right)\left(1 - x^2\right). \tag{2.53}$$

Taking its derivative and simplifying the resulting expressions yield

$$\frac{\mathrm{d}^2x}{\mathrm{d}t^2} + (1 + m)\,x - 2mx^3 = 0. \tag{2.54}$$

Finally, a similar calculation for $x=\mathrm{dn}(u\,|m\,)$ with $u = t$ and Eq. (2.50) results in

$$\frac{\mathrm{d}^2x}{\mathrm{d}t^2} - (2 - m)\,x + x^3 = 0. \tag{2.55}$$

All these equations can be found in Appendix B.

Equations (2.52), (2.54) and (2.55) can be related to the form of Eq. (2.22) and four types of oscillators listed below it (HDO, PCO, SDO, BDO). These solutions are presented in detail in the forthcoming subsections.

2.3.2 Hardening Duffing Oscillators (HDOs)

It is assumed here that $c_1 > 0$ and $c_3 > 0$, i.e. the oscillator is of a hardening type (Hardening Duffing Oscillator (HDO)). The diagram of the corresponding potential energy (single-well potential),

$$E_p = \frac{1}{2}c_1\, x^2 + \frac{1}{4}c_3\, x^4, \tag{2.56}$$

is shown in Fig. 2.9. The equilibrium $x = 0$ is stable (centre) and it is surrounded by closed phase trajectories, which imply that the corresponding motion is periodic. Analogous to Fig. 2.1, the phase trajectories associated with different energy levels are also shown in Fig. 2.9.

Its free response can be expressed as closed form in terms of the Jacobi cn elliptic function [4] in the form:

$$x = a\mathrm{cn}\left(u\left|m\right.\right), \tag{2.57}$$

$$u = \omega t + b, \tag{2.58}$$

where ω, m, a and b need to be obtained. By using Eqs. (2.48)–(2.50), one can differentiate (2.57) with respect to time and substitute it into Eq. (2.22), equating to zero coefficients next to cn and cn^3 to derive (note that this algorithm corresponds actually to elliptic balancing):

$$a\left(2\omega^2 m - \omega^2 + c_1\right) = 0, \tag{2.59}$$

$$a\left(-2\omega^2 m + a^2 c_3\right) = 0. \tag{2.60}$$

This system of equation can be solved for ω and m. For the initial conditions corresponding to the ZIV case, one can obtain $a = A$, $b = 0$. Thus, solving Eqs. (2.59), (2.60), the solution for the response of such free HDOs is

$$x_{\mathrm{HDO,\,ZIV}} = A\mathrm{cn}\left(\omega_{\mathrm{HDO,\,ZIV}}t\left|m_{\mathrm{HDO,\,ZIV}}\right.\right), \tag{2.61}$$

$$\omega_{\mathrm{HDO,\,ZIV}} = \sqrt{c_1 + c_3 A^2}, \tag{2.62}$$

$$m_{\mathrm{HDO,\,ZIV}} = \frac{c_3 A^2}{2\left(c_1 + c_3 A^2\right)}. \tag{2.63}$$

The insight into the content of this response with respect to its harmonics is given by Eq. (2.36) and also in Appendix C. This closed-form response includes odd multiplications of the first harmonic (third, fifth, etc.), which is one of the basic differences of this response with respect to the previously discussed SHO. In addition, both the frequency of the elliptic function (2.62) ω and its elliptic parameter m (2.63) are amplitude-dependent. Note that for $c_3 = 0$, Eq. (2.63) gives $m = 0$, so that the

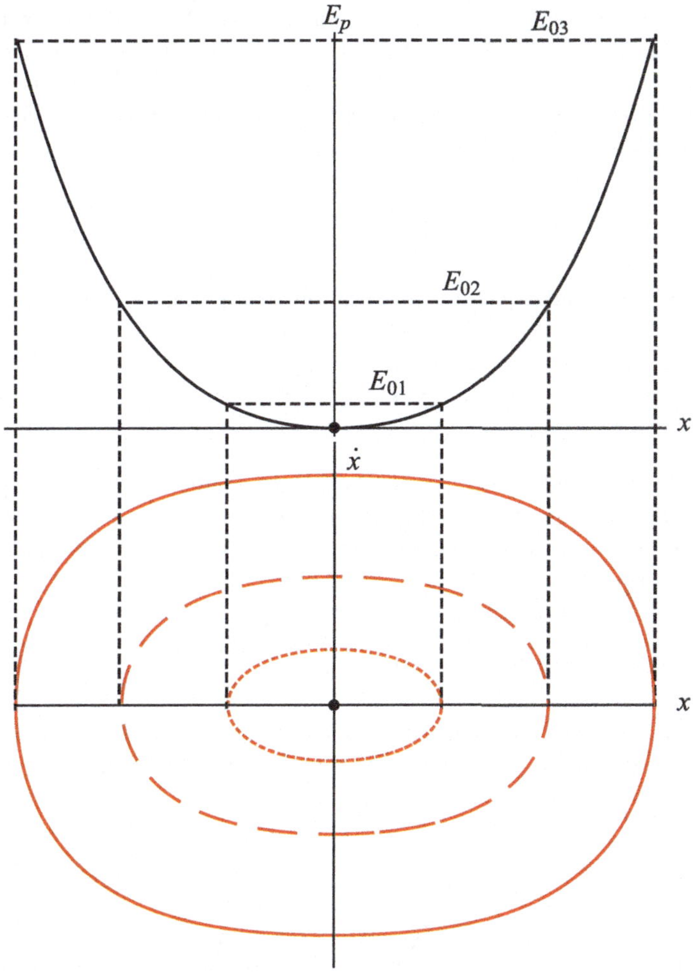

Fig. 2.9 General form of a diagram of the potential energy of the HDO and the phase trajectories corresponding to different levels of the initial energy E_0

cn function turns into the cosine function [7] and the solution of motion for the SHO is obtained, Eq. (2.8). Note also that when $c_3 A^2 \to \infty$, one has $m \to 1/2$.

Given the fact that the period of the cn and sn functions is $T = 4K(m)/\omega$ (2.41), one can now write the period for the HDO as follows:

$$T_{\mathrm{HDO}} = \frac{4K\left(\frac{c_3 A^2}{2(c_1+c_3A^2)}\right)}{\sqrt{c_1 + c_3 A^2}}, \tag{2.64}$$

which clearly shows that the period is amplitude-dependent and not constant as in the case of the SHO. This period is plotted in Fig. 2.10 as a function of the product $\bar{A}=\sqrt{c_3}\,|A|$ for $c_1 = 1$ to show how it differs from the period of the SHO, which is equal to 2π. Further, this period can be approximated by using the expansion for K given by Eq. (2.40)

$$T_{\mathrm{HDO}} \approx \frac{4}{\omega}\frac{\pi}{2}\left[1+\frac{1}{4}m+\frac{9}{64}m^2+\frac{25}{256}m^3+O\left(m^4\right)\right]. \tag{2.65}$$

By using Eq. (2.63) and expanding the resulting expression for small c_3A^2, one can derive

$$T_{\mathrm{HDO}} \approx \frac{2\pi}{\sqrt{c_1}}\left(1-\frac{3}{8}\frac{c_3A^2}{c_1}+\cdots\right), \tag{2.66}$$

which is convenient analytically as it shows how the coefficients from the equation of motion and the initial amplitude influence the period of vibrations, making it overall smaller than the one of the SHO. The approximate frequency–amplitude relation now has the form:

$$\omega_{\mathrm{HDO}} \approx \sqrt{c_1}\left(1+\frac{3}{8}\frac{c_3A^2}{c_1}+\cdots\right). \tag{2.67}$$

This is plotted as a backbone curve together with the one for the SHO in Fig. 2.11, where it is clearly seen how it bends from it to the right-hand side.

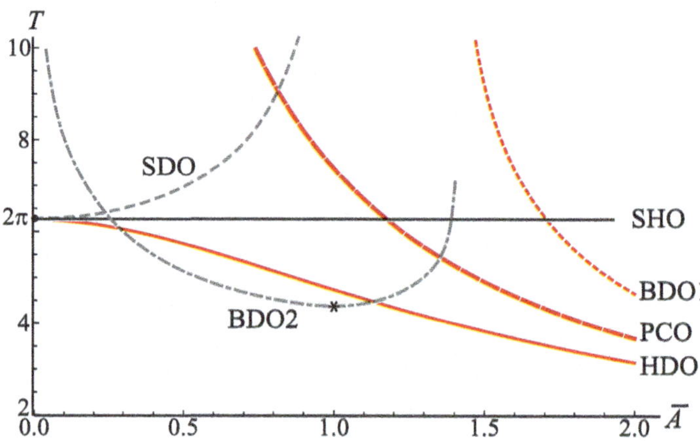

Fig. 2.10 Periods of the HDO, SDO, PCO, BDO1 and BDO2 for $|c_1| = 1$ as a function of the product $\bar{A}=\sqrt{|c_3|}\,|A|$ and with respect to the period of the SHO equal to 2π

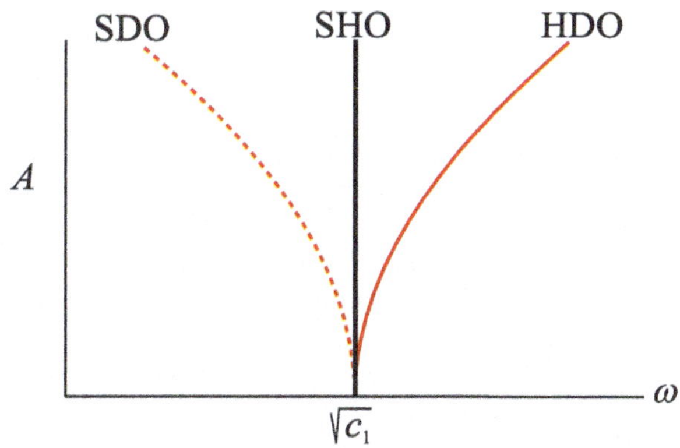

Fig. 2.11 Backbone curves of the HDO, SHO and SDO

For the ZID case, following the elliptic balancing algorithm with respect to the sn function, the solution can be derived as

$$x_{\text{HDO, ZID}} = \frac{v_0}{\omega_{\text{HDO, ZID}}} \operatorname{sn}\left(\omega_{\text{HDO, ZID}} t \left| m_{\text{HDO, ZID}} \right.\right), \tag{2.68}$$

$$\omega_{\text{HDO, ZID}} = \sqrt{\frac{c_1 + \sqrt{c_1^2 + 2c_3 v_0^2}}{2}}, \tag{2.69}$$

$$m_{\text{HDO, ZID}} = -\frac{c_3 v_0^2}{c_1^2 + c_3 v_0^2 + c_1\sqrt{c_1^2 + 2c_3 v_0^2}}. \tag{2.70}$$

It is seen that for $c_3 = 0$, the solution gets the form for the SHO, Eq. (2.9).

In a general case when neither the initial displacement nor the initial velocity is zero, the superposition principle does not hold and one cannot sum up the solution for the ZIV and ZID cases as for the SHO. In this general case, the solution is

$$x_{\text{HDO}} = a \operatorname{sn}\left(\omega_{\text{HDO}} t + b \left| m_{\text{HDO}} \right.\right), \tag{2.71}$$

$$\omega_{\text{HDO}} = \sqrt{c_1 + \frac{1}{2} c_3 a^2}, \tag{2.72}$$

$$m_{\text{HDO}} = -\frac{c_3 a^2}{2\left(c_1 + \frac{1}{2} c_3 a^2\right)}, \tag{2.73}$$

and one needs to calculate a and b based on:

$$A = a \operatorname{sn}\left(b \left| m_{\text{HDO}} \right.\right), \quad v_0 = a\omega_{\text{HDO}} \operatorname{cn}\left(b \left| m_{\text{HDO}} \right.\right) \operatorname{dn}\left(b \left| m_{\text{HDO}} \right.\right). \tag{2.74}$$

By squaring and summing up these two expressions, and using the identities (2.44) and (2.45), one can derive the quartic equation in the amplitude a:

$$\frac{1}{2}c_3a^4 + (c_1 - c_3A^2)a^2 + \frac{1}{2}c_3A^4 - c_1A^2 = 0. \tag{2.75}$$

Then, obtaining a from it, one needs to calculate ω_{HDO} and m_{HDO} so as to finally obtain the value b either in a pure numerical procedure or by using the inverse Jacobi elliptic functions.

Thus, four main characteristics of the response of the free conservative HDOs can be emphasized (and compared to four main characteristics of the SHO listed in Sect. 2.2) as follows:

HDO1. There is an exact solution for its response and it can be expressed in terms of Jacobi elliptic cn or sn function.

HDO2. The frequency and the period of vibrations are amplitude-dependent, as is the elliptic modulus of the cn/sn function.

HDO3. The backbone curve is bent to the right-hand side.

HDO4. The response contains multiple harmonics, which are odd multipliers of the first one.

Let us go back to the energy considerations. The conservation law reads as

$$\frac{1}{2}\dot{x}^2 + \frac{1}{2}c_1\,x^2 + \frac{1}{4}c_3\,x^4 = E_0. \tag{2.76}$$

This expression can be used as an implicit one for plotting phase trajectories corresponding to different levels of the initial energy E_0 and can be transformed into the expression analogous to the one defining the phase trajectories of the SHO (2.19) to point out the difference between the SHO and the HDO:

$$\frac{\left(x^2 + \frac{c_1}{c_3}\right)^2}{\frac{4E_0}{c_3} + \frac{c_1^2}{c_3^2}} + \frac{\dot{x}^2}{2E_0 + \frac{c_1^2}{2c_3}} = 1. \tag{2.77}$$

By calculating the sectorial velocity Eq. (2.20) for the case defined by Eqs. (2.61)–(2.63),

$$\mathbf{v}_s = -\frac{1}{4}A^2\left[2c_1 + c_3A^2 + c_3A^2\mathrm{cn}^4\left(\omega_{\mathrm{HDO,\,ZIV}}t\,\middle|\,m_{\mathrm{HDO,\,ZIV}}\right)\right](\mathbf{i}\times\mathbf{j})\,, \tag{2.78}$$

one can conclude that this parameter is time-dependent. Its extremal values are equal to $|v_s| = |E_0| = c_1A^2/2 + c_3A^4/4$ and $|v_s| = c_1A^2/2 + c_3A^4/2 > |E_0|$.

2.3.3 Pure Cubic Oscillators (PCOs)

Considerations related to the response of the PCO can be done in a straightforward way, treating it as a special case of the HDO. Its potential well corresponding to the potential energy,

$$E_p = \frac{1}{4} c_3 x^4, \tag{2.79}$$

is also single-welled (Fig. 2.12) as in the case of the HDO. Based on Eqs. (2.61)–(2.63) and for the ZIV conditions, the solution for its response is found to be

$$x_{\mathrm{PCO}} = A\mathrm{cn}\left(\omega_{\mathrm{PCO}} t \,|m_{\mathrm{PCO}}\right), \tag{2.80}$$

$$\omega_{\mathrm{PCO}} = \sqrt{c_3} A, \tag{2.81}$$

$$m_{\mathrm{PCO}} = \frac{1}{2}. \tag{2.82}$$

It is seen that the frequency is amplitude-dependent, while the elliptic parameter is constant. The backbone curve in the $\omega - A$ plane is a straight line, whose slope is $1/\sqrt{c_3}$.

By using the Fourier series defined by Eqs. (2.36)–(2.39), which is also given in Appendix C, one can first calculate that for $m = 1/2$ and one has $K\,(1/2) = 1.85407$. Further, the Nome q is

$$q\left(m = \frac{1}{2}\right) = \exp\left(-\pi \frac{K\,(1-m)}{K\,(m)}\right) = \exp\left(-\pi\right). \tag{2.83}$$

This yields the following Fourier series with the harmonics that are odd multiplications of the first one:

$$\mathrm{cn}\left(\omega t \left| \frac{1}{2} \right.\right) \approx 0.955006\cos\left(0.847213\omega t\right) + 0.0430495\cos\left(2.54164\omega t\right) + 0.00186049\cos\left(4.23607\omega t\right) + 0.0000803976\cos\left(5.93049\omega t\right) + \cdots .$$

The period is defined by

$$T_{\mathrm{PCO}} = \frac{4K\left(\frac{1}{2}\right)}{\sqrt{c_3 A^2}} = \frac{7.4163}{\sqrt{c_3 A^2}}, \tag{2.84}$$

which is plotted in Fig. 2.10 as a function of the product $\bar{A} = \sqrt{c_3} A$.

The main characteristics of the response of the free conservative PCO are as follows:

PCO1. There is an exact solution for its response and it can be expressed in terms of Jacobi elliptic cn function.

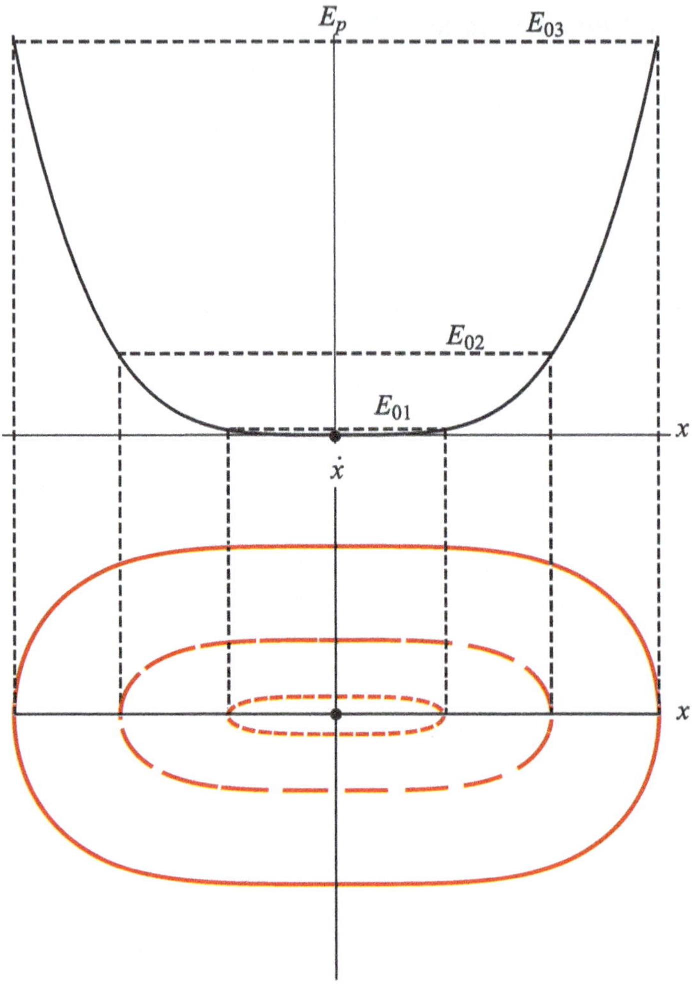

Fig. 2.12 General form of a diagram of the potential energy of the PCO and the phase trajectories corresponding to different levels of the initial energy E_0

PCO2. The frequency and the period of vibrations are amplitude-dependent, while the elliptic modulus is amplitude-independent.

PCO3. The backbone curve is a straight-line curve, whose slope is related to the square root of the coefficient of the cubic term.

PCO4. The response is multi-frequency, where the higher harmonics are odd multiplications of the lowest one.

The conservation law reads as

$$\frac{1}{2}\dot{x}^2 + \frac{1}{4}c_3\, x^4 = E_0, \tag{2.85}$$

yielding

$$\frac{x^4}{\frac{4E_0}{c_3}} + \frac{\dot{x}^2}{2E_0} = 1. \tag{2.86}$$

This type of a plane curve is a special class of a generalized superellipse

$$\left|\frac{x}{a}\right|^p + \left|\frac{y}{b}\right|^q = 1, \tag{2.87}$$

where $y = \dot{x}$, $a = 4E_0/c_3$, $b = 2E_0$, $p = 4$, $q = 2$. It is known that the corresponding parametric equations are

$$x\,(t) = |\cos\vartheta|^{\frac{2}{p}}\, a\mathrm{sgn}\,(\cos\vartheta)\,, \tag{2.88}$$

$$y\,(t) = |\sin\vartheta|^{\frac{2}{q}}\, a\mathrm{sgn}\,(\sin\vartheta)\,, \tag{2.89}$$

where ϑ is a parameter.

The sectorial velocity is now:

$$\mathbf{v}_s = -\frac{1}{4}c_3A^4\left[1+\mathrm{cn}^4\left(\omega t\left|\frac{1}{2}\right.\right)\right](\mathbf{i}\times\mathbf{j}) = -E_0\left[1+\mathrm{cn}^4\left(\omega t\left|\frac{1}{2}\right.\right)\right](\mathbf{i}\times\mathbf{j})\,. \tag{2.90}$$

It changes during time as the system oscillates, and its magnitude has two extremal values $|v_s| = c_3A^4/2 = 2E_0$ and $|v_s| = c_3A^4/4 = E_0$.

2.3.4 *Softening Duffing Oscillator (SDO)*

The negative sign in front of the cubic term in Eq. (2.22) ($c_3 = -|c_3|$) corresponds to a Softening Duffing oscillator (SDO), and to emphasize this fact, Eq. (2.22) is written in the form

$$\ddot{x} + c_1 x - |c_3|\; x^3 = 0. \tag{2.91}$$

It has three equilibria: a trivial one $x_1 = 0$ and two non-trivial ones $x_{2/3} = \pm\sqrt{c_1/|c_3|}$. They are indicated on the plot of the corresponding potential energy

$$E_p = \frac{1}{2}c_1x^2 - \frac{1}{4}|c_3|\,x^4, \tag{2.92}$$

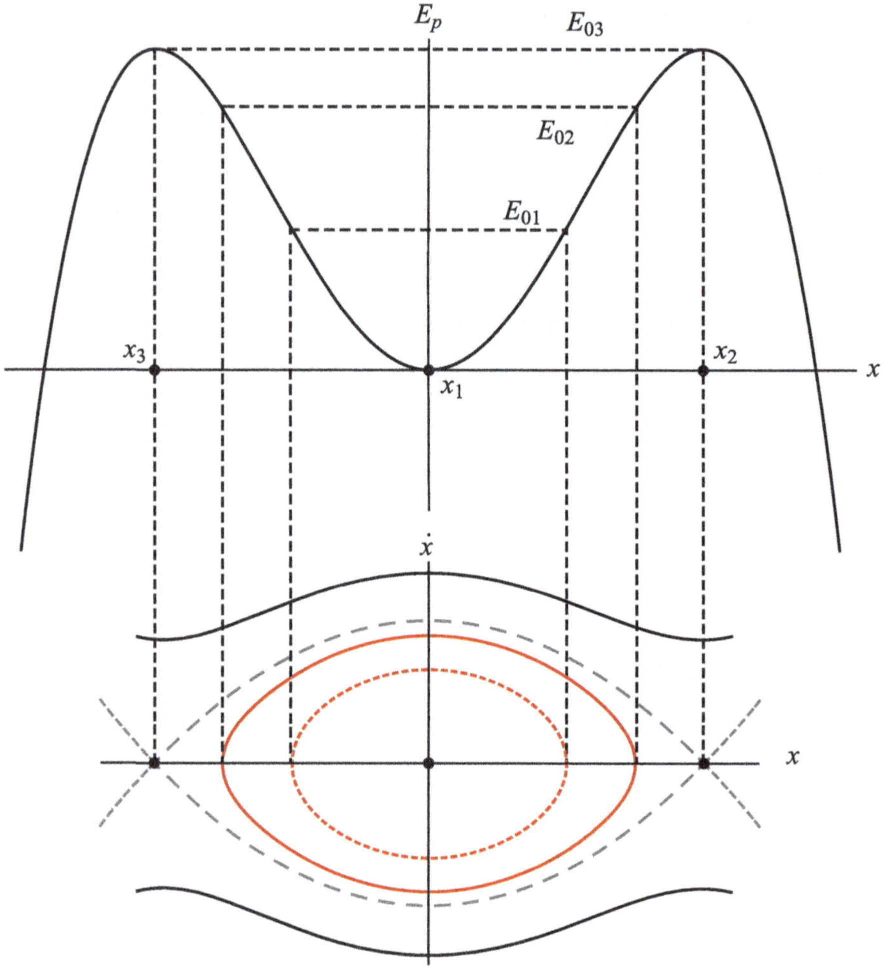

Fig. 2.13 General form of a diagram of the potential energy of the SDO and the phase trajectories corresponding to different levels of the initial energy E_0

in Fig. 2.13. By examining the second derivative of the potential energy and the sign of these values corresponding to the equilibria, one can determine that this sign is positive for x_1 and negative for $x_{2/3}$. Therefore, the origin is stable (it represents a centre), while non-trivial equilibria are unstable (they represent two saddles). The phase trajectories surrounding the centre and the separatrices passing through the saddles are plotted in Fig. 2.13. There is another type of trajectory that corresponds to motion that grows unbounded. Note that in the case of the SDO, one requires $|c_3|\, A^2 < c_1$ for the system to exhibit closed orbits around the stable equilibrium.

Its free response can be expressed in terms of the Jacobi sn elliptic function [4]. The response is assumed in the form:

$$x = a\text{sn}\,(u\,|m\,)\,, \tag{2.93}$$

$$u = \omega t + b, \tag{2.94}$$

where ω, m, a and b need to be calculated. By using Eqs. (2.48)–(2.50), one can differentiate (2.93) with respect to time and substitute it into Eq. (2.91), equating to zero coefficients next to sn and sn^3 to derive:

$$a\left(-\omega^2 m - \omega^2 + c_1\right) = 0, \tag{2.95}$$

$$-a\left(-2\omega^2 m + a^2\,|c_3|\right) = 0. \tag{2.96}$$

This system of equation can be solved for ω and m. For the initial condition corresponding to the ZIV case, one can obtain $a = A, b = K(m)$. Thus, the solution for the response of the free SDO is

$$x_{\text{SDO, ZIV}} = A\text{sn}\left(\omega_{\text{SDO, ZIV}}t + K(m_{\text{SDO, ZIV}})\,\middle|m_{\text{SDO, ZIV}}\right), \tag{2.97}$$

$$\omega_{\text{SDO, ZIV}} = \sqrt{c_1 - \frac{1}{2}\,|c_3|\,A^2}, \tag{2.98}$$

$$m_{\text{SDO, ZIV}} = \frac{c_3 A^2}{2\left(c_1 - \frac{1}{2}\,|c_3|\,A^2\right)}. \tag{2.99}$$

The insight into the content of this response with respect to its harmonics is given by Eq. (2.42) and also in Appendix C. This closed-form response contains sine terms with odd harmonics (first, third, fifth, etc.). In addition, as in the case of the HDO, both the frequency of the elliptic function (2.98) $\omega_{\text{SDO, ZIV}}$ and its elliptic parameter $m_{\text{SDO, ZIV}}$ (2.99) are amplitude-dependent. Note that for $c_3 = 0$, Eq. (2.99) gives $m = 0$, so that the sn function turns into the sine function [7] and the solution of motion for the SHO is obtained, Eq. (2.8).

Note that alternatively, one can assume the solution in the form of the cn function, follow an analogous procedure, deriving $a = A, b = 0$ and

$$x_{\text{SDO, ZIV}} = A\text{cn}\left(\omega_{\text{SDO, ZIV}}t\,\middle|m_{\text{SDO, ZIV}}\right), \tag{2.100}$$

$$\omega_{\text{SDO, ZIV}} = \sqrt{c_1 - |c_3|\,A^2}, \tag{2.101}$$

$$m_{\text{SDO, ZIV}} = -\frac{c_3 A^2}{2\left(c_1 - |c_3|\,A^2\right)}. \tag{2.102}$$

Based on Eq. (2.98) and knowing that the period of the sn function is $T = 4K\,(m)\,/\omega$ (2.41), one can now write the period for the SDO as follows:

$$T_{\text{SDO}} = \frac{4K\left(\frac{c_3 A^2}{2\left(c_1 - c_3\frac{A^2}{2}\right)}\right)}{\sqrt{c_1 - |c_3|\,\frac{A^2}{2}}} = \frac{4K\left(\frac{c_3 A^2}{2\left(c_1 - |c_3|A^2\right)}\right)}{\sqrt{c_1 - |c_3|\,A^2}}.$$

This period is plotted in Fig. 2.10 as a function of the product $\bar{A}=\sqrt{|c_3|}A$ to show how it differs from the period of the SHO.

Further, the period can be approximated analogously as done for the HDO to derive

$$T_{\mathrm{SDO}} \approx \frac{2\pi}{\sqrt{c_1}}\left(1+\frac{3}{8}\frac{c_3 A^2}{c_1}+\cdots\right), \tag{2.103}$$

while the approximate frequency–amplitude relation has the form:

$$\omega_{\mathrm{SDO}} \approx \sqrt{c_1}\left(1-\frac{3}{8}\frac{c_3 A^2}{c_1}+\cdots\right). \tag{2.104}$$

This is plotted as a backbone curve next to the one for the SHO and the HDO in Fig. 2.11, where it is clearly seen how it bends to the left-hand side. Note that the same result for the approximate expressions for T_{SDO} and ω_{SDO} can analogously be derived from Eqs. (2.101) and (2.102).

Thus, four main characteristics of the response of the free conservative SDO can be summarized as follows:

SDO1. There is an exact solution for its response and it can be expressed in terms of Jacobi elliptic sn or cn function.

SDO2. The frequency and the period of vibrations are amplitude-dependent, as is the elliptic modulus.

SDO3. The backbone curve is bent to the left-hand side.

SDO4. The response is a multi-frequency response.

The conservation law reads as

$$\frac{1}{2}\dot{x}^2+\frac{1}{2}c_1x^2-\frac{1}{4}\left|c_3\right|\,x^4=E_0, \tag{2.105}$$

yielding:

$$\frac{\left(\frac{c_1}{|c_3|}-x^2\right)^2}{\frac{4E_0}{|c_3|}+\frac{c_1^2}{c_3^2}}+\frac{\dot{x}^2}{2E_0+\frac{c_1^2}{2|c_3|}}=1. \tag{2.106}$$

By calculating the sectorial velocity Eq. (2.20) for the ZID case defined by Eqs. (2.100)–(2.102), one can conclude that this parameter becomes time-dependent

$$\mathbf{v}_s=\frac{1}{4}A^2\left[-2c_1+c_3A^2+c_3A^2\mathrm{cn}^4\left(\sqrt{c_1-|c_3|\,A^2}t\,\left|-\frac{c_3A^2}{2\left(c_1-|c_3|\,A^2\right)}\right.\right)\right](\mathbf{i}\times\mathbf{j}), \tag{2.107}$$

changing between two extreme values: $|v_s|=|E_0|=\left|c_1A^2/2-|c_3|\,A^4/4\right|$ and $|v_s|=\left|c_1A^2/2-|c_3|\,A^4/2\right|<|E_0|$.

2.3.5 Bistable Duffing Oscillators (BDOs)

The negative sign of the linear coefficient ($c_1 = -|c_1|$) and a positive cubic one in Eq. (2.22) correspond to a Bistable Duffing oscillator (BDO). Thus, this equation can be written as follows:

$$\ddot{x} - |c_1|\, x + c_3\, x^3 = 0. \tag{2.108}$$

There are three equilibria:a trivial one $x_1 = 0$ and two non-trivial ones $x_{2/3} = \pm\sqrt{|c_1|/c_3}$. They are indicated on the plot of the corresponding potential energy

$$E_p = -\frac{1}{2}|c_1|\, x^2 + \frac{1}{4}c_3 x^4, \tag{2.109}$$

in Fig. 2.14. By examining the second derivative of the potential energy and the sign of these values corresponding to the equilibria, one can determine that this sign is negative for x_1 (so, this corresponds to a saddle) and positive for $x_{2/3}$ (so, these correspond to centres). Consequently, there are two types of periodic motion that are separated by the separatrix, which passes through the saddle (Fig. 2.14).

For the ZIV case and the parameters $c_3A^2 > 2|c_1|$, phase trajectories corresponding to large amplitude vibrations occur, surrounding all three equilibria (Fig. 2.14). This is the so-called 'full-swing mode' or 'out-of-well motion'. The solution for this motion can be obtained based on the analogy with the HDO and is labelled by BDO1 here:

$$x_{\mathrm{BDO1,\,ZIV}} = A\mathrm{cn}\left(\omega_{\mathrm{BDO1,\,ZIV}}t\,\middle|\, m_{\mathrm{BDO1,\,ZIV}}\right), \tag{2.110}$$

$$\omega_{\mathrm{BDO1,\,ZIV}} = \sqrt{c_3A^2 - |c_1|}, \tag{2.111}$$

$$m_{\mathrm{BDO1,\,ZIV}} = \frac{c_3A^2}{2\left(c_3A^2 - |c_1|\right)}. \tag{2.112}$$

Another bounded periodic response takes place around one of the two stable equilibria $x^2 = |c_1|/c_3$ (Fig. 2.14).This is the so-called 'half-swing mode' or 'in-well motion' that is exhibited for $0 < c_3A^2 < 2|c_1|$, excluding $c_3A^2 = |c_1|$, as it correspond to the centre. To obtain the exact solution for this type of the response, one can relate Eqs. (2.108) and (2.55), assuming the solution as $x = A\mathrm{dn}(\omega t\,|m)$. Differentiating it twice, substituting in Eq. (2.108), using Eqs. (2.48)–(2.50) and equating to zero coefficients next to dn and dn^3, one can derive

$$-\omega^2 m + 2\omega^2 - |c_1| = 0, \tag{2.113}$$

$$-2\omega^2 + c_3A^2 = 0. \tag{2.114}$$

Solving these two equations for ω and m, one can obtain the exact solution for the BDO2

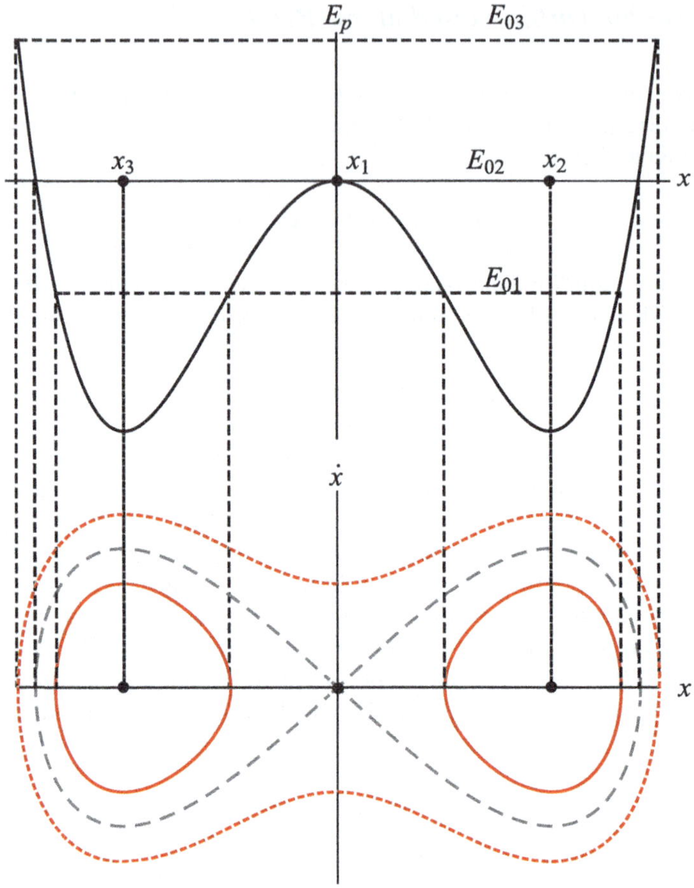

Fig. 2.14 General form of a diagram of the potential energy of the BDO and the phase trajectories corresponding to different levels of the initial energy E_0

$$x_{\text{BDO2, ZIV}} = A\text{dn}\left(\omega_{\text{BDO2, ZIV}}t \left| m_{\text{BDO2, ZIV}}\right.\right), \tag{2.115}$$

$$\omega_{\text{BDO2, ZIV}} = \sqrt{\frac{c_3}{2}}A, \tag{2.116}$$

$$m_{\text{BDO2, ZIV}} = \frac{2\left(c_3 A^2 - |c_1|\right)}{c_3 A^2}. \tag{2.117}$$

As seen from the Fourier expansion for the dn function given by Eq. (2.43), this solution contains an offset and odd and even harmonics. Note that again the case $c_3 A^2 = |c_1|$ should be excluded as it corresponds to the centre and not the trajectories.

Given the fact that the periods of the cn function are $T = 4K(m)/\omega$ and of the dn function $T = 2K(m)/\omega$, one can now write the periods for the BDO1 and BDO2 as

$$T_{\mathrm{BDO1}} = \frac{4K\left(\frac{c_3A^2}{2(c_3A^2-|c_1|)}\right)}{\sqrt{c_3A^2-|c_1|}}, \tag{2.118}$$

$$T_{\mathrm{BDO2}} = \frac{2K\left(\frac{2(c_3A^2-|c_1|)}{c_3A^2}\right)}{\sqrt{\frac{c_3}{2}}A}. \tag{2.119}$$

These two expressions are plotted in Fig. 2.10 together with the periods of all other Duffing-type oscillators discussed above as a function of the product $\bar{A}=\sqrt{|c_3|}A$ (for positive A) to show how much they differ from the period of a linear oscillator 2π as $\bar{A}$ increases from zero. As it can be seen, for small values of $\bar{A}$, this difference for the HDO and SDO is not large, but it increases as $\bar{A}$ becomes higher. The period of the PCO is monotonous: when $\bar{A}\to 0$, then $T_{\mathrm{PCO}} \to \infty$; when $\bar{A}\to \infty$, then $T_{\mathrm{PCO}} \to 0$. The periods corresponding to the BDO1 and BDO2 are also shown (the case $c_3A^2 = |c_1|$ is excluded for the BDO2 and it is labelled by the asterix). It is interesting to note that there are a few intersections of the curves plotted in Fig. 2.10, which imply that some of these oscillators, although being phenomenologically different, can have the response characterized by the same amplitude and the same period (for example, the HDO and the BDO, or the BDO2 and PCO). Note also that the period of BDO1 and BDO2 expressed in terms of the product of the cubic coefficient and the initial energy is given in [8].

The sectorial velocity Eq. (2.20) for the case of out-of-well oscillations (2.110)–(2.112) changes between two extreme values $|v_s| = \left||c_1|A^2/2 - c_3A^4/2\right| < E_0$ and $|v_s| = |E_0| = \left||c_1|A^2/2 - c_3A^4/4\right|$, which is similar to the SDO.

For the case of in-well oscillations (2.115)–(2.117), the extreme values of the sectorial velocity are the same.

2.4 Quadratic Oscillators (QOs)

Let us show now how to obtain the exact solution for motion for the undamped oscillators whose equation of motion contains quadratic nonlinearity

$$\ddot{x} + c_1x + c_2x^2 = 0, \tag{2.120}$$

where, again, one can rescale the equation to get $c_1 = 1$.

The corresponding potential energy,

$$E_p = \frac{1}{2}c_1x^2 + \frac{1}{3}c_2x^3, \tag{2.121}$$

is plotted in Fig. 2.15 for a positive and negative value of c_2. In both cases, the existence of two fixed points $x_1 = 0$ and $x_2 = -c_1/c_2$ is seen. It should be pointed out that closed and bounded phase trajectories surrounding the origin exist for the

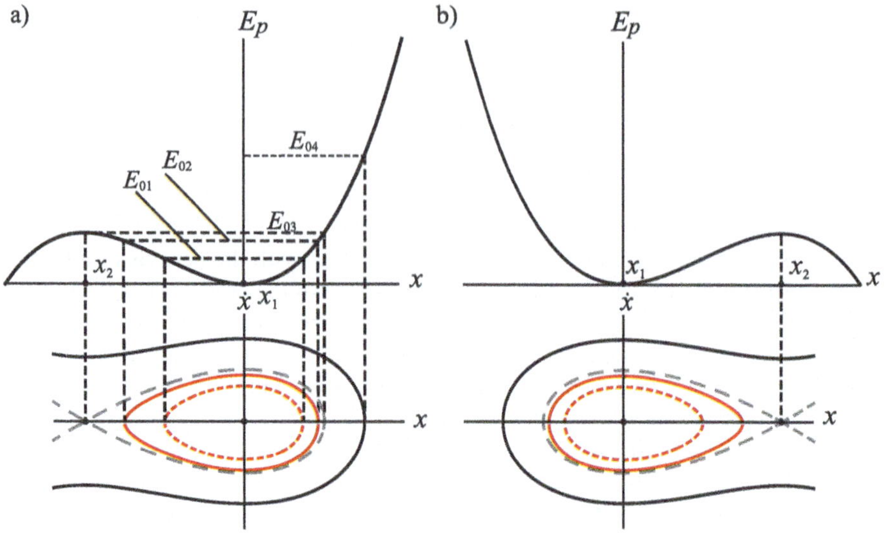

Fig. 2.15 General form of a diagram of the potential energy of the QO and the phase trajectories corresponding to different levels of the initial energy E_0: **a** $c_2 > 0$; **b** $c_2 < 0$

initial energy levels $0 < E_0 < c_1^3/(6c_2^2)$. The asymmetry of the shapes of the potential energy and phase trajectories is apparent.

The solution for motion can be taken in the form [9]

$$x_{QO} = A_{0QO} + A_{2QO}\,\mathrm{sn}^2\left(\omega_{QO}t\,\middle|m_{QO}\right). \tag{2.122}$$

Differentiating this solution twice, using Eqs. (2.48)–(2.50), substituting it into Eq. (2.120) and collecting free terms and those next to cn^2 and cn^4 lead to

$$A_{0QO} = \frac{c_1}{2c_2}\frac{m+1-\sqrt{m^2-m+1}}{\sqrt{m^2-m+1}}, \tag{2.123}$$

$$A_{2QO} = -\frac{3c_1}{2c_2}\frac{m}{\sqrt{m^2-m+1}}, \tag{2.124}$$

$$\omega_{QO} = \frac{\sqrt{c_1}}{2}\frac{1}{(m^2-m+1)^{1/4}}. \tag{2.125}$$

The ZIV initial conditions imply $A_{0QO} = A$, and the elliptic parameter is calculated from the implicit equation

$$A = \frac{c_1}{2c_2}\frac{m_{QO,\,ZIV}+1-\sqrt{m_{QO,\,ZIV}^2-m_{QO,\,ZIV}+1}}{\sqrt{m_{QO,\,ZIV}^2-m_{QO,\,ZIV}+1}}. \tag{2.126}$$

Given the fact that the period of the sn^2 is twice as small as the of the sn function, one can conclude that the period of (2.122) with these conditions is

$$T_{\mathrm{QO}} = \frac{2K\left(m_{\mathrm{QO,ZIV}}\right)}{\omega_{\mathrm{QO}}\left(m_{\mathrm{QO,ZIV}}\right)}. \tag{2.127}$$

Note also that the exact solution for motion can be taken in the form given by Eq. (2.122), but with the square of the cn or dn functions instead of the square of the sn function [10], i.e. a general form of the solution is

$$x = \bar{A}_{0\mathrm{ep}} + \bar{A}_{2\mathrm{ep}}\,\mathrm{ep}^2\left(\bar{\omega}_{\mathrm{ep}}t\,|m\right), \tag{2.128}$$

where $\mathrm{ep}\left(\bar{\omega}_{\mathrm{ep}}t\,|m\right)$ stands for all three of them. The case $\mathrm{ep}\left(\bar{\omega}_{\mathrm{ep}}t\,|m\right) \equiv \mathrm{cn}(\omega t\,|m)$ can easily be derived based on the identity (2.44) to get

$$\bar{A}_{0\mathrm{cn}} = A_0 + A_2, \tag{2.129}$$

$$\bar{A}_{2\mathrm{cn}} = -A_2. \tag{2.130}$$

2.5 Purely Nonlinear Oscillators (PNOs)

Purely nonlinear oscillators are characterized by the restoring (elastic) term G_r that is a power function of the displacement x:

$$G_r \propto \mathrm{sgn}\,(x)\,|x|^{\alpha}, \tag{2.131}$$

where α is any positive real number. The sign and absolute value functions are used to assure that G_r is an odd function of the displacement for all the values of α defined.

This section provides an overview on how one can use a variety of special functions to find and express characteristics and a general exact solution for the motion of free conservative oscillators whose restoring terms are purely nonlinear, governed by

$$\ddot{x} + c_{\alpha}\mathrm{sgn}\,(x)\,|x|^{\alpha} = 0, \tag{2.132}$$

or non-dimensionalized with respect to time to

$$\ddot{x} + \mathrm{sgn}\,(x)\,|x|^{\alpha} = 0. \tag{2.133}$$

The oscillators modelled by Eq. (2.133) are called truly nonlinear as the restoring force has no linear approximation in any neighbourhood of $x = 0$ [11]. As stated in Chapter 1, instead of $\mathrm{sgn}(x)\,|x|^{\alpha}$ one can alternatively use the expression $x\,|x|^{\alpha-1}$.

2.5.1 On the Period of Oscillations

The energy conservation law corresponding to the oscillators governed by Eq. (2.133) reads as

$$\frac{1}{2}\dot{x}^2 + \frac{|x|^{\alpha+1}}{\alpha+1} = E_0, \tag{2.134}$$

where E_0 is the initial total mechanical energy, as before.

If the amplitude of oscillations is labelled by A, the exact period of oscillations can be calculated as follows:

$$T = 4\int_0^A \frac{\mathrm{d}x}{|\dot{x}|} = 4\int_0^A \frac{\mathrm{d}x}{\sqrt{2E_0 - \frac{2|x|^{\alpha+1}}{\alpha+1}}}, \tag{2.135}$$

or as

$$T = 4\sqrt{\frac{\alpha+1}{2}}\left[E_0(\alpha+1)\right]^{(1-\alpha)/(2\alpha+2)} \int_0^{A[E_0(\alpha+1)]^{-1/(\alpha+1)}} \frac{\mathrm{d}X}{\sqrt{1-|X|^{\alpha+1}}}. \tag{2.136}$$

It should be noted that the integral in Eq. (2.135) has been recognized by many researchers as crucial for defining the period or frequency of truly nonlinear oscillators (Lyapunov [12], Gelb and Vander Velde [13], Nayfeh and Mook [14], to name just a few). Herein, the solution of the integral is expressed in terms of a hypergeometric function ${}_2F_1$ (see Appendix A) as follows:

$$T = 4\sqrt{\frac{\alpha+1}{2}}\frac{A}{\sqrt{E_0(\alpha+1)}}\ {}_2F_1\left[1, \frac{1}{\alpha+1}, 1+\frac{1}{\alpha+1}, \frac{A^{\alpha+1}}{E_0(\alpha+1)}\right]. \tag{2.137}$$

The maximal displacement, i.e. the amplitude A can be calculated by going back to the energy conservation law given by Eq. (2.134), knowing that the velocity corresponding to this position is zero, which gives

$$A = \left[E_0(\alpha+1)\right]^{1/(\alpha+1)}. \tag{2.138}$$

Equation (2.137) now becomes

$$T = 4\sqrt{\frac{\alpha+1}{2}}A^{(1-\alpha)/2}\ {}_2F_1\left[1, \frac{1}{\alpha+1}, 1+\frac{1}{\alpha+1}, 1\right]. \tag{2.139}$$

By expressing the hypergeometric function in terms of the Euler gamma function Γ (see Appendix A)

$${}_2F_1\left[p, q, r, 1\right] = \frac{\Gamma(r)\Gamma(r-p-q)}{\Gamma(r-p)\Gamma(r-q)}, \tag{2.140}$$

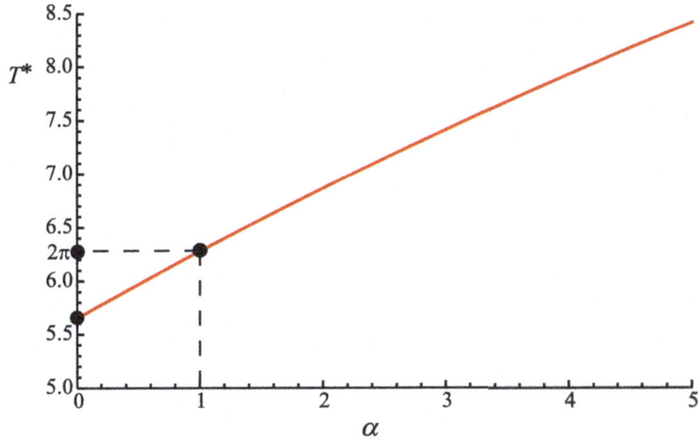

Fig. 2.16 Change of the coefficient T^* from Eq. (2.143) with the power of nonlinearity α for $c_\alpha = 1$

and performing some transformations, one obtains

$$T = 4\sqrt{\frac{\alpha+1}{2}}\frac{\sqrt{\pi}\Gamma\left(1+\frac{1}{\alpha+1}\right)}{\Gamma\left(\frac{1}{2}+\frac{1}{\alpha+1}\right)}A^{(1-\alpha)/2}, \tag{2.141}$$

which due to $\Gamma\left(1+\frac{1}{\alpha+1}\right) = \frac{1}{\alpha+1}\Gamma\left(\frac{1}{\alpha+1}\right)$ (see Appendix A) can also be written as follows:

$$T = \sqrt{\frac{8\pi}{\alpha+1}}\frac{\Gamma\left(\frac{1}{\alpha+1}\right)}{\Gamma\left(\frac{\alpha+3}{2(\alpha+1)}\right)}A^{(1-\alpha)/2} = T^*\, A^{(1-\alpha)/2}, \tag{2.142}$$

where

$$T^* = \sqrt{\frac{8\pi}{\alpha+1}}\frac{\Gamma\left(\frac{1}{\alpha+1}\right)}{\Gamma\left(\frac{\alpha+3}{2(\alpha+1)}\right)} = 4\sqrt{\frac{\alpha+1}{2}}\frac{\sqrt{\pi}\Gamma\left(1+\frac{1}{\alpha+1}\right)}{\Gamma\left(\frac{1}{2}+\frac{1}{\alpha+1}\right)}. \tag{2.143}$$

Note that when the initial velocity is zero $\dot{x}(0)=0$, the amplitude A corresponds to the initial displacement $x(0)=A$. Then, one can always non-dimensionalize the generalized coordinate with respect to A and end up with the initial displacement and the amplitude being equal to unity (see, for example, [11]). Then $T \equiv T^*$, and the period depends on the power α only, which is plotted in Fig. 2.16: the higher the power of nonlinearity, the longer the period. Although being restricted to positive α, the previous derivation also holds for $\alpha = 0$, giving the period is $4\sqrt{2} \approx 5.65685$ (see Appendix A for the values of the gamma function). For the case $\alpha = 1$, the period is equal to 2π (both of these values are labelled in Fig. 2.16).

Table 2.1 The expressions for the approximate period $T_{\text{app}}^{\text{HBF}}$ for different α

Power $\alpha = \alpha(p, q)$; $p, q \in Z^{+}$	$T_{\text{app}}^{\text{HBF}}$
Inverse odd integers $\alpha = \frac{1}{2p+1}$	$\pi 2^{\frac{p+1}{2p+1}} \left[\frac{(p+1)!n!}{(2p+1)!}\right]^{-\frac{1}{2(2p+1)}} A^{\frac{p}{2p+1}}$
Inverse even integers $\alpha = \frac{1}{2p}$	$\pi^{\frac{4p-1}{4p}} 2^{\frac{2p-1}{4p}} \left[\frac{(p!)^2}{(2p)!}\right]^{-\frac{1}{4p}} A^{\frac{1-2p}{4p}}$
Ratios of odd integers $\alpha = \frac{2q+1}{2p+1}$	$\pi 2^{\frac{p+q+1}{2p+1}} \left[\frac{(2q+1)!(p+1)!p!}{(2p\pm1)!(q+1)!q!}\right]^{-\frac{1}{2(2p+1)}} A^{\frac{q-p}{2p+1}}$
Ratios of odd and even integers $\alpha = \frac{2q+1}{2p}$	$\pi^{\frac{4n-1}{4n}} 2^{\frac{2n-1}{4n}} \left[\frac{(2q)!!(n!)^2}{(2q+1)!!(2n)!}\right]^{-\frac{1}{4n}} A^{\frac{1-2n+2q}{4n}}$
Ratios of even and odd integers $\alpha = \frac{2q}{2p+1}$	$\pi^{\frac{4n+1}{2(2n+1)}} 2^{\frac{2n}{2(2n+1)}} \left[\frac{(2q)!!(n+1)!n!}{(2q+1)!!(2n+1)!}\right]^{-\frac{1}{2(2n+1)}} A^{\frac{1-2n+2q}{2(2n+1)}}$

Cveticanin [15] expressed the period in Eq. (2.135) by means of the complete beta function. Knowing that the complete beta function is defined as (see Appendix A)

$$B(u, v) \equiv B(v, u) = \int_{0}^{1} (1 - |z|)^{v-1} z^{u-1} \mathrm{d}z, \tag{2.144}$$

one can write down that

$$T = \sqrt{\frac{8}{\alpha+1}} B\left(\frac{1}{\alpha+1}, \frac{1}{2}\right) A^{(1-\alpha)/2}. \tag{2.145}$$

This expression can further be transformed to Eq. (2.142) by using

$$B(u, v) = \frac{\Gamma(u)\Gamma(v)}{\Gamma(u+v)}, \tag{2.146}$$

and knowing that $\Gamma(1/2) = \sqrt{\pi}$ (see Appendix A).

Gottlieb [16] applied the first-order Harmonic Balance Method (HBF) to calculate the approximate expressions for the frequency $\omega_{\text{app}}^{\text{HBF}}$ of the response for different values of the positive power $\alpha < 1$, writing them down in terms of the factorial and double factorial function. By using $T_{\text{app}}^{\text{HBF}} = 2\pi/\omega_{\text{app}}^{\text{HBF}}$, his results are summarized in Table 2.1. As shown in [16], these approximate results are of a reasonably good accuracy for the powers relatively close to unity, but their accuracy worsens as the power becomes very small and approaches zero.

In order to solve the integral of a rational type analogous to the one in Eq. (2.136), Cveticanin and Pogany [17] expanded the integrand into a binomial series

$$\int \frac{\mathrm{d}X}{\sqrt{1-|X|^{\alpha+1}}} = \int \sum_{n=0}^{\infty} (-1)^n \binom{-\frac{1}{2}}{n} X^{(\alpha+1)n} dX. \tag{2.147}$$

Using the Pochhammer symbol $(b)_n = b\,(b+1)\cdots(b+n-1)\,,\, n \in \mathbb{N}$, one has

$$\binom{-\frac{1}{2}}{n} = (-1)^n \frac{\left(\frac{1}{2}\right)_n}{n!}, \tag{2.148}$$

$$\frac{1}{(\alpha+1)\,n+1} = \frac{\frac{1}{\alpha+1}}{n+\frac{1}{\alpha+1}} = \frac{\left(\frac{1}{\alpha+1}\right)_n}{\left(\frac{\alpha+2}{\alpha+1}\right)_n}. \tag{2.149}$$

Thus, Eq. (2.147) becomes

$$\int \frac{\mathrm{d}X}{\sqrt{1-|X|^{\alpha+1}}} = |X| \sum_{n=0}^{\infty} \frac{\left(\frac{1}{2}\right)_n \left(\frac{1}{\alpha+1}\right)_n}{n!\left(\frac{\alpha+2}{\alpha+1}\right)_n} X^{(\alpha+1)n} = |X|\ {}_2F_1\left[\frac{1}{2}, \frac{1}{\alpha+1}, \frac{\alpha+2}{\alpha+1}, X^{\alpha+1}\right]. \tag{2.150}$$

By putting this last expression into the integral in Eq. (2.136) and using the transformation given by Eq. (2.140), one can derive the expression (2.142). To sum up, either by using the direct integration of Eq. (2.136) via hypergeometric, beta and gamma functions, or expanding it into a binomial series using the Pochhammer symbol, the same expression for the period in terms of the power of nonlinearity and the amplitude (2.142) can be calculated.

2.5.2 *On the Motion of Conservative Oscillators*

In this section, several approaches to find exact and approximate solutions for PNOs are presented. What is common for all of them is that they give a closed-form expression in terms of special functions.

2.5.2.1 Exact Solution via Lyapunov's Function

Lyapunov [12] seems to have been the first to provide a general solution for motion of a conservative purely nonlinear oscillator with a single-term power-form nonlinearity. He considered the restoring term (2.131) with the powers $\alpha = 2n-1$, where $n \in \mathbb{N}$, $n \geq 2$. He did not use the sign and absolute function as in Eq. (2.133); to make the restoring term odd, he focused on odd powers of the restoring term. His starting point was also the energy conservation law as above, but instead of the initial mechanical energy, he expressed the right-hand side in Eq. (2.134) in terms of the constant c, which is

$$c = [2n E_0]^{1/(2n)}\,. \tag{2.151}$$

Given Eq. (2.138), it follows that $c \equiv A$.

The solution for motion and velocity is then defined by the two Cs ad Sn Lyapunov functions

$$x = c\,\mathrm{Cs}\vartheta, \tag{2.152}$$
$$\dot{x} = -c^{n}\,\mathrm{Sn}\vartheta, \tag{2.153}$$

where

$$\vartheta = c^{n-1}t + \gamma, \tag{2.154}$$

with γ being a constant.

Equations (2.134) and (2.151) imply that these two functions are related mutually by

$$(\mathrm{Cs}\vartheta)^{2n} + n\ (\mathrm{Sn}\vartheta)^{2} = 1, \tag{2.155}$$

and by

$$\frac{d\,\mathrm{Cs}\vartheta}{d\vartheta} = -\mathrm{Sn}\vartheta, \quad \frac{d\mathrm{Sn}\vartheta}{d\vartheta} = (\mathrm{Cs}\vartheta)^{2n-1}. \tag{2.156}$$

In accordance with Eq. (2.155), these Lyapunov functions are defined as having the following values for the zero argument:

$$\mathrm{Cs}\,(0) = 1, \quad \mathrm{Sn}\,(0) = 0. \tag{2.157}$$

The function $\mathrm{Cs}\vartheta$ is even, while the function $\mathrm{Sn}\vartheta$ is odd. Lyapunov stated that the expression for the half-period

$$\frac{T}{2} = 2\sqrt{n}\int_0^1 \frac{\mathrm{d}x}{\sqrt{1-x^{2n}}} = 2\int_0^{1/\sqrt{n}} \left(1-nx^2\right)^{1-2n/(2n)}\mathrm{d}x \tag{2.158}$$

can be expressed by means of the Euler gamma function Γ as

$$\frac{T}{2} = \sqrt{\frac{\pi}{n}}\frac{\Gamma\left(\frac{1}{2n}\right)}{\Gamma\left(\frac{n+1}{2n}\right)}. \tag{2.159}$$

It is easy to show that for $\alpha = 2n-1$, i.e $n = (\alpha+1)/2$, Eq. (2.159) gives the result coinciding with the period T^* described in Eq. (2.143).

In order to give the expressions by means of which the two new functions will be defined completely for all real values of the argument ϑ, a new variable φ is introduced

$$\vartheta = \sqrt{n}\int_0^{\varphi} \frac{\mathrm{d}\varphi}{\sqrt{1+\cos^2\varphi+\cdots+\cos^{2n-2}\varphi}}, \tag{2.160}$$

so that

$$\mathrm{Cs}\vartheta = \cos\varphi, \quad \mathrm{Sn}\vartheta = \frac{\sin\varphi}{\sqrt{n}}\sqrt{1+\cos^2\varphi+\cdots+\cos^{2n-2}\varphi}. \tag{2.161}$$

Thus, when ϑ increases for $T/2$, one has that φ increases for π, which gives

$$\mathrm{Cs}\left(\vartheta+\frac{T}{2}\right) = -\mathrm{Cs}\vartheta, \quad \mathrm{Sn}\left(\vartheta+\frac{T}{2}\right) = -\mathrm{Sn}\vartheta, \tag{2.162}$$

as well as

$$\mathrm{Cs}\,(\tau-\vartheta) = -\mathrm{Cs}\vartheta, \quad \mathrm{Sn}\,(\tau-\vartheta) = \mathrm{Sn}\vartheta. \tag{2.163}$$

If ϑ increases from zero to $T/4$, then $\mathrm{Cs}\vartheta$ decreases and $\mathrm{Sn}\vartheta$ increases, having the following values at the fourth of the period:

$$\mathrm{Cs}\frac{T}{4} = 0, \quad \mathrm{Sn}\frac{T}{4} = \frac{1}{\sqrt{n}}. \tag{2.164}$$

It is seen that the Lyapunov functions are analogous to the cosine and sine functions and that they reduce to them for $n = 1$. Moreover, when $n = 2$, Lyapunov pointed out that his functions transform to the elliptic functions (see Sect. 2.3.1) in the following way:

$$\mathrm{Cs}\vartheta = \mathrm{cn}\left(\vartheta\left|\frac{1}{2}\right.\right), \tag{2.165}$$

$$\mathrm{Sn}\vartheta = \mathrm{sn}\left(\vartheta\left|\frac{1}{2}\right.\right)\mathrm{dn}\left(\vartheta\left|\frac{1}{2}\right.\right). \tag{2.166}$$

2.5.2.2 Exact Solution via Ateb Functions

To solve Eq. (2.132) with $\alpha > 0$, Rosenberg [18, 19] first introduced the following variables and parameters:

$$\tau = \sqrt{\frac{c_\alpha}{n}}A^{n-1}t, \quad \xi = \frac{x}{A}, \quad n = \frac{\alpha+1}{2}, \tag{2.167}$$

getting the governing equation

$$\xi'' + n\xi\,|\xi|^{2(n-1)} = 0, \tag{2.168}$$

and then considering the first integral for the ZIV conditions

$$\xi' = \pm\sqrt{1 - |\xi|^{2n}}, \tag{2.169}$$

where the primes in Eqs. (2.168) and (2.169) denote differentiation with respect to τ. Note also that the original time is rescaled in a such a way that new time τ is amplitude-dependent (non-isochronous). Note also that the ZIV initial conditions now became $\xi(0) = 1, \xi'(0) = 0$ (this will be referred to as Case 1 and the corresponding subscript will be used to indicate it).

Unlike the Lyapunov case, which was solved for the power corresponding to odd integer numbers, this Rosenberg's consideration holds for all real powers higher than 1/2.

Equation (2.169) yields

$$\tau_1 = -\int_1^{0 \le \xi_1 \le 1} \frac{d\xi}{\sqrt{1 - |\xi|^{2n}}}, \tag{2.170}$$

which can be written as

$$\tau_1 = \int_0^1 \frac{d\xi}{\sqrt{1 - |\xi|^{2n}}} - \int_0^{0 \le \xi_1 \le 1} \frac{d\xi}{\sqrt{1 - |\xi|^{2n}}}. \tag{2.171}$$

The first integral in Eq. (2.171) can be related to the complete beta function $B\left(\frac{1}{2n}, \frac{1}{2}\right)$ (see Appendix A):

$$\tau_1 = \int_0^1 \frac{d\xi}{\sqrt{1 - |\xi|^{2n}}} = \frac{1}{2n} \int_0^1 s^{(1-2n)/2n} (1 - s)^{-1/2} ds = \frac{1}{2n} B\left(\frac{1}{2n}, \frac{1}{2}\right). \tag{2.172}$$

The second integral can be expressed in terms of the incomplete beta function, which is a generalization of the beta function with the same integrand and the same lower limit, but with an unspecified upper limit:

$$\int_0^{0 \le \xi_1 \le 1} \frac{d\xi}{\sqrt{1 - |\xi|^{2n}}} = \frac{1}{2n} B_{\xi^{1/2n}}\left(\frac{1}{2n}, \frac{1}{2}\right), \tag{2.173}$$

where the subscript indicates the upper limit.

Therefore, Eq. (2.171) becomes

$$\tau_1 = \frac{1}{2n}\left[B\left(\frac{1}{2n}, \frac{1}{2}\right) - B_{\xi^{1/2n}}\left(\frac{1}{2n}, \frac{1}{2}\right)\right]. \tag{2.174}$$

Rosenberg also recognized the expressions for the period T_ξ of ξ and the period T of x, respectively, as

$$T_\xi = \frac{2}{n} B\left(\frac{1}{2n}, \frac{1}{2}\right), \tag{2.175}$$

$$\sqrt{\frac{2c_\alpha}{\alpha+1}} A^{(\alpha-1)/2} T = \frac{4}{\alpha+1} B\left(\frac{1}{2n}, \frac{1}{2}\right), \tag{2.176}$$

where Eq. (2.175) gives the period T coinciding with the one in Eq. (2.142).

The transformation

$$\xi_1 = \cos^{1/n}(\varphi_1), \tag{2.177}$$

can be introduced into Eq. (2.170) to get

$$n\tau_1 = \int\limits_0^{0\le\varphi_1\le\pi/2} \cos^{(1-n)/n}(\varphi)\,\mathrm{d}\varphi. \tag{2.178}$$

The upper limit φ_1 of the integral in Eq. (2.178) is the amplitude of $n\tau$

$$\varphi_1 = \operatorname{amp}(n\tau_1), \tag{2.179}$$

so that Eq. (2.177) becomes the Ateb cam function

$$\xi_1 = \cos^{1/n}(\operatorname{amp}(n\tau_1)) \equiv \operatorname{cam}(n\tau_1). \tag{2.180}$$

The counterpart for the sine function was developed while considering the case $x(0) = 0$, i.e. $\xi(0) = 0$ (this is Case 2 herein and the corresponding subscript is used to indicate it). Now, Eq. (2.169) gives

$$\tau_2 = \int\limits_0^{0\le\xi\le1} \frac{\mathrm{d}\xi}{\sqrt{1-|\xi|^{2n}}}. \tag{2.181}$$

By using the substitution

$$\xi_2 = \sin^{1/n}(\varphi_2), \tag{2.182}$$

the integral in Eq. (2.181) becomes

$$n\tau_2 = \int\limits_0^{0\le\varphi_2\le\pi/2} \sin^{(1-n)/n}(\varphi)\,\mathrm{d}\varphi. \tag{2.183}$$

As previously, the upper limit φ_2 of the integral in Eq. (2.183) corresponds to the amplitude of $n\tau$

$$\varphi_2 = \text{amp}\,(n\tau_2)\,, \tag{2.184}$$

so that Eq. (2.182) can be written as the Ateb sam function

$$\xi_2 = \sin^{1/n}(\text{amp}\,(n\tau_2)) \equiv \text{sam}(n\tau_2). \tag{2.185}$$

Thus, one can conclude that the Ateb can and sam functions are the inversion of the incomplete beta function, and this is where their name comes from. Let us show some of their characteristic values and properties.

The following values are easily calculated: when amp(0) $= 0$, then

$$\text{cam}(0) = 1. \tag{2.186}$$

$$\text{sam}(0) = 0, \tag{2.187}$$

$$\text{cam}\left(\frac{\pi}{2}\right) = 0, \tag{2.188}$$

$$\text{sam}\left(\frac{\pi}{2}\right) = 1. \tag{2.189}$$

The amplitudes $\varphi_{1,2}$ and the sam function are odd, while the cam function is even.

To derive the identities that relate the cam and the sam functions mutually, one should notice that this can be done when $\tau_1 = \tau_2$, i.e. when

$$\text{amp}\,(n\tau_2) = \text{amp}\,(n\tau_2) = \text{amp}\left(n\tau^*\right). \tag{2.190}$$

One can easily show that

$$\text{cam}^{2n}(n\tau^*) + \text{sam}^{2n}(n\tau^*) = 1, \tag{2.191}$$

while differentiation results in

$$\frac{d}{d\tau^*}\left[\text{cam}(n\tau^*)\right] = -\text{sam}^n(n\tau^*), \tag{2.192}$$

$$\frac{d}{d\tau^*}\left[\text{sam}(n\tau^*)\right] = \text{cam}^n(n\tau^*). \tag{2.193}$$

It is easy to see that for $n = 1$, the cam function turns into the cosine function and the sam function turns into the sine function. In addition, when $n = 2$, one can, for example, use Eq. (2.181) with the substitution $\xi = \cos\theta$ to derive

$$u = -\frac{1}{\sqrt{2}}\int \frac{d\theta}{\sqrt{1 - \frac{1}{2}\sin^2(\theta)}}. \tag{2.194}$$

Thus, the inversion is the Jacobi elliptic function with the elliptic parameters $m = 1/2$ (see Appendices A and B), which is the case presented in Sect. 2.3.3 for the PCO.

Senik's Contribution

Senik [20] extended Rosenberg's work [18], noting that Eq. (2.133) belongs to a wide class of equations which can be solved by utilizing Rosenber's Ateb functions. This class of equations is described by

$$\dot{x} + \beta y^n = 0, \tag{2.195}$$

$$\dot{y} - \gamma x^m = 0, \tag{2.196}$$

where $\beta, \gamma > 0$, while m, n are odd natural numbers or ratios of odd natural numbers. So, this case may be related to the case (2.168) studied by Rosenberg for particular values of β, γ, m and n. Keeping the same names for the three Ateb functions (amp, cam and sam), Senik [20] defined them as being the functions of amplitude that depends on both power n and m.

In his subsequent work [21], Senik demonstrated that the system,

$$\frac{\mathrm{d}u}{\mathrm{d}t} + \frac{2}{m+1} v^n = 0, \tag{2.197}$$

$$\frac{\mathrm{d}v}{\mathrm{d}t} - \frac{2}{n+1} u^m = 0, \tag{2.198}$$

has the exact solution in the form of the two basic Ateb functions, which he defined as three-parameter ca and sa functions:

$$u = \mathrm{ca}\,(m, n, t)\,, \tag{2.199}$$

$$v = \mathrm{sa}\,(n, m, t)\,, \tag{2.200}$$

where t is defined in terms of the incomplete beta function (see Appendix A):

$$t = \frac{1}{2} B_x \left(\frac{1}{n+1}, \frac{1}{m+1} \right) = \frac{1}{2} \int_0^{0 \le x \le 1} z^{\frac{-n}{n+1}} (1-z)^{\frac{-m}{m+1}} \mathrm{d}z. \tag{2.201}$$

Thus, the ca and sa functions are inversions of the half of the incomplete beta function. They both have the period $2B\left(\frac{1}{n+1}, \frac{1}{m+1}\right)$, where the ca function is even and the sa function is odd, i.e.

$$\mathrm{ca}\,(m, n, t) = \mathrm{ca}\,(m, n, -t)\,, \tag{2.202}$$

$$\mathrm{sa}\,(n, m, t) = -\mathrm{sa}\,(n, m, -t)\,. \tag{2.203}$$

Senik [21] derived that

$$\mathrm{ca}^{m+1}(m,n,t)+\mathrm{sa}^{n+1}(n,m,t)=1, \tag{2.204}$$

$$\frac{\mathrm{d}}{\mathrm{d}t}\mathrm{ca}(m,n,t)=-\frac{2}{m+1}\mathrm{sa}^{n}(n,m,t), \tag{2.205}$$

$$\frac{\mathrm{d}}{\mathrm{d}t}\mathrm{sa}(n,m,t)=\frac{2}{n+1}\mathrm{ca}^{m}(m,n,t). \tag{2.206}$$

In order to achieve the same power of nonlinearity as in Eq. (2.133), one can select $n=1$, $m\equiv\alpha$, so that Eqs. (2.197), (2.198) become

$$\frac{\mathrm{d}u}{\mathrm{d}t}+\frac{2}{\alpha+1}v=0, \tag{2.207}$$

$$\frac{\mathrm{d}v}{\mathrm{d}t}-u^{\alpha}=0, \tag{2.208}$$

yielding

$$\frac{\mathrm{d}^2u}{\mathrm{d}t^2}+\frac{2}{\alpha+1}u^{\alpha}=0. \tag{2.209}$$

Senik's solutions are

$$u=\mathrm{ca}(\alpha,1,t), \tag{2.210}$$

$$v=\mathrm{sa}(1,\alpha,t), \tag{2.211}$$

and their period is

$$T_{\mathrm{ca,\,sa}}=2B\left(\frac{1}{2},\frac{1}{\alpha+1}\right)=2B\left(\frac{1}{\alpha+1},\frac{1}{2}\right). \tag{2.212}$$

The following identity holds

$$\mathrm{ca}^{\alpha+1}(\alpha,1,t)+\mathrm{sa}^{2}(1,\alpha,t)=1.$$

The first derivatives of these two functions are

$$\frac{\mathrm{d}}{\mathrm{d}t}\mathrm{ca}(\alpha,1,t)=-\frac{2}{\alpha+1}\mathrm{sa}(1,\alpha,t), \tag{2.213}$$

$$\frac{\mathrm{d}}{\mathrm{d}t}\mathrm{sa}(1,\alpha,t)=\mathrm{ca}^{\alpha}(\alpha,1,t). \tag{2.214}$$

2.5.2.3 Summary for the Use of Ateb Functions

Let us now utilize the previously presented results in terms of Ateb function for PNOs governed by Eq. (2.133) and for the initial conditions corresponding to the ZIV case defined by Eq. (2.3). This utilization will be based on Rosenberg's and Senik's contribution, which relates to the governing equation whose coefficient of the nonlinear term is dependent of the power of nonlinearity, while the one in Eq. (2.133) is labelled by c_α. Although the equation of motion can be non-dimensionalized to exclude it (or just set $c_\alpha = 1$, it will be done later on in numerical calculations), this latter coefficient is left in the equation for the further use and Senik's approach is utilized for this purpose as well. Thus, it should be noted that their equations are analogous to Eq. (2.133), but one may consider that the coefficient c_α depends on the power of nonlinearity or rescale the time in it. Besides this, as pointed out by Mickens [22], Senik's Eq. (2.209) will be odd parity to yield oscillatory response if $\alpha = (2n+1)/(2m+1), n, m = 0, 1, 2, 3, \ldots$ or if the term u^α is replaced by $|u|^\alpha \mathrm{sgn}(u)$, so certain adjustments are needed. They are presented subsequently and will be consistently used throughout this book.

The period of vibrations T of PNOs can be obtained in an exact form, and is given by (see Eqs. (2.142) and (2.212)):

$$T = \sqrt{\frac{8}{c_\alpha (\alpha+1)}} B\left(\frac{1}{\alpha+1}, \frac{1}{2}\right) |A|^{(1-\alpha)/2} = \sqrt{\frac{8\pi}{c_\alpha (\alpha+1)}} \frac{\Gamma\left(\frac{1}{\alpha+1}\right)}{\Gamma\left(\frac{\alpha+3}{2(\alpha+1)}\right)} |A|^{(1-\alpha)/2}, \tag{2.215}$$

where B stands for the beta function and Γ for the Euler gamma function (see Appendix A).

The solution for motion can be expressed as closed form in terms of the three-argument Ateb ca function:

$$x = A\, \mathrm{ca}\left(\alpha, 1, \omega_{\mathrm{ca}} t\right), \tag{2.216}$$

$$\omega_{\mathrm{ca}} = |A|^{(\alpha-1)/2} \sqrt{\frac{c_\alpha (\alpha+1)}{2}}. \tag{2.217}$$

Note that in the expression for the ca solution (2.216), the first argument is the power of nonlinearity α, the second one is always unity to follow Senik's notation in this respect [21], while the frequency ω_{ca} existing in the third argument has been calculated from the fact that the Ateb function $\mathrm{ca}(\alpha, 1, z)$ has the following exact period (see Eq. (2.212)):

$$T_{\mathrm{ca}(\alpha,1,z)} = 2B\left(\frac{1}{\alpha+1}, \frac{1}{2}\right) = 2\frac{\sqrt{\pi}\,\Gamma\left(\frac{1}{\alpha+1}\right)}{\Gamma\left(\frac{\alpha+3}{2(\alpha+1)}\right)}. \tag{2.218}$$

Thus, by changing the variable $z = \omega_{ca}t$, one has $\omega_{ca} = T_{ca(\alpha,1,z)}/T$, which together with Eq. (2.215) yields Eq. (2.217).

In case of the linear oscillator one has $\alpha = 1$, and Eqs. (2.216), (2.217) simplify to the well-known form $x = A\ \mathrm{ca}\big(1, 1, \sqrt{c_1}t\big) = A\cos\big(\sqrt{c_1}t\big)$.

In case of the pure cubic nonlinearity $\alpha = 3$, the Ateb ca function turns into the Jacobi cn function, and the solution has the form presented in Sect. 2.3.3:

$$x_{\mathrm{PCO}} = A\ \mathrm{cn}\left(\omega_{\mathrm{PCO}}t \,\middle|\, \frac{1}{2}\right), \tag{2.219}$$

$$\omega_{\mathrm{PCO}} = \sqrt{c_3}\,|A|\,. \tag{2.220}$$

The second argument of the cn function in Eq. (2.219) is its elliptic parameter $m = 1/2$. The frequency of the cn function ω_{PCO} is given by Eq. (2.220). It can be verified as such by using the fact that the function $\mathrm{cn}\big(z\,\big|\,\frac{1}{2}\big)$ has the exact period $T_{\mathrm{cn}(z|\frac{1}{2})} = 4K\,(1/2)$, where K stands for the complete elliptic integral of the first kind, as well as by calculating from Eq. (2.215) $T = |A|^{-1}\,\Gamma^2\,(1/4)\,/\sqrt{\pi c_3} = |A|^{-1}\,4K\,(1/2)\,/\sqrt{c_3}$. Thus, one has $\omega_{\mathrm{PCO}} = T_{\mathrm{PCO}(z|\frac{1}{2})}/T = \sqrt{c_3}\,|A|$, which agrees with the solution obtained in Sect. 2.3.3.

Representations of Ateb Function

There have been different representations or approximations of the Ateb function constructed or suggested to enable one to calculate it more easily or present/visualize it more conveniently. The approximated form of the cam function presented in [23] is expressed as the natural logarithm of the hyperbolicus cosine. Pilipchuk [24] proposed the solution as a series in certain time-powered functions. Andrianov approximated it as a power-form function of the sine function whose argument is also a powered time function [25, 26].

However, to provide an insight into the harmonic character of the response, a Fourier series can be used as follows:

$$\mathrm{ca}\,(\alpha, 1, t) = \sum_{N=1}^{\infty} C_{2N-1}\cos\left[(2N-1)\,\frac{2\pi}{T}t\right], \tag{2.221}$$

where the Fourier coefficients are defined by

$$C_{2N-1}\,(\alpha) = \frac{8}{T}\int_0^{T/4} \mathrm{ca}\,(\alpha, 1, t)\cos\left[(2N-1)\,\frac{2\pi}{T}t\right], \tag{2.222}$$

with T being its period. These Fourier coefficients can further be simplified [27] and calculated numerically from the expression (see Appendix C for details):

$$C_{2N-1}\,(\alpha) = \frac{2\Gamma\left(\frac{\alpha+3}{2(\alpha+1)}\right)}{\sqrt{\pi}\Gamma\left(\frac{1}{\alpha+1}\right)}\int_0^1 \frac{(1-y)^{(1-\alpha)/(1+\alpha)}}{\sqrt{y}}\cos\left[\frac{(2N-1)\,\pi}{2}I\left(y, \frac{1}{2}, \frac{1}{\alpha+1}\right)\right]\mathrm{d}y, \tag{2.223}$$

where $I\left(y, \frac{1}{2}, \frac{1}{\alpha+1}\right)$ stands for the regularized incomplete beta function (see Appendix A). Given the fact that the Ateb ca function is odd, the Fourier series encompasses odd harmonics (first, third, fifth, etc). First four Fourier coefficients are calculated numerically using Eq. (2.223) and plotted in Fig. 2.17 as a function of the power α. It is seen that their sign and values change mainly with respect to $\alpha = 1$: C_1 is higher than unity for $\alpha < 1$, and then smaller than unity for $\alpha > 1$; C_3 and C_7 are negative for $\alpha < 1$, and then positive for $\alpha > 1$; C_5 is negative for $1 < \alpha < 2.34$, and positive otherwise.

To provide further explanations about the solution for motion described by the Ateb ca function, this solution is written for several powers of nonlinearity by using Eqs. (2.216), (2.217) and developed into the Fourier series defined by Eqs. (2.221) and (2.223). The solutions corresponding to $\alpha = 1/2; 2/3; 2, 3$ (two under-linear cases and two over-linear cases) are

$$x\,(\alpha = 1/2) = \text{ca}\left(\frac{1}{2}, 1, \frac{\sqrt{3}}{2}t\right) \tag{2.224}$$

$$\approx 1.01511\text{cos}\,(1.05164t) - 0.01729\text{cos}\,(3.15491t) + 0.00279\text{cos}\,(5.25818t) - 0.00085\text{cos}\,(7.36146t)\,, \tag{2.225}$$

$$x\,(\alpha = 2/3) = \text{ca}\left(\frac{2}{3}, 1, \sqrt{\frac{5}{6}}t\right) \tag{2.226}$$

$$\approx 1.00987\text{cos}\,(1.03365t) - 0.01114\text{cos}\,(3.10096t) + 0.00161\text{cos}\,(5.16826t) - 0.00046\text{cos}\,(7.23556t)\,, \tag{2.227}$$

$$x\,(\alpha = 2) = \text{ca}\left(2, 1, \sqrt{\frac{3}{2}}t\right) \tag{2.228}$$

$$\approx 0.97480\text{cos}\,(0.91468t) + 0.02572\text{cos}\,(2.74404t) - 0.00064\text{cos}\,(4.57341t) + 0.00014\text{cos}\,([6.40277t)\,, \tag{2.229}$$

$$x\,(\alpha = 3) = \text{ca}\left(3, 1, \sqrt{2}t\right) = \text{cn}\left(t \,\middle|\, \frac{1}{2}\right) = \tag{2.230}$$

$$\approx 0.95501\text{cos}\,(0.84721t) + 0.04305\text{cos}\,(2.54164t) + 0.00186\text{cos}\,(4.23607t) + 0.00008\text{cos}\,(5.93049t)\,. \tag{2.231}$$

It is seen that these representations give a clear insight into the meaning of the Ateb ca function in terms of the harmonics that it encompasses and their share in the overall response. It also provides an idea how this function can be obtained practically—as a combination of certain harmonics with mutually dependent amplitudes and angular frequencies. It is easy to check that the frequency of the first harmonic Ω_1 satisfies the equality of the period of the cosine and the Ateb ca function: $2\pi/\Omega_1 =$

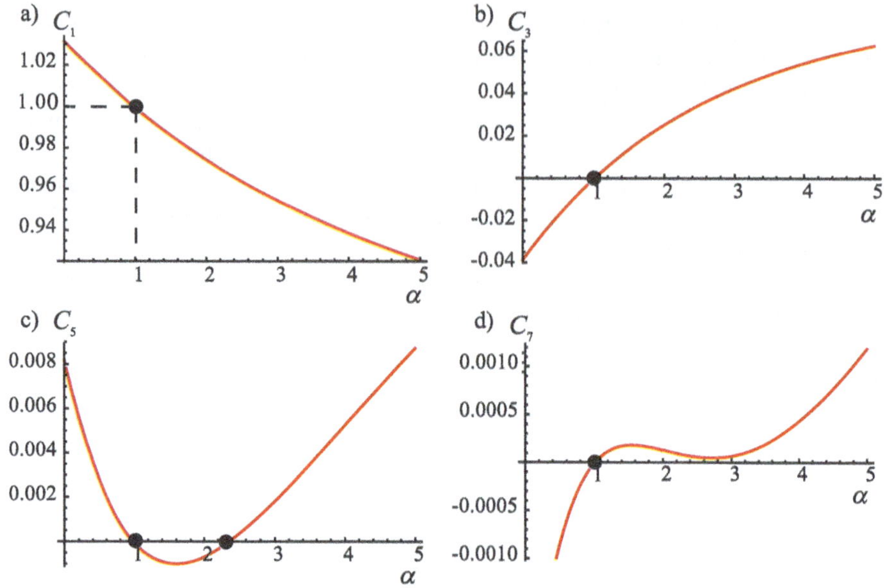

Fig. 2.17 Fourier coefficients given by Eq. (2.222) as a function of the power of nonlinearity α for $c_\alpha = 1$, $A = 1$: **a** C_1; **b** C_3; **c** C_5; **d** C_7

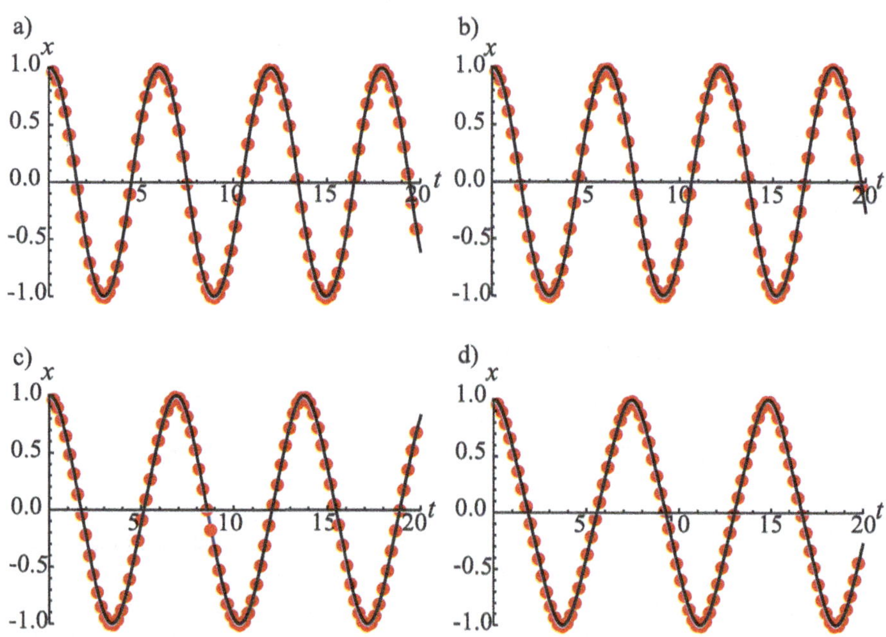

Fig. 2.18 Time response corresponding to $c_\alpha = 1$, $A = 1$ and: **a** $\alpha = 1/2$; **b** $\alpha = 2/3$; **c** $\alpha = 2$; **d** $\alpha = 3$. The numerical solution of Eq. (2.132)—red dots, Fourier series approximations, Eqs. (2.225)–(2.231)—black solid line

$2B\left(\frac{1}{\alpha+1}, \frac{1}{2}\right)/\omega_{ca}$, which also means that Ω_1 can be calculated in advance from the expression $\Omega_1 = \omega_{ca}\pi/B\left(\frac{1}{\alpha+1}, \frac{1}{2}\right)$.

To illustrate their shape and period, all these solutions are shown in Fig. 2.18 both for the case of the exact solution obtained numerically from the equation of motion and the Fourier series presented. At this level of truncation, the latter agree well with the former. The increase of the period with the increase of α is apparent, which is in accordance with Fig. 2.16.

2.5.2.4 Approximate Solutions via Jacobi Elliptic Functions

As demonstrated previously, both Lyapunov functions and Ateb functions give exact solutions for motion for a general model given by Eqs. (2.132) or (2.133), covering a wide class of oscillators as α can take different values that are specified in Sects. 2.5.2.1 and 2.5.2.2. However, they are not built in contemporary computer algebra software, such as Mathematica or Maxima, unlike Jacobi elliptic functions, for example. The question that naturally arises is if one can use Jacobi elliptic functions to express the solution for motion for a range of values of α. In addition, a great deal of publications (handbooks, textbooks, papers, websites) contain the theory related to Jacobi elliptic functions, while the theory related to Lyapunov functions and Ateb functions is neither easily attainable nor widely available. Being motivated by these facts, the approach presented below is to find the solution for motion for the oscillator modelled by Eq. (2.133) with α being any non-negative real number, as done in [28]. This solution is assumed in the form

$$x(t) = \mathrm{cn}\left(\omega(\alpha)\, t \,|m(\alpha)\right), \tag{2.232}$$

where the frequency ω of the elliptic function and the elliptic parameter m are unknown functions of the power α and need to be specified.

To find the frequency ω, the known fact that the period of the cn Jacobi elliptic function is $4K(m)$ is to be used (see Sect. 2.3.1.2). This frequency $\omega(\alpha)$ can then be related to this period by using the expression for the period of the response given by Eq. (2.142) as follows:

$$\omega(\alpha) = \frac{4K(m(\alpha))}{T} = \frac{4K(m(\alpha))}{T^*} A^{(\alpha-1)/2}. \tag{2.233}$$

By choosing the ratio in Eq. (2.233) to be equal to unity and using the expression T^* based on Eq. (2.141), the following is derived:

$$K(m(\alpha)) = \sqrt{\frac{\alpha+1}{2}} \frac{\sqrt{\pi}\Gamma\left(1+\frac{1}{1+\alpha}\right)}{\Gamma\left(\frac{1}{2}+\frac{1}{1+\alpha}\right)}. \tag{2.234}$$

Equation (2.234) represents the implicit equation for calculating the elliptic parameter m as a function of the power α, which is plotted in Fig. 2.19. For the

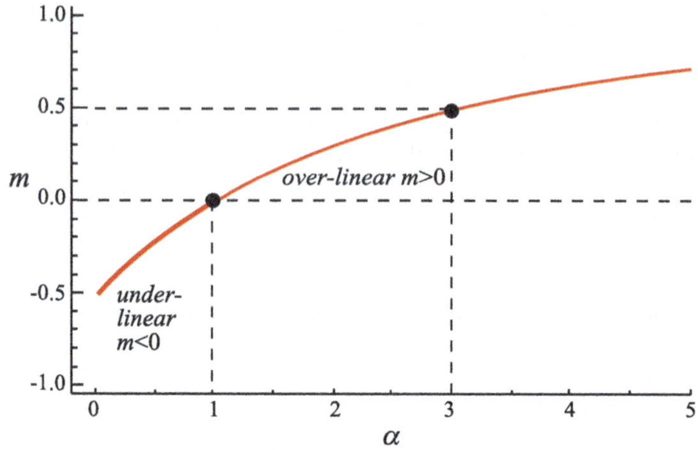

Fig. 2.19 The elliptic parameter m versus the power α, Eq. (2.234)

Table 2.2 Values of the elliptic parameter m calculated from Eq. (2.234) for various α

α	m, Eq. (2.234)	α	m, Eq. (2.234)
0	−0.506161	1.1	0.0376767
1/10	−0.442696	3/2	0.171103
2/9	−0.369589	2	0.305831
2/7	−0.333449	5/2	0.413276
1/3	−0.307134	3	0.5
2/5	−0.271391	7/2	0.570772
1/2	−0.220082	4	0.629107
2/3	−0.140326	9/2	0.677625
1	0	5	0.718309

convenience of the reader, the values of the elliptic parameter m are also calculated numerically for certain values of the power α from Eq. (2.234) and given in Table 2.2

It is seen from Fig. 2.19 and Table 2.2 that as the power α increases, the values of the elliptic parameter increase, too. When α is less than unity (under-linear case), the parameter m is negative. As the parameter m represents the square of the modulus k, i.e. $m = k^2$ (see Sect. 2.3.1.1), it follows that the elliptic modulus k is imaginary in the under-linear case. For α higher than unity (over-linear case), the parameter m is positive. In the case of a linear oscillator, i.e. when $\alpha = 1$, one has $m = 0$. For the pure cubic oscillator $\alpha = 3$, this approach yields $m = 0.5$, which is the well-known results (see Sect. 2.3.3).

Assuming that the initial velocity is zero and taking into account the fact that the oscillators modelled by Eq. (2.133) can always be non-dimensionalized with respect to the initial non-zero amplitude, it follows from Eq. (2.233) that A can always be made equal to unity. Therefore, $\omega_{\mathrm{cn}} = 1$. The solution for motion in Eq. (2.232) now becomes

$$x(t) = \mathrm{cn}\left(t \,|m\, (\alpha)\right), \tag{2.235}$$

where m follows from Eq. (2.234).

In order to check the accuracy of the approximate solutions obtained, Eqs. (2.234), (2.235) are used to plot the resulting time histories as a solid line in Fig. 2.20 for an under-linear oscillator ($\alpha = 1/10$) and in Fig. 2.21 for an over-linear oscillator ($\alpha = 4$). In addition, numerical solutions obtained by integrating Eq. (2.133) directly are also shown and depicted by dots connected by a dashed line. These comparisons illustrate excellent accuracy of the approximate solution obtained, which characterizes also the response after a longer period of time, implying that there is no accu-

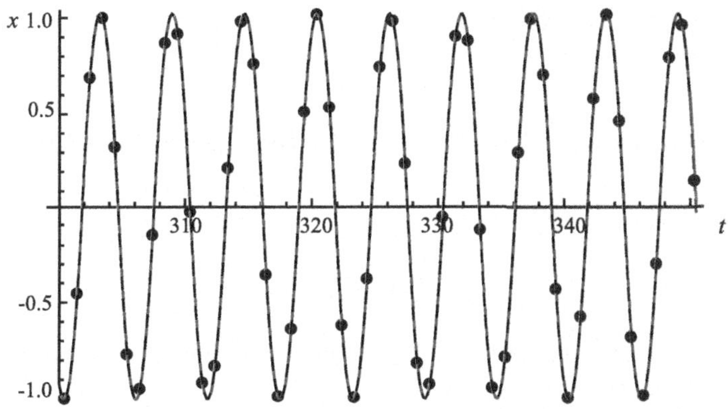

Fig. 2.20 Time history for an under-linear oscillator after a long period of time: approximate solutions, Eqs. (2.234), (2.235) (solid line) and numerical solutions of Eq. (2.133) (dots and a dashed line) for $x(0) = 1$, $\dot{x}(0) = 1$ and $\alpha = 1/10$

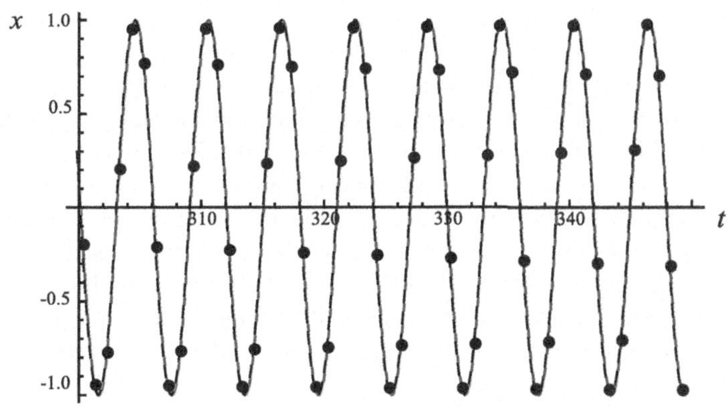

Fig. 2.21 Time history for an over-linear oscillator after a long period of time: approximate solutions, Eqs. (2.234), (2.235) (solid line) and numerical solutions of Eq. (2.133) (dots and a dashed line) for $x(0) = 1$, $\dot{x}(0) = 1$ and $\alpha = 4$

mulating error that would yield an increasing delay with respect to the numerical solution.

This approach, in which the approximation for conservative truly nonlinear oscillators (2.133) with the power of nonlinearity being any non-negative real number is expressed in terms of the Jacobi elliptic functions, is also utilized by Rakaric and Kovacic in [29]. Therein, the elliptic parameter is calculated by using Hamilton's variation principle and the stationary condition of the action integral. Besides free oscillators, forced truly nonlinear oscillators have also been treated by expressing their response in terms of the Jacobi elliptic function with the elliptic parameter being defined by Eq. (2.234) [30, 31]. In [30], externally forced truly nonlinear non-conservative oscillators are considered, while in [31] parametrically forced truly nonlinear oscillators are examined.

2.6 Oscillators with Constant Restoring Force (CRFO)

The oscillators governed by

$$\ddot{x} + \operatorname{sgn}(x) = 0 \tag{2.236}$$

are referred to as antisymmetric oscillators with a constant [11, 32] or piece-wise constant restoring force [33, 34].

Lipscomb and Mickens [32] emphasized that this is one of few nonlinear oscillators for which the exact solution can be expressed in terms of elementary functions: their solution is a piece-wise continuous time-dependent function composed of two parts defined separately over two half-period intervals. The corresponding Fourier series is presented in their work, too. The same solution was also derived by Govindan Potti et al. [35], who proposed the restoring term function $\operatorname{sgn}(x)$ in the form of Fourier series and integrated the equation of motion directly. Pilipchuk [33] obtained a closed-form analytical solution using the saw-tooth transformation of the time proposed in [36]. Pilipchuk wrote the following remark [36]: "It is known that the trigonometric series appear to be "bad working" around the discontinuities due to the Gibbs phenomenon... In terms of acceleration, the series performs an oscillating error near those points of time t at which the acceleration $\ddot{x}(t)$ has step-wise discontinuities switching its value from -1 to 1 or back ...". Referring to this, Awrejcewicz and Andrianov [37] remarked that this is true if one applies a simple summation of the Fourier series and pointed out that one can utilize the regularization properties of the Padé approximants as it possess the self-correction properties, as a result of which this drawback is omitted. Awrejcewicz and Andrianov [38] also considered a system with a restoring force close to a zero power and adjusted the so-called small δ-method to this problem, obtaining the solution for motion as a logarithmic and polynomial function of time. Beléndez et al. [39] utilized a generalized harmonic balance method to calculate the analytical approximate solutions having a rational form. Rakaric [40] applied a modified harmonic balance method to obtain higher order approximate solutions for conservative oscillators with a quasi-constant restoring force, i.e. oscillators with a low fractional-order restoring force, including antisymmetric oscillators. The assumed solution is taken in the form of the truncated Fourier expansion and represented as a sum of the so-called modified solution and a small unknown correction part solution, yielding simple and explicit expressions for the frequency and amplitude of harmonics. Kovacic [41] considered harmonically forced purely nonlinear oscillators and developed two perturbation approaches (the Lindstedt–Poincaré method and the method of multiple scales) for the cases of weak and strong nonlinearities, deriving the frequency–amplitude equation and analytical expressions for the steady-state response at the frequency of excitation, and extending it to various PNOs in [42]. Frequency response curves of the antisymmetric oscillators are presented and examined as well. The response has been also modelled by using elliptic functions [30, 43]. In [43], performance characteristics of the isolator in a base excited system are determined and discussed, while in [30] this type of oscillator has been considered as a class of harmonically excited purely nonlinear systems.

In addition to finding solutions for motion, some of the authors have focused on obtaining the period/frequency of free conservative antisymmetric oscillations. Considering the oscillators with a zero initial amplitude and a non-zero initial velocity v_0, Lipscomb and Mickens [32] noted that the period of oscillations is $T = 4v_0$. Gottlieb [16] considered the oscillator with a non-zero initial amplitude A and a zero initial velocity and calculated the exact frequency $\Omega = 1.110721/\sqrt{A}$, while his first-order harmonic balance approach yielded $\Omega = 1.128379/\sqrt{A}$. Van Horssen [44] analysed the period of a non-integer order oscillator whose power is a reciprocal of odd numbers $1/(2n+1)$, and found that the period corresponding to $n \to \infty$ is $T = 4\sqrt{2}\,\sqrt{A}$.

The oscillators governed by Eq. (2.236) has the following energy conservation law:

$$|x| + \frac{1}{2}\dot{x}^2 = E_0, \tag{2.237}$$

where E_0 is its initial total mechanical energy ($E_0 > 0$).

If the amplitude of oscillations is labelled by a, the exact period of oscillations is given by

$$T = 4\int_0^a \frac{\mathrm{d}x}{|\dot{x}|} = 4\int_0^a \frac{\mathrm{d}x}{\sqrt{2E_0 - 2x}}. \tag{2.238}$$

The maximal displacement, i.e. the amplitude a can be calculated by going back to the energy conservation law given by Eq. (2.237), knowing that the velocity corresponding to this position is zero, which gives $a = E_0$. The integral in Eq. (2.238) now becomes

$$T = 4\sqrt{2}\,\sqrt{a}. \tag{2.239}$$

For the initial conditions given by Eq. (2.2) and in accordance with the energy conservation law (2.237), the amplitude a can be calculated as follows:

$$a = A + \frac{v_0^2}{2}. \tag{2.240}$$

For the ZID oscillator, one holds $a_{\text{ZID}} \equiv (v_0)^2 /2$ and $T_{\text{ZID}} = 4v_0$, which is the result obtained by Lipscomb and Mickens [32]. For the ZIV oscillator, one has $a_{\text{ZIV}} \equiv A$ and $T_{\text{ZIV}} = 4\sqrt{2}\sqrt{A}$, which is the result obtained by van Horssen [44].

The exact solution for motion of the oscillator under consideration can be achieved by starting from the expression for acceleration, as Lipscomb and Mickens [32] did, but expressing it in the following form for general initial conditions:

$$\ddot{x} = -\text{sgn}\left(\sin\left(\hat{\omega}t + \hat{\varphi}\right)\right), \tag{2.241}$$

where

$$\hat{\omega} = \frac{2\pi}{T} = \frac{\pi}{2\sqrt{2}\sqrt{a}}, \tag{2.242}$$

$$\hat{\varphi} = -\hat{\omega}v_0 + \frac{\pi}{2}. \tag{2.243}$$

This can be integrated to derive the exact solution for the corresponding velocities

$$\dot{x}(t) = \frac{1}{\hat{\omega}}\arcsin\left(\sin\left(\hat{\omega}t + \hat{\varphi} + \frac{\pi}{2}\right)\right). \tag{2.244}$$

By using the energy conservation law (2.237), the exact solution for motion can be obtained as follows:

$$x = \text{sgn}\left(\sin\left(\hat{\omega}t + \hat{\varphi}\right)\right)\left\{a - \frac{1}{2}\left(\dot{x}(t)\right)^2\right\}. \tag{2.245}$$

These solutions coincide with the numerical solutions plotted in Figs. 2.22, 2.23 and 2.24. Figures 2.22a, 2.23a and 2.24a show, respectively, the acceleration of the oscillator with general initial conditions, and for the ZID and ZIV oscillators. Figures 2.22b, 2.23b and 2.24b show their velocity, and Figs. 2.22c, 2.23c and 2.24c their displacement response. Analytical solutions are plotted as lines, while the results of numerical integration of the equation of motion are labelled by dots.

It should be noted that the expression given on the right-hand side of Eq. (2.241) is related to the square wave (pulse train, pulse wave, Rademacher function). One of the analytic formulas for the square wave SW(t) with amplitude A, period T and offset t_0 is

$$\text{SW}(t) = A\,\text{sgn}\left(\sin\left(\frac{2\pi}{T}t - \frac{2\pi}{T}t_0\right)\right). \tag{2.246}$$

The square wave is implemented in Mathematica as SquareWave[t], which gives a square wave that alternates between +1 and -1 with unit period. Using this notation, the solution (2.241) corresponds to

$$\ddot{x} = -\mathrm{SW}\left(\frac{t}{4\sqrt{2}\sqrt{a}} - \frac{v_0}{4\sqrt{2}\sqrt{a}} + \frac{1}{4}\right). \tag{2.247}$$

In terms of frequency and phase, it reads as

$$\ddot{x}_{\mathrm{CRFO}} = -\mathrm{SW}\left(\omega_{\mathrm{CRFO}} t + \varphi_{\mathrm{CRFO}}\right), \tag{2.248}$$

$$\omega_{\mathrm{CRFO}} = \frac{1}{4\sqrt{2}\sqrt{a}}, \tag{2.249}$$

$$\varphi_{\mathrm{CRFO}} = -\omega v_0 + \frac{1}{4}. \tag{2.250}$$

Further, the solution for the velocity (2.244) is associated with the triangle wave TW, whose analytic representations with period 2 and variation between -1 and 1 are

$$\mathrm{TW}(\ddot{x}) = \frac{2}{\pi}\arcsin\left(\sin\left(\pi\ddot{x}\right)\right). \tag{2.251}$$

The triangle wave is also implemented in Mathematica as TriangleWave[t], giving a triangle wave that varies between -1 and $+1$ but with the unit period. Thus, if one wants to use the solution (2.244) in Mathematica, its representation should be

$$\dot{x}_{\mathrm{CRFO}} = 8a\omega\,\mathrm{TW}\left(\omega_{\mathrm{CRFO}}\, t + \varphi_{\mathrm{CRFO}} + \frac{1}{4}\right), \tag{2.252}$$

or

$$\dot{x} = \sqrt{2}\sqrt{a}\mathrm{TW}\left(\omega_{\mathrm{CRFO}} t + \varphi_{\mathrm{CRFO}} + \frac{1}{4}\right). \tag{2.253}$$

The solution for motion (2.245) can be expressed in this way as

$$x_{\mathrm{CRFO}} = a\,\mathrm{SW}\left(\omega_{\mathrm{CRFO}} t + \varphi_{\mathrm{CRFO}}\right)\left[1 - \mathrm{TW}^2\left(\omega_{\mathrm{CRFO}} t + \varphi_{\mathrm{CRFO}} + \frac{1}{4}\right)\right]. \tag{2.254}$$

Table 2.3 contains the simplified expressions (2.247)–(2.254) corresponding to the ZID and ZIV oscillators. As the expressions (2.247)–(2.254) are the exact one, their graphical presentations coincide with those corresponding to Eqs. (2.241)–(2.245) plotted as lines in Figs. 2.22, 2.23 and 2.24.

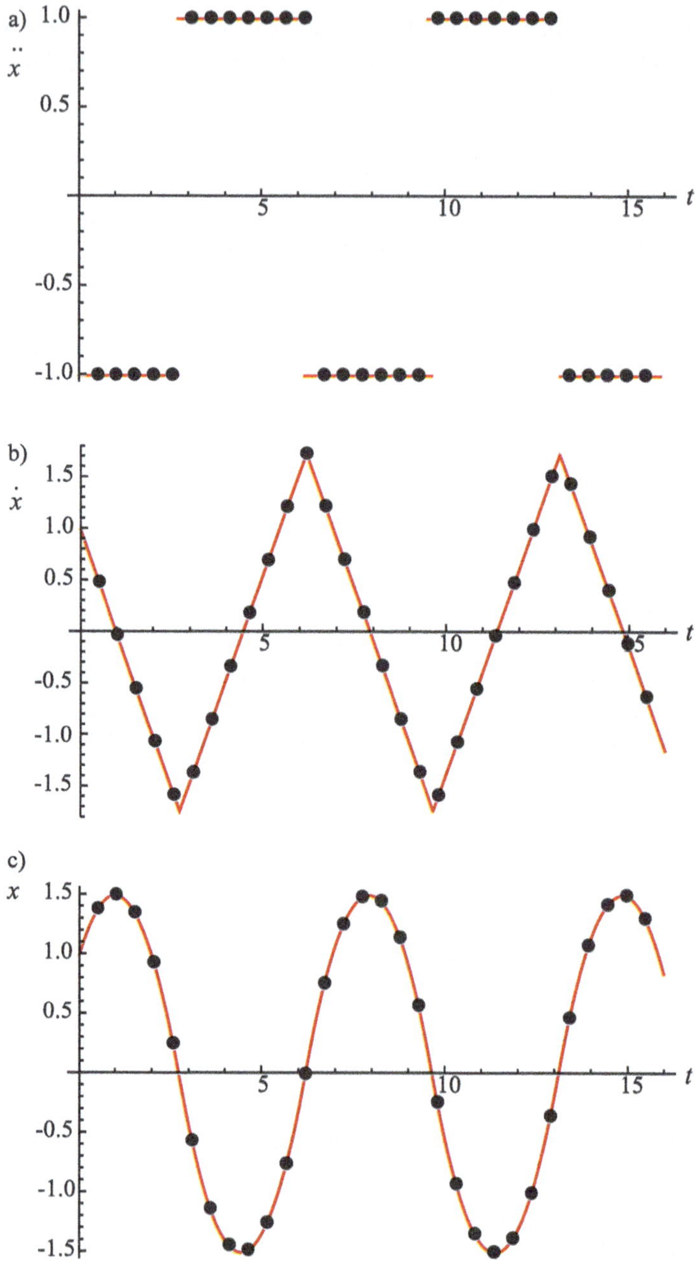

Fig. 2.22 Exact analytical solutions (red solid lines) and numerical solutions (black dots) for the antisymmetric oscillator with $A = 1$, $v_0 = 1$: **a** time history of acceleration; **b** time history of velocity; **c** displacement response

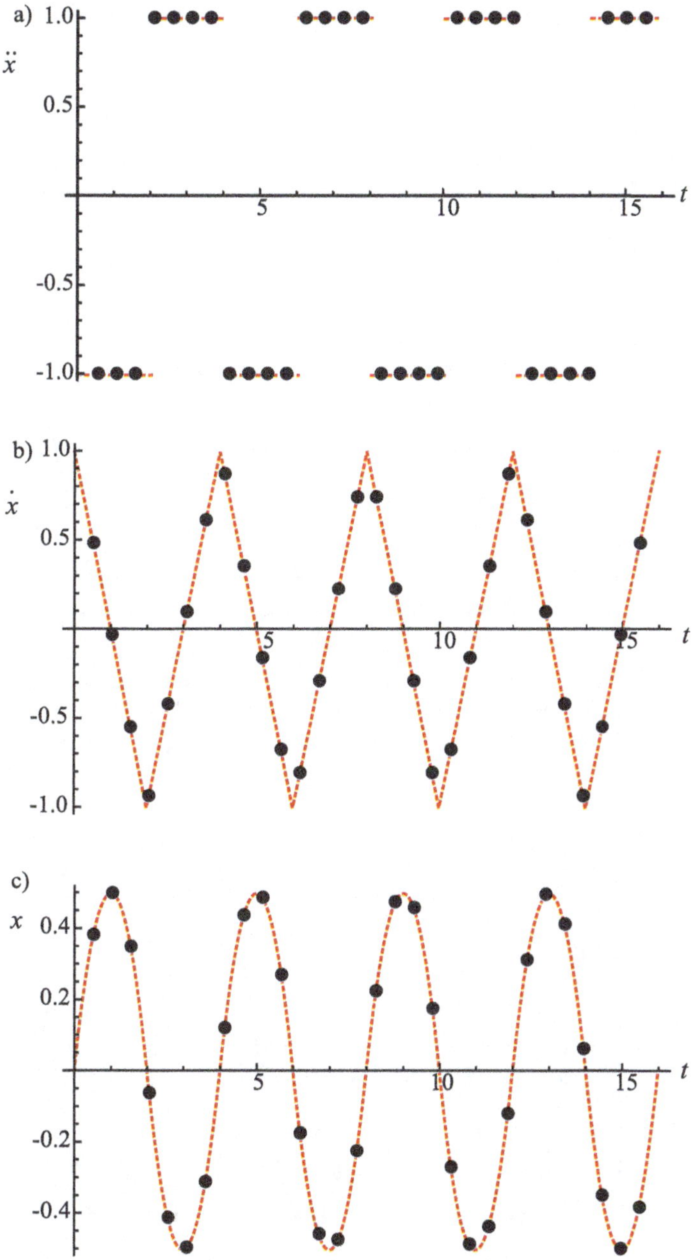

Fig. 2.23 Exact analytical solutions (red dotted lines) and numerical solutions (black dots) for the antisymmetric ZID oscillator with $v_0 = 1$: **a** time history of acceleration; **b** time history of velocity; **c** displacement response

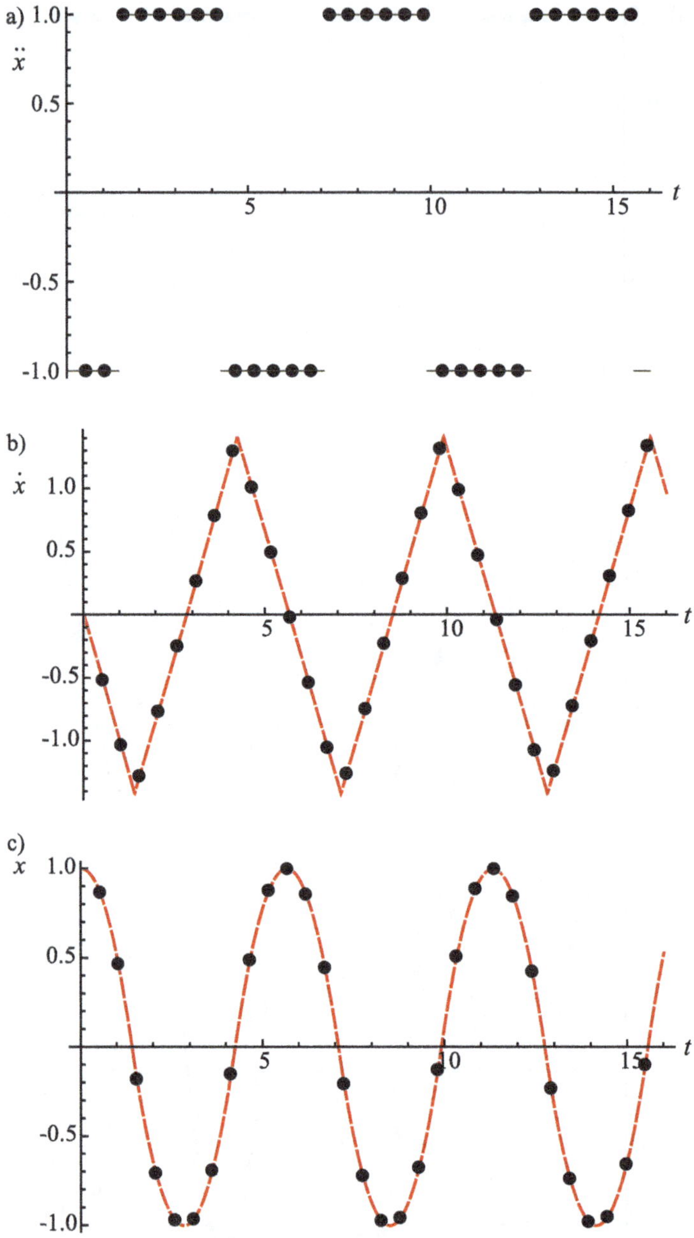

Fig. 2.24 Analytical solutions (red dashed lines) and numerical solutions (black dots) for the antisymmetric ZIV oscillator with $A = 1$: **a** time history of acceleration; **b** time history of velocity; **c** displacement response

Table 2.3 Simplified expressions for the exact solution for acceleration, velocity and motion of the ZID and ZIV oscillators

Exact solutions	ZID oscillator	ZIV oscillator
Acceleration	$-\mathrm{SW}\left(\frac{t}{4v_0}\right)$	$-\mathrm{SW}\left(\frac{t}{4\sqrt{2A}}+\frac{1}{4}\right)$
Velocity	$v_0\mathrm{TW}\left(\frac{t}{4v_0}+\frac{1}{4}\right)$	$\sqrt{2A}\mathrm{TW}\left(\frac{t}{4\sqrt{2A}}+\frac{1}{2}\right)$
Displacement	$\frac{v_0^2}{2}\,\mathrm{SW}\left(\frac{t}{4v_0}\right)\left[1-\mathrm{TW}^2\left(\frac{t}{4v_0}+\frac{1}{4}\right)\right]$	$A\,\mathrm{SW}\left(\frac{t}{4\sqrt{2A}}+\frac{1}{4}\right)$ $\left[1-\mathrm{TW}^2\left(\frac{t}{4\sqrt{2A}}+\frac{1}{2}\right)\right]$

References

1. Tenenbaum M, Pollard H (1985) Ordinary differential equations: an elementary textbook for students of mathematics, engineering, and the sciences. Dover Publications, New York
2. Arnold VI (1978) Ordinary differential equations, 1st edn. MIT Press, Boca Raton
3. Jacobi CGJ (2012) New foundations of the theory of elliptic functions, Konigsberg, Borntraeger 1829, Reprinted by Cambridge University Press (In Latin)
4. Byrd P, Friedman M (1954) Handbook of elliptic integrals for engineers and scientists. Springer, Berlin
5. Abramowitz M, Stegun I (1965) Handbook of mathematical functions. Dover Publications, New York
6. Gradshteyn IS, Ryzhik IM (2000) Tables of integrals, series and products. Academic, New York
7. Kovacic I, Cveticanin L, Zukovic M, Rakaric Z (2016) Jacobi elliptic functions: a review of nonlinear oscillatory application problems. J Sound Vib 380:1–36
8. Lenci S, Rega G (2011) Forced harmonic vibration in a Duffing oscillator with negative linear stiffness and linear viscous damping. In: Kovacic I, Brennan MJ (eds) The Duffing equation: nonlinear oscillators and their behaviour. Wiley, New York
9. Rand RH (1990) Using computer algebra to handle elliptic functions in the method of averaging. In: Noor AK, Elishakoff I, Hulbert G (eds) Symbolic computations and their impact on mechanics. American Society of Mechanical Engineers, PVP-vol 205, pp 311–326
10. Chen SH, Yang XM, Cheung YK (1999) Periodic solutions of strongly quadratic non-linear oscillators by the elliptic Lindstedt-Poincar é method. J Sound Vib 227:1109–1118
11. Mickens RE (2010) Truly nonlinear oscillations: harmonic balance, parametric Expansions, Iteration, and Averaging Methods. World Scientific, Singapore
12. Lyapunov AM (1950) Stability of motion. GITTL, Moscow (in Russian)
13. Gelb A, Vander Velde WE (1968) Multiple-input describing functions and nonlinear system design. McGraw-Hill, New York
14. Nayfeh AH, Mook DT (1979) Nonlinear oscillations. Wiley, New York
15. Cveticanin L (2009) Oscillator with fraction order restoring force. J Sound Vib 320:1064–1077
16. Gottlieb HPW (2003) Frequencies of oscillators with fractional-power nonlinearities. J Sound Vib 261:557–56
17. Cveticanin L, Pogány T (2012) Oscillator with a sum of noninteger-order nonlinearities. J Appl Math Art. No. 649050
18. Rosenberg RM (1963) The Ateb(h)-functions and their properties. Q Appl Math 21:37–47
19. Rosenberg RM, Hsu CS (1963) On the geometrization of normal vibrations of nonlinear systems having many degrees of freedom. In: Analytical methods of the theory of nonlinear oscillations-papers from the international symposium on nonlinear oscillations (Kiev, 12–18 September 1961), Academy of Science USSR, Kiev, pp 380-415
20. Senik PM (1968) On the Ateb functions, DAN URSR Ser. A 21:23–26 (In Ukranian)

21. Senik PM (1969) Inversions of the incomplete beta function. Ukr Math J 21:325–333 (In Russian)
22. Mickens RE (2019) Generalized trigonometric and hyperbolic functions. CRC Press, Boca Raton
23. Gendelman O, Vakakis AF (2000) Transitions from localization to nonlocalization in strongly nonlinear damped oscillators. Chaos Solitons Fract 11:1535–1542
24. Pilipchuk VN (2010) Nonlinear dynamics: between linear and impact limits. Springer, New York
25. Andrianov IV (2002) Asymptotics of nonlinear dynamical systems with a high degree of nonlinearity. Dokl Math 66:270–273
26. Andrianov I, Olevskyi V, Olevska Y (2016) Analytic approximation of periodic Ateb functions via elementary functions in nonlinear dynamics. In: AIP conference proceedings, vol 1773, Art. No. 040001
27. Beléndez A, Francés J, Beléndez T, Bleda S, Pascual C, Arribas E (2015) Nonlinear oscillator with power-form elastic-term: Fourier series expansion of the exact solution. Commun Nonlinear Sci Numer Simul 22:134–148
28. Kovacic I (2014) On the use of special functions for studying truly nonlinear conservative oscillators. In: Gummel A (ed) Mathematics of discrete and continuous dynamical systems conference organizer (contemporary mathematics book series celebrating the contributions of Professor Ronald Mickens (in conjunction with his 70th birthday). Contemporary mathematics, vol 618, pp 281–298
29. Rakaric Z, Kovacic I (2011) Approximations for motion of the oscillators with a non-negative real-power restoring force. J Sound Vib 330:321–336
30. Rakaric Z, Kovacic I (2013) An elliptic averaging method for harmonically excited oscillators with a purely nonlinear non-negative real-power restoring force. Commun Nonlinear Sci Numer Simul 18:1888–1901
31. Zukovic M, Kovacic I (2012) On the behaviour of parametrically excited purely nonlinear oscillators. Nonlinear Dyn 70:2117–2128
32. Lipscomb T, Mickens RE (1994) Exact solution to the antisymmetric, constant force oscillator equation. J Sound Vib 169:138–140
33. Pilipchuk VN (1999) An explicit form general solution for oscillators with a non-smooth restoring force, $\ddot{x}+\text{sign}(x)\, f(x) = 0$. J Sound Vib 226:795–798
34. Araki Y, Asai T, Masui T (2009) Vertical vibration isolator having piecewise-constant restoring force. Earthq Eng Struct Dyn 38:1505–1523
35. Govindan Potti PK, Sarma MS, Nageswara Rao B (1999) On the exact periodic solution for $\ddot{x}+\text{sign}(x) = 0$. J Sound Vib 220:378–381
36. Pilipchuk VN (1996) Analytical study of vibrating systems with strong non-linearities by employing saw-tooth time transformations. J Sound Vib 192:43–64
37. Andrianov I, Awrejcewicz J (2001) Solutions in the Fourier series form, Gibbs phenomena and Padé approximations. J Sound Vib 24:753–756
38. Awrejcewicz J, Andrianov IV (2002) Oscillations of non-linear with restoring force close to $\text{sign}(x)$. J Sound Vib 252:962–966
39. Beléndez A, Gimeno E, Alvarez ML, Méndez DI (2009) Nonlinear oscillator with discontinuity by generalized harmonic balance method. Comput Math Appl 58:2117–2123
40. Rakaric Z (2011) Oscillators with a quasi-constant restoring force: approximations for motion. Meccanica 46:1047–1053
41. Kovacic I (2011) Forced vibrations of oscillators with a purely nonlinear power-form restoring force. J Sound Vib 330:4313–4327
42. Kovacic I (2010) The method of multiple scales for forced oscillators with some real-power nonlinearities in the stiffness and damping force. Chaos Solitons Fract 44:891–901
43. Kovacic I (2014) On some performance characteristics of base excited oscillatory systems with a purely nonlinear restoring force. Int J Nonlinear Mech 65:44–52
44. van Horssen WT (2003) On the periods of the periodic solutions of the non-linear oscillator equation $\ddot{x} + x^{1/(2n+1)} = 0$. J Sound Vib 260:961–964

Chapter 3
Free Damped Oscillators

3.1 Introduction

In this chapter, the focus is on damped oscillators. Nonlinear oscillators are mainly dealt with, but linear oscillators are also considered or referred to for the sake of comparison or clear extension of the related methodology. The equation of motion that encompasses the cases covered has the following general form:

$$\ddot{x} + f\left(x, \dot{x}, D^Q x\right) + c_1\, x + c_3\, x^3 + c_\alpha \mathrm{sgn}\,(x)\, |x|^\alpha = 0, \tag{3.1}$$

with c_1, c_3, c_α and α being constants and $f\left(x, \dot{x}, D^Q x\right)$ standing, in general, for a non-conservative term that can stem from different types of damping considered: linear viscous damping, quadratic damping, van der Pol damping or fractional derivative damping.

Section 3.2 deals with Lagrangians and conservation laws of viscously damped ($f\,(x, \dot{x}) \propto \dot{x}$) linear oscillators ($c_3 = c_\alpha = 0$) as well as Duffing oscillators ($c_\alpha = 0$).

Section 3.3 is focused on quadratically damped ($f\,(x, \dot{x}) \propto \mathrm{sgn}(\dot{x})\, \dot{x}^2$) purely nonlinear oscillators ($c_1 = c_3 = 0$). Exact expressions for their energy-displacement functions are derived based on energy considerations. In addition, the expressions for phase trajectories are also obtained.

Section 3.4 deals with fractional derivative damping, i.e. f is taken as a fractional derivative term $D^Q x$ of order Q, where $0 < Q < 1$. The approximate solutions for motion and some physical insights are obtained based on an approximate solution expressed in terms of trigonometric functions.

Section 3.5 deals with purely nonlinear oscillators ($c_1 = c_3 = 0$) perturbed with f being dependent on the generalized velocity and/or generalized coordinate: $f\,(\dot{x})$, $f\,(x, \dot{x})$ or $f\,(x)$. The approximate solutions for motion are obtained based on an approximate solution expressed in terms of Jacobi elliptic functions.

Section 3.6 is focused on constant restoring force oscillators ($c_1 = c_3 = 0, \alpha = 0$) with linear or quadratic viscous damping. Approximate solutions for motion are derived based on an approximate solution expressed in terms of wave functions.

It should be pointed out that only quantitative techniques are presented, without ex ante qualitative analyses. It goes without saying that for a complete overview of the dynamics of nonlinear oscillators, one can carry out such examinations as this

I. Kovacic, *Nonlinear Oscillations*,
https://doi.org/10.1007/978-3-030-53172-0_3

completes the overview, including, for example, the type of responses that appear for certain values of the system parameters. In addition, the physical insight provided is less developed, and the approach that the reader should take is to use this chapter to get the theoretical foundations, in more or less complete detail, and then impose a physical context on top, to give the reality needed to apply this new knowledge.

3.2 Oscillators with Linear Viscous Damping: Lagrangians and Conservation Laws

This section is dedicated to linearly viscously damped linear and nonlinear oscillators. It will be shown that despite the fact that their overall mechanical energy changes in time, there are certain quantities that stay constant and can be expressed in the exact form. The aim of this section is to derive and interpret some of the expressions for such quantities.

3.2.1 Linear Oscillators

Let us start from the case when a linear oscillator is viscously damped, i.e. when $f(x, \dot{x})$ in Eq. (3.1) is proportional to the generalized velocity, i.e. $f(x, \dot{x}) \propto \dot{x}$, while $c_3 = c_\alpha = 0$ and $c_1 = \omega^2$. The equation of motion is given by

$$\ddot{x} + 2\delta\dot{x} + \omega^2 x = 0, \tag{3.2}$$

where δ and ω are constants and the dots stand for the derivatives with respect to time t.

The first interesting fact about this type of oscillators is that they can completely be represented by the time-dependent Lagrangian. This Lagrangian is

$$L = \left(\frac{1}{2}\dot{x}^2 - \frac{1}{2}\omega^2 x^2\right) e^{2\delta t}. \tag{3.3}$$

Indeed, if one would use Eq. (3.3) in conjunction with the Lagrange equation of the second kind:

$$\frac{\mathrm{d}}{\mathrm{d}t}\frac{\partial L}{\partial \dot{x}} - \frac{\partial L}{\partial x} = 0, \tag{3.4}$$

one would derive the equation of motion (3.2).

The second interesting fact regards the existence of conservation laws of this oscillator. Let us start from the expression:

$$\left(\frac{1}{2}\dot{x}^2 + \frac{1}{2}\omega^2 x^2 + \delta x\dot{x}\right) e^{2\delta t} = const. \tag{3.5}$$

It is seen that when $\delta = 0$, this expression corresponds to the energy conservation law of an undamped simple harmonic oscillator. If one would differentiate Eq. (3.5) with respect to time t, the following equation can be derived:

$$(\dot{x} + \delta x)\left(\ddot{x} + 2\delta\dot{x} + \omega^2 x\right) e^{2\delta t} = 0, \tag{3.6}$$

which is satisfied by Eq. (3.2). Thus, the expression given by Eq. (3.5) represents its conservation law. This conservation law can also be written as follows:

$$\left[\frac{1}{2}(\dot{x} + \delta x)^2 + \frac{1}{2}\varkappa^2 x^2\right] e^{2\delta t} = const, \tag{3.7}$$

where $\varkappa = \sqrt{\omega^2 - \delta^2}$.

Let us assume now that the velocity can be presented as a field $\Phi = \Phi(t, x)$ depending on time t and the generalized coordinate x:

$$\Phi(t, x) \equiv \dot{x} = g(t)\, x, \tag{3.8}$$

where the time-dependent function $g(t)$ needs to be determined. The partial differential equation for the field has the form

$$\frac{\partial\Phi}{\partial t} + \frac{\partial\Phi}{\partial x}\Phi + 2\delta\Phi + \omega^2 x = 0. \tag{3.9}$$

Substituting Eq. (3.8) into Eq. (3.9) yields

$$\dot{g} + g^2 + 2\delta g + \omega^2 = 0. \tag{3.10}$$

Separating variables, its integration gives

$$\frac{g + \delta}{\varkappa} = -\cot(\varkappa t + C), \tag{3.11}$$

where C is an integration constant. Going back to Eq. (3.8), one can obtain the following conservation law:

$$\operatorname{arc\,cot}\left(\frac{\dot{x}}{\varkappa x} + \frac{\delta}{\varkappa}\right) - \varkappa t = C = const., \tag{3.12}$$

i.e.

$$\dot{x} = \left[-\delta + \varkappa\cot(\varkappa t + C)\right] x. \tag{3.13}$$

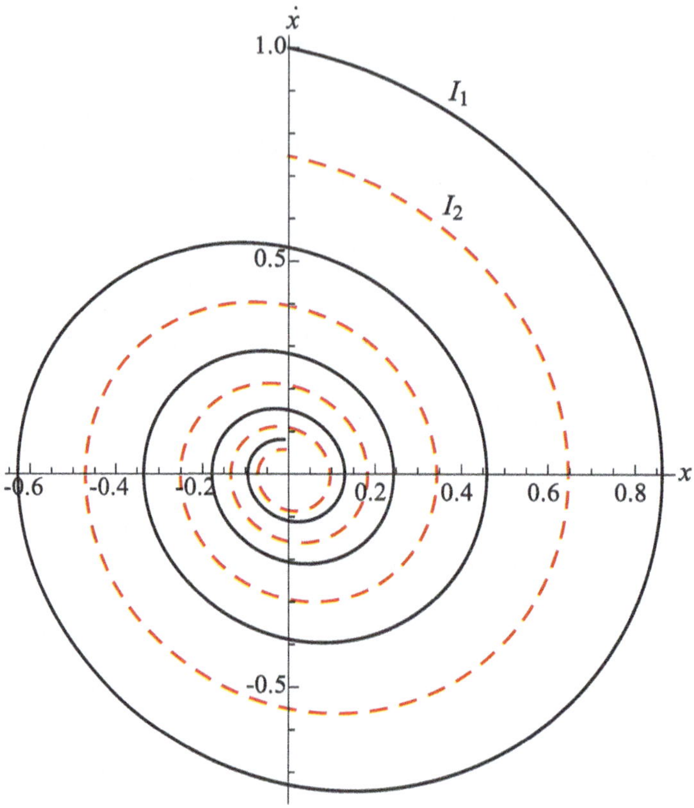

Fig. 3.1 Phase trajectories obtained from Eq. (3.14)

The value of the constant C can be calculated by using the initial conditions. Taking them in the most general form: $x(0) = x_0$ and $\dot{x}(0) = v_0$, it can be derived that the value of this constant actually corresponds to the initial phase $C =$arctan $(\varkappa x_0/(v_0 + \delta x_0))$.

It is interesting to note that one can express t from Eq. (3.12) and substitute it back to the conservation law (3.5) to derive the expression in x and $\dot{x}$ only:

$$\left[(\dot{x} + \delta x)^2 + \varkappa^2 x^2\right] e^{\frac{2\delta}{\varkappa}\operatorname{arc\,cot}\left(\frac{\dot{x}+\delta x}{\varkappa x}\right)} = I = const. \tag{3.14}$$

Equation (3.14) can be plotted to derive spirals in the phase plane, as shown in Fig. 3.1 for two values of the constant I: I_1 and I_2.

3.2.2 Duffing Oscillators

This part follows the concept presented previously for linear oscillators in terms of the Lagrangian and conservation laws, but the concept is related to damped Duffing oscillators:

$$\ddot{x} + 2\delta\dot{x} + c_1 x + c_3 x^3 = 0. \tag{3.15}$$

This equation of motion can be derived from the time-dependent Lagrangian:

$$L = \left(\frac{1}{2}\dot{x}^2 - \frac{1}{2}c_1 x^2 - \frac{1}{4}c_3 x^4\right) e^{2\delta t}. \tag{3.16}$$

Based on the Noether identity, it is derived in [1] that the conservation law of the oscillator governed by Eq. (3.15) is of the form:

$$\frac{1}{2\delta}\left(\frac{1}{2}\dot{x}^2 + \frac{1}{2}c_1 x^2 + \frac{1}{4}c_3 x^4\right) e^{2\delta t} + \int L\mathrm{d}t = const, \tag{3.17}$$

with L being defined by Eq. (3.16).

Another conservation law analogous to the one used to plot Fig. 3.1 can be derived herein again by introducing the field of velocity $\Phi \equiv \dot{x} = \Phi(t, x)$. The partial differential equation for the field has the form

$$\frac{\partial \Phi}{\partial t} + \frac{\partial \Phi}{\partial x}\Phi + 2\delta\Phi + c_1 x + c_3 x^3 = 0. \tag{3.18}$$

The solution for the field is assumed in the form [2]:

$$\Phi \equiv \dot{x} = Rx + \sqrt{SF(t) + g(x)}, \tag{3.19}$$

where R and S are the constants, while $F(t)$ and $g(x)$ are unknown functions. Substituting Eq. (3.19) into Eq. (3.18), one obtains

$$S\left(\dot{F} + 4\delta F + 2RF\right) + Rxg' + 2Rg + 4\delta g + \\ \sqrt{SF(t) + g(x)}\left[2x\left(R^2 + 2\delta R + c_1\right) + g' + 2c_3 x^3\right] = 0.$$

This equation can be separated into four equations, as follows:

$$\dot{F} + 4\delta F + 2RF = 0,$$

$$Rxg' + 2Rg + 4\delta g = 0, \tag{3.20}$$

$$R^2 + 2\delta R + c_1 = 0, \tag{3.21}$$

$$g' + 2c_3 x^3 = 0. \tag{3.22}$$

The last one gives

$$g = -\frac{1}{2}c_3x^4. \tag{3.23}$$

Now, for this g, the second one yields

$$\delta = -\frac{3}{2}R, \tag{3.24}$$

while the third results in

$$R = -\sqrt{\frac{c_1}{2}}. \tag{3.25}$$

Note that the minus sign is retained in R since δ in Eq. (3.24) must be positive. This implies that the following relationship should exist between the damping coefficient and the natural frequency:

$$2\delta = 3\sqrt{\frac{c_1}{2}}. \tag{3.26}$$

Finally, integrating the first equation from the set above, one has

$$F = e^{-2\sqrt{2c_1}t}. \tag{3.27}$$

Thus, it is concluded that the Duffing oscillator,

$$\ddot{x} + 3\sqrt{\frac{c_1}{2}}\dot{x} + c_1x + c_3x^3 = 0, \tag{3.28}$$

has the following conservation law:

$$\dot{x} = -\sqrt{\frac{c_1}{2}}x + \sqrt{Be^{-2\sqrt{2c_1}t} - \frac{1}{2}c_3x^4}, \tag{3.29}$$

which can also be written as:

$$\left[\left(\dot{x} + \sqrt{\frac{c_1}{2}}x\right)^2 + \frac{1}{2}c_3x^4\right]e^{2\sqrt{2c_1}t} = const. \tag{3.30}$$

3.3 Purely Nonlinear Oscillators with Quadratic Viscous Damping: Exact Solution Based on Energy Considerations and Approximations via Trigonometric Functions

The oscillators considered in this section have the equation of motion corresponding to Eq. (3.1) with $f(\dot{x}) = \mu\,\mathrm{sgn}(\dot{x})\dot{x}^2$, $c_1 = c_3 = 0$, $c_\alpha = 1$:

$$\ddot{x} + \mu\,\mathrm{sgn}(\dot{x})\dot{x}^2 + \mathrm{sgn}(x)\,|x|^\alpha = 0, \tag{3.31}$$

where μ is a constant parameter.

The physics behind it stems from one of the basic damping mechanisms, quadratic damping, which occurs, for example, when an immersed object moves through a fluid at relatively high Reynolds numbers [3, 4]. The corresponding magnitude of the drag force F_d is proportional to the square of the velocity

$$F_d(\dot{x}) = b\,\mathrm{sgn}(\dot{x})\,\dot{x}^2, \tag{3.32}$$

where the constant b depends, in general, on the object geometry and the fluid properties.

The nonlinearity stems from a purely nonlinear restoring force whose magnitude changes in accordance with

$$F_r(x) = k_\alpha\,\mathrm{sgn}(x)\,|x|^\alpha, \tag{3.33}$$

where k_α and α are non-negative real constants.

Thus, the equation of motion of the system considered having the mass m is

$$m\ddot{x} + b\,\mathrm{sgn}(\dot{x})\,\dot{x}^2 + k_\alpha\,\mathrm{sgn}(x)\,|x|^\alpha = 0. \tag{3.34}$$

The initial conditions are assumed in the zero-initial velocity form

$$x(0) = X_0,\ \dot{x}(0) = 0, \tag{3.35}$$

where X_0 is assumed to be a positive real constant.

Numerous authors have performed quantitative analyses of oscillators with restoring force nonlinearities and small quadratic damping by obtaining their approximate solution for motion. Some of the methods used have been: the method of multiple scales [3], the averaging method (Krylov–Bogoliubov method) [5] and elliptic Krylov–Bogoliubov method [6]. In addition, some techniques have been developed for converting the given equation with a certain type of nonlinearity both in the restoring force and damping force into equivalent linear systems (see, for example, [7], and the references cited therein). On the other hand, a qualitative analysis given in [8] for this type of nonlinear damping includes an analytical proof that the energy of the system continuously decreases along the phase path, i.e. the energy is continuously withdrawn from the system, causing the corresponding phase path to spiral in towards the equilibrium $x = \dot{x} = 0$. In [9], the system with quadratic damping and a linear restoring force is studied qualitatively by analysing the phase plane. It is shown that the phase trajectories in the upper half of the phase plane are symmetric to the phase trajectories in the lower half with respect to the equilibrium point. Klotter [10] considered the oscillators with purely nonlinear integer power or linear-plus-nonlinear integer power restoring forces. He transformed the equations of motion into linear first-order differential equations for the square of the velocity, which actually corresponds to double kinetic energy, and derived the transcendental relationship between two consecutive maximal displacement amplitudes.

The aim of this section is to carry out the analysis of the nonlinear oscillator governed by Eq. (3.34) for the initial conditions (3.35) focusing on the influence of an arbitrary non-negative real power α on the energy-displacement function, as done in [11] by Rakaric and Kovacic. The energy-displacement function obtained is used to find the values of the amplitudes of motion. These values are exact, unlike the results of the application of the methods listed above, which are mainly approximate. In addition, the expressions for phase trajectories are also derived. Several examples are presented to illustrate the findings.

The following non-dimensional variables are introduced:

$$\bar{x} = \frac{x}{X_0}, \quad \bar{t} = \frac{t}{\sqrt{\frac{m}{k_\alpha}} x_0^{\frac{1-\alpha}{2}}}. \tag{3.36}$$

Omitting the overbars, Eq. (3.34) can be expressed in non-dimensional form given by Eq. (3.31), where the parameter μ stands for:

$$\mu = \frac{ck_\alpha}{m} X_0. \tag{3.37}$$

Thus, instead of having five parameters as in Eqs. (3.34) and (3.35), the system behaviour will depend on two parameters only: the coefficient μ and the power α. The initial conditions (3.35) are now

$$x\,(0) \equiv x_0^+ = 1, \;\; x'\,(0) = 0, \tag{3.38}$$

where the superscript '+' is used to emphasize the positive sign of the displacement, which is of importance for the procedure developed below.

3.3.1 Energy-Displacement Function

As shown analytically in [8], the total mechanical energy of the system (3.31) decreases with time. However, instead of analysing the change of this energy in time, it is also possible and useful to consider how it changes with the displacement.

The non-dimensional equation of motion (3.31) can be written down in the form

$$\frac{\mathrm{d}}{\mathrm{d}t}[\frac{1}{2}\dot{x}^2 + \frac{1}{\alpha+1}|x|^{\alpha+1}] = -\mu \operatorname{sgn}(\dot{x})\,(\dot{x})^3\,. \tag{3.39}$$

Recognizing the total mechanical energy

$$E = E_k + E_p, \quad E_k = \frac{1}{2}\dot{x}^2, \quad E_p = \frac{|x|^{\alpha+1}}{\alpha+1}, \tag{3.40}$$

as well as changing the variable of differentiation from non-dimensional time to non-dimensional displacement, Eq. (3.39) becomes

$$\frac{\mathrm{d}E}{\mathrm{d}x} = -\mu \operatorname{sgn}(\dot{x}) \left[2\left(E - \frac{1}{\alpha+1}|x|^{\alpha+1}\right)\right]. \tag{3.41}$$

Based on Eq. (3.40), the following substitution is introduced:

$$E_k = E - \frac{1}{\alpha+1}|x|^{\alpha+1}, \tag{3.42}$$

so that Eq. (3.41) transforms into

$$\frac{\mathrm{d}E_k}{dx} + 2\mu \operatorname{sgn}(\dot{x}) E_k = -\operatorname{sgn}(x)\,|x|^{\alpha}\,. \tag{3.43}$$

In this way, the problem is transformed from a second-order differential equation to a first-order differential equation. In order to solve Eq. (3.43), two cases are to be considered separately, for the power α being odd or even.

3.3.1.1 Case I: Odd-Power Restoring Force

Let us assume that the motion starts from the positive initial displacement with respect to the equilibrium position and with the zero velocity as described in Eq. (3.35), being directed to the left side ($\leftarrow$), so that $\operatorname{sgn}(\dot{x}) < 0$. For the odd-parity restoring force, Eq. (3.43) can be expressed as

$$\frac{\mathrm{d}E_{k\leftarrow}}{\mathrm{d}x} - 2\mu E_{k\leftarrow} = -x^{\alpha}. \tag{3.44}$$

It should be noted that Eq. (3.44) holds both for $x > 0$ and $x < 0$, due to the fact that the restoring force is odd.

The general solution of Eq. (3.44) is

$$E_{k\leftarrow}(x) = C_{\mathrm{I}_1} e^{2\mu x} - e^{2\mu x} \int e^{-2\mu x} x^{\alpha} \mathrm{d}x, \tag{3.45}$$

with C_{I_1} being a constant of integration. The subscript I indicates the case considered and the subscript 1 denotes the first interval of motion.

In order to find the solution for $E_{k\leftarrow}(x)$, the integral

$$J = \int e^{-2\mu x} x^{\alpha} \mathrm{d}x, \tag{3.46}$$

needs to be solved. Therefore, the following substitution is introduced:

$$2\mu x = u, \tag{3.47}$$

which leads to

$$J = -\left(\frac{1}{2\mu}\right)^{\alpha+1} \int e^{-u} u^{\alpha} \mathrm{d}u. \tag{3.48}$$

This integral can be related to the definition of the upper incomplete gamma function (see Appendix A or [12, 13]):

$$\Gamma[s, y] = \int_{y}^{\infty} e^{-u} u^{s-1} \mathrm{d}u. \tag{3.49}$$

In addition, taking $\alpha = s - 1$, the integral (3.48) becomes

$$J = -\left(\frac{1}{2\mu}\right)^{\alpha+1} \Gamma[\alpha+1, 2\mu x], \tag{3.50}$$

and the solution for $E_{k\leftarrow}$ in Eq. (3.45) is found to be

$$E_{k\leftarrow} = C_{\mathrm{I}_1} e^{2\mu x} + e^{2\mu x} \left(\frac{1}{2\mu}\right)^{\alpha+1} \Gamma[\alpha+1, 2\mu x]. \tag{3.51}$$

On the basis of Eq. (3.42), the energy-displacement function is

$$E_{\mathrm{I}_1\leftarrow} = \frac{1}{\alpha+1} |x|^{\alpha+1} + C_{\mathrm{I}_1} e^{2\mu x} + \left(\frac{1}{2\mu}\right)^{\alpha+1} e^{2\mu x} \Gamma[\alpha+1, 2\mu x]. \tag{3.52}$$

The constant C_{I_1} can be obtained by using the initial conditions (3.38)

$$C_{\mathrm{I}_1} = -\left(\frac{1}{2\mu}\right)^{\alpha+1} \Gamma\left[\alpha+1, 2\mu x_0^+\right], \tag{3.53}$$

so that the energy during the first interval changes in accordance with

$$E_{\mathrm{I}_1\leftarrow} = \frac{|x|^{\alpha+1}}{\alpha+1} + \left(\frac{1}{2\mu}\right)^{\alpha+1} e^{2\mu x} \left(\Gamma\left[\alpha+1, 2\mu x\right] - \Gamma\left[\alpha+1, 2\mu x_0^+\right]\right). \tag{3.54}$$

The motion changes direction when the velocity and the kinetic energy are zero. Based on Eq. (3.54) this is satisfied at $x = x_1^-$, when

$$\Gamma\left[\alpha+1, 2\mu x_1^-\right] - \Gamma\left[\alpha+1, 2\mu x_0^+\right] = 0. \tag{3.55}$$

When the system moves from the left side to the right side (labelled by '→'), one holds $\mathrm{sgn}(x^{'}) > 0$, and Eq. (3.43) becomes

$$\frac{dE_{k\rightarrow}}{dx} + 2\mu E_{k\rightarrow} = -x^{\alpha}. \tag{3.56}$$

Its general solution is

$$E_{k\rightarrow} = C_{\mathrm{I}_2} e^{-2\mu x} + e^{-2\mu x}\left(\frac{1}{2\mu}\right)^{\alpha+1} \Gamma\left[\alpha+1, -2\mu x\right]. \tag{3.57}$$

The corresponding energy-displacement function can be obtained analogously as before and is given by

$$E_{\mathrm{I}_2\rightarrow} = \frac{|x|^{\alpha+1}}{\alpha+1} + C_{\mathrm{I}_2} e^{-2\mu x} + \left(\frac{1}{2\mu}\right)^{\alpha+1} e^{-2\mu x} \Gamma\left[\alpha+1, -2\mu x\right]. \tag{3.58}$$

The constant of integration C_{I_2} can be obtained from the fact that at x_1^- the energy is equal to the potential one. Thus, one has

$$C_{\mathrm{I}_2} = -\left(\frac{1}{2\mu}\right)^{\alpha+1} \Gamma\left[\alpha+1, -2\mu x_1^-\right]. \tag{3.59}$$

The expression for the energy-displacement function is

$$E_{\mathrm{I}_2\rightarrow} = \frac{|x|^{\alpha+1}}{\alpha+1} + \left(\frac{1}{2\mu}\right)^{\alpha+1} e^{-2\mu x} \left(\Gamma\left[\alpha+1, -2\mu x\right] - \Gamma\left[\alpha+1, -2\mu x_1^-\right]\right). \tag{3.60}$$

Proceeding in the same way, the following recurrent expressions can be found for the energy-displacement functions for the motion to the left and right sides, respectively

Table 3.1 Values of the negative and positive amplitudes of a pure cubic oscillator for $\mu = 0.5$

x_0	x_1^-	x_2^+	x_3^-	x_4^+	x_5^-
1	–0.71149	0.55294	–0.45240	0.38288	–0.33193

$$E_{I_{i\leftarrow}} = \frac{|x|^{\alpha+1}}{\alpha+1} + \left(\frac{1}{2\mu}\right)^{\alpha+1} e^{2\mu x}\left(\Gamma\left[\alpha+1, 2\mu x\right] - \Gamma\left[\alpha+1, 2\mu x_{i-1}^+\right]\right),$$
$$x_i^- \leq x \leq x_{i-1}^+, \tag{3.61}$$
$$E_{I_{i+1\rightarrow}} = \frac{|x|^{\alpha+1}}{\alpha+1} + \left(\frac{1}{2\mu}\right)^{\alpha+1} e^{-2\mu x}\left(\Gamma\left[\alpha+1, -2\mu x\right] - \Gamma\left[\alpha+1, -2\mu x_i^-\right]\right),$$
$$x_i^- \leq x \leq x_{i+1}^+, i = 2j-1, \quad j \in \mathbb{N}. \tag{3.62}$$

Extremal displacements, i.e. local minima and maxima x_i^- and x_{i+1}^+, correspond to $E_k = 0$. They can be found by equating the second term in Eqs. (3.61) and (3.62) to zero. These equations give the subsequent amplitude as the functions of the previous one

$$\Gamma\left[\alpha+1, 2\mu x_i^-\right] - \Gamma\left[\alpha+1, 2\mu x_{i-1}^+\right] = 0 \Rightarrow \quad x_i^-, \tag{3.63}$$
$$\Gamma\left[\alpha+1, -2\mu x_{i+1}^+\right] - \Gamma\left[\alpha+1, -2\mu x_i^-\right] = 0 \Rightarrow \quad x_{i+1}^+. \tag{3.64}$$

The procedure developed and the results obtained will be illustrated by considering a pure cubic oscillator.

Example 3.1. Pure cubic oscillator On the basis of Eqs. (3.63) and (3.64), local minima and maxima of the amplitude of a pure cubic oscillator, which correspond to $\alpha = 3$, are found to be

$$\Gamma\left[4, 2\mu x_1^-\right] - \Gamma\left[4, 2\mu\right] = 0, \tag{3.65}$$
$$\Gamma\left[4, -2\mu x_2^+\right] - \Gamma\left[4, -2\mu x_1^-\right] = 0, \tag{3.66}$$
$$\Gamma\left[4, 2\mu x_3^-\right] - \Gamma\left[4, 2\mu x_2^+\right] = 0, \tag{3.67}$$
$$\Gamma\left[4, -2\mu x_4^+\right] - \Gamma\left[4, -2\mu x_3^-\right] = 0, \tag{3.68}$$
$$\Gamma\left[4, 2\mu x_5^-\right] - \Gamma\left[4, 2\mu x_4^+\right] = 0. \tag{3.69}$$

Solving these equations numerically successively for $\mu = 0.5$, starting from Eq. (3.65), the values given in Table 3.1 are obtained.

Then, the energy-displacement function in the first interval of motion (3.61), starting from the initial position x_0^+ and lasting until the position x_1^-, is

$$E_{I_{1\leftarrow}} = \frac{x^4}{4} + \left(\frac{1}{2\mu}\right)^4 e^{2\mu x}\left(\Gamma\left[4, 2\mu x\right] - \Gamma\left[4, 2\mu\right]\right), -0.71149 \leq x \leq 1. \tag{3.70}$$

This function in plotted in the upper part of Fig. 3.2, where the potential well $E_p = x^4/4$ is shown, too. By using Eq. (3.62), the energy-displacement function in the next interval, when the system moves to the right side, is obtained

$$E_{\mathrm{I}_2\rightarrow} = \frac{x^4}{4} + \left(\frac{1}{2\mu}\right)^4 e^{-2\mu x} \left(\Gamma\left[4, -2\mu x\right] - \Gamma\left[4, 2\mu 0.71149\right]\right), \quad -0.71149 \leq x \leq 0.55294. \tag{3.71}$$

Further use of Eqs. (3.61) and (3.62) gives the following energy-displacement functions in the subsequent intervals of motion

$$E_{\mathrm{I}_3\leftarrow} = \frac{x^4}{4} + \left(\frac{1}{2\mu}\right)^4 e^{2\mu x} \left(\Gamma\left[4, 2\mu x\right] - \Gamma\left[4, 2\mu 0.55294\right]\right), \quad -0.45240 \leq x \leq 0.55294, \tag{3.72}$$

$$E_{\mathrm{I}_4\rightarrow} = \frac{x^4}{4} + \left(\frac{1}{2\mu}\right)^4 e^{-2\mu x} \left(\Gamma\left[4, -2\mu x\right] - \Gamma\left[4, 2\mu 0.45240\right]\right), \quad -0.45240 \leq x \leq 0.38288, \tag{3.73}$$

$$E_{\mathrm{I}_5\leftarrow} = \frac{x^4}{4} + \left(\frac{1}{2\mu}\right)^4 e^{2\mu x} \left(\Gamma\left[4, 2\mu x\right] - \Gamma\left[4, 2\mu 0.38288\right]\right), \quad -0.33193 \leq x \leq 0.38288. \tag{3.74}$$

The energy-displacement functions (3.71)–(3.74) are also plotted in the upper part of Fig. 3.2, clearly indicating how the energy decreases. The positions at which the motion changes direction, i.e. the amplitudes $x_0^+ - x_5^-$, are depicted as well. The time evolution $x(t)$ obtained by integrating the equation of motion numerically is shown in the lower part of Fig. 3.2 The link between the amplitudes $x_0^+ - x_5^-$ in the energy-displacement function and the time evolution is also shown to verify the results obtained. It is seen that the amplitudes obtained above completely coincide with the ones calculated numerically from the equation of motion.

3.3.1.2 Case II: Even-Power Restoring Force

When the restoring force is an even-power function, the initial motion to the left side needs to be divided into two parts, depending on the sign of the displacement. When the displacement is positive, the analysis given above for the motion to the left side holds and, consequently, the energy changes in accordance with

$$E_{\mathrm{II}_{1\leftarrow+}} = \frac{|x|^{\alpha+1}}{\alpha+1} + \left(\frac{1}{2\mu}\right)^{\alpha+1} e^{2\mu x} \left(\Gamma\left[\alpha+1, 2\mu x\right] - \Gamma\left[\alpha+1, 2\mu x_0^+\right]\right), \tag{3.75}$$

where the subscript II indicates Case II and '+' stands for the sign of the displacement.

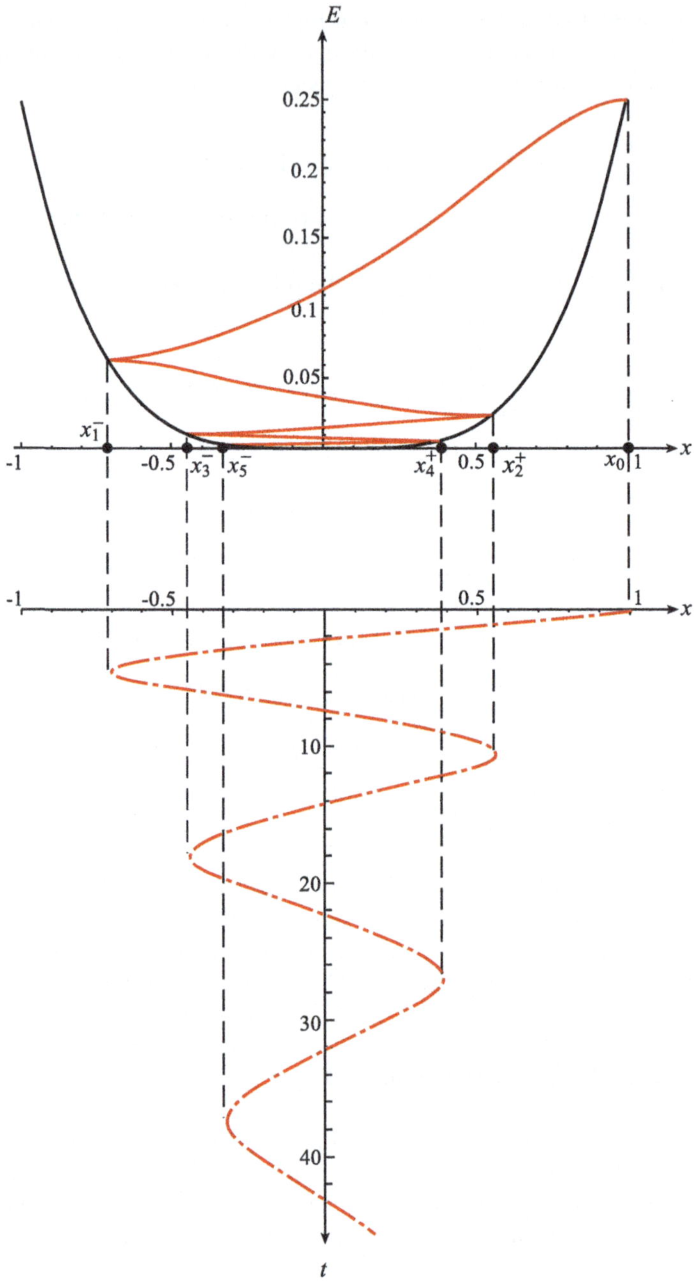

Fig. 3.2 Energy-displacement curves and time response for a pure cubic oscillator $\alpha = 3$ and $\mu = 0.5$

However, when the displacement is negative, which will be labelled by '−', Eq. (3.43) can be written down as

$$\frac{\mathrm{d}E_{k\leftarrow-}}{\mathrm{d}x} - 2\mu E_{k\leftarrow-} = x^{\alpha}. \tag{3.76}$$

By carrying out the procedure similar to the one described above, the following expression for the energy-displacement function can be found

$$E_{\mathrm{II}_{1\leftarrow-}} = \frac{|x|^{\alpha+1}}{\alpha+1} + C_{\mathrm{II}_1} e^{2\mu x} - e^{2\mu x}\left(\frac{1}{2\mu}\right)^{\alpha+1} \Gamma\left[\alpha+1, 2\mu x\right], \tag{3.77}$$

where the constant C_{II_1} can be obtained from the condition

$$E_{\mathrm{II}_{1\leftarrow+}}\left(x=0\right) = E_{\mathrm{II}_{1\leftarrow-}}\left(x=0\right). \tag{3.78}$$

Knowing that $\Gamma\left[\alpha+1, 0\right] \equiv \Gamma\left[\alpha+1\right]$, i.e. that the incomplete gamma function turns into the Euler gamma function when the second argument is equal to zero, one obtains

$$C_{\mathrm{II}_1} = \left(\frac{1}{2\mu}\right)^{\alpha+1} \left(2\Gamma\left[\alpha+1\right] - \Gamma\left[\alpha+1, 2\mu x_0^+\right]\right). \tag{3.79}$$

Thus, the energy-displacement function is

$$E_{\mathrm{II}_{1\leftarrow-}} = \frac{|x|^{\alpha+1}}{\alpha+1} + \left(\frac{1}{2\mu}\right)^{\alpha+1} e^{2\mu x}\left(2\Gamma\left[\alpha+1\right] - \Gamma\left[\alpha+1, 2\mu x_0^+\right] - \Gamma\left[\alpha+1, 2\mu x\right]\right). \tag{3.80}$$

The motion changes direction at the position x_1^- when

$$2\Gamma\left[\alpha+1\right] - \Gamma\left[\alpha+1, 2\mu x_0^+\right] - \Gamma\left[\alpha+1, 2\mu x_1^-\right] = 0. \tag{3.81}$$

When the motion is directed towards the right side and the displacement is negative, Eq. (3.43) becomes

$$\frac{\mathrm{d}E_{k\rightarrow-}}{\mathrm{d}x} + 2\mu E_{k\rightarrow-} = x^{\alpha}, \tag{3.82}$$

which results in the following solution for the energy-displacement function

$$E_{\mathrm{II}_{2\rightarrow-}} = \frac{1}{\alpha+1}|x|^{\alpha+1} + C_{\mathrm{II}_2} e^{-2\mu x} - \left(\frac{1}{-2\mu}\right)^{\alpha+1} e^{-2\mu x}\Gamma\left[\alpha+1, -2\mu x\right], \tag{3.83}$$

with the constant C_{II_2} being defined by the condition

$$E_{\mathrm{II}_{1\leftarrow+}}\left(x_1^-\right) = E_{\mathrm{II}_{2\rightarrow-}}\left(x_1^-\right). \tag{3.84}$$

Its value is then

$$C_{\mathrm{II}_2}=\left(\frac{1}{-2\mu}\right)^{\alpha+1}\Gamma\left[\alpha+1,-2\mu x_1^-\right], \tag{3.85}$$

and the energy-displacement function is

$$E_{\mathrm{II}_{2\to-}}=\frac{1}{\alpha+1}|x|^{\alpha+1}+\left(\frac{1}{2\mu}\right)^{\alpha+1}e^{-2\mu x}\left(\Gamma\left[\alpha+1,-2\mu x\right]-\Gamma\left[\alpha+1,-2\mu x_1^-\right]\right). \tag{3.86}$$

Finally, when the displacement is positive, Eq. (3.43) corresponds to Eq. (3.56). Its solution (3.58) can be used then together with the condition $E_{\mathrm{II}_{2\to-}}(0)=E_{\mathrm{II}_{2\to+}}(0)$ to derive the constant C_{II_3}

$$C_{\mathrm{II}_3}=\left(\frac{1}{-2\mu}\right)^{\alpha+1}\left(\Gamma\left[\alpha+1,-2\mu x_1^-\right]-2\Gamma\left[\alpha+1\right]\right), \tag{3.87}$$

which completes the expression for the energy-displacement function

$$E_{\mathrm{II}_{2\to+}}=\frac{|x|^{\alpha+1}}{\alpha+1}+\left(\frac{1}{2\mu}\right)^{\alpha+1}e^{-2\mu x}\left(2\Gamma\left[\alpha+1\right]-\Gamma\left[\alpha+1,-2\mu x_1^-\right]-\Gamma\left[\alpha+1,-2\mu x\right]\right). \tag{3.88}$$

In the subsequent intervals of motion, the energy-displacement functions are defined by

$$E_{\mathrm{II}_{i\leftarrow+}}=E_p+\left(\frac{1}{2\mu}\right)^{\alpha+1}e^{2\mu x}\left(\Gamma\left[\alpha+1,2\mu x\right]-\Gamma\left[\alpha+1,2\mu x_{i-1}^+\right]\right),\ 0\le x\le x_{i-1}^+, \tag{3.89}$$

$$E_{\mathrm{II}_{i\leftarrow-}}=E_p+\left(\frac{1}{2\mu}\right)^{\alpha+1}e^{2\mu x}\left(2\Gamma\left[\alpha+1\right]-\Gamma\left[\alpha+1,2\mu x\right]-\Gamma\left[\alpha+1,2\mu x_{i-1}^+\right]\right),\quad x_i^-\le x\le 0, \tag{3.90}$$

$$E_{\mathrm{II}_{i+1\to-}}=E_p+\left(\frac{1}{2\mu}\right)^{\alpha+1}e^{-2\mu x}\left(\Gamma\left[\alpha+1,-2\mu x\right]-\Gamma\left[\alpha+1,-2\mu x_i^-\right]\right),\quad x_i^-\le x\le 0, \tag{3.91}$$

$$E_{\mathrm{II}_{i+1\to+}}=E_p+\left(\frac{1}{2\mu}\right)^{\alpha+1}e^{-2\mu x}\left(2\Gamma\left[\alpha+1\right]-\Gamma\left[\alpha+1,-2\mu x\right]-\Gamma\left[\alpha+1,-2\mu x_i^-\right]\right),\quad 0\le x\le x_{i+1}^+, \tag{3.92}$$

$$i=2j-1,\quad j\in\mathbb{N}, \tag{3.93}$$

where E_p is defined by the third expression in Eq. (3.40), while the minima and maxima x_i^- and x_{i+1}^+ can be calculated, respectively, from

Table 3.2 Values of the negative and positive amplitudes of a purely nonlinear oscillator for $\alpha = 4/3$, $\mu = 0.5$

x_0	x_1^-	x_2^+	x_3^-	x_4^+	x_5^-
1	–0.619508	0.450347	–0.354093	0.291847	–0.248257

$$2\Gamma\left[\alpha+1\right]-\Gamma\left[\alpha+1,2\mu x_i^-\right]-\Gamma\left[\alpha+1,2\mu x_{i-1}^+\right]=0\Rightarrow x_i^-, \quad (3.94)$$

$$2\Gamma\left[\alpha+1\right]-\Gamma\left[\alpha+1,-2\mu x_{i+1}^+\right]-\Gamma\left[\alpha+1,-2\mu x_i^-\right]=0\Rightarrow x_{i+1}^+. \quad (3.95)$$

Thus, it is seen that, unlike in Case I with an even restoring case, in the case with an odd restoring force twice more expressions for the energy-displacement functions have been derived due to the fact that the motion in one direction needed to be split into two parts: one for positive and the other one for negative displacements.

The use of the expressions (3.89)–(3.95) will be illustrated on the example of a restoring force the power of which is a ratio involving an even integer.

Example 3.2. Purely nonlinear oscillator with the power involving an even integer A purely nonlinear restoring force with $\alpha = 4/3$ is considered here. The amplitudes of motion are defined by Eqs. (3.94) and (3.95), which give the values presented in Table 3.2. Starting from $i = 1$ in Eq. (3.89) and knowing the value of x_0^+, the energy-displacement function (3.89) can be plotted (Fig. 3.3). Then, using Eq. (3.94), the value of x_1^- can be obtained numerically and it is given in Table 3.2. Further, for the known x_1^-, Eq. (3.90) can be used to plot the energy-displacement diagram from the zero position until $x_1^- \leq x \leq 0$. Then, for the known x_1^- and $i = 1$, Eq. (3.95) yields the value of x_2^+ and Eq. (3.92) for the energy-displacement function is completed. By proceeding in the analogous way, the energy-displacement function shown in Fig. 3.3 can be obtained. The accuracy of the obtained amplitudes is confirmed by comparing them with the numerically obtained amplitudes on the time response plotted in the lower part of Fig. 3.3.

3.3.2 Phase Trajectories and Some Characteristics of Motion

In this subsection, the phase trajectories of the system (3.31) will be obtained analytically. To that end, Eq. (3.31) is written down in the following form:

$$\frac{\mathrm{d}E_k}{\mathrm{d}x}+\operatorname{sgn}(x)\left|x\right|^{\alpha}=-\mu\operatorname{sgn}(\dot{x})\,\dot{x}^2, \quad (3.96)$$

so that the velocity can be expressed as

$$\dot{x}=\pm\sqrt{-\frac{1}{\mu}\frac{dE_k}{dx}-\frac{1}{\mu}\operatorname{sgn}(x)\left|x\right|^{\alpha}}. \quad (3.97)$$

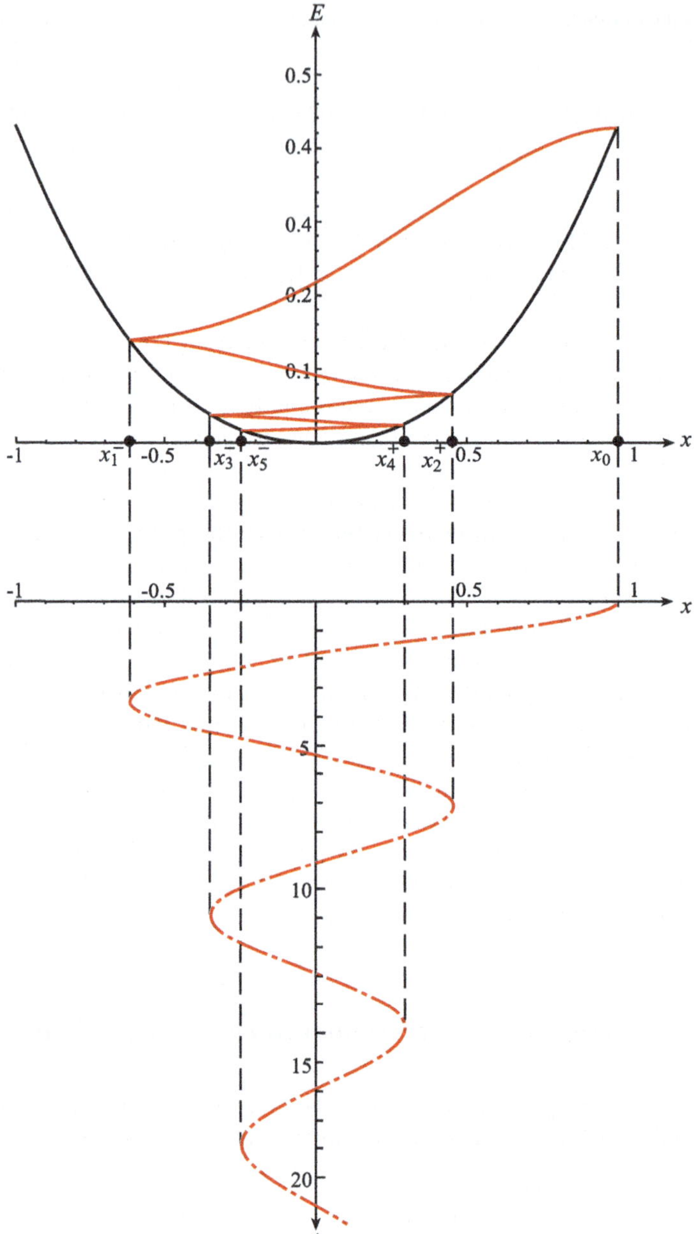

Fig. 3.3 Energy-displacement curves and time response for a purely nonlinear oscillator $\alpha = 4/3$ and $\mu = 0.5$

Now, the expressions for the energy-displacement function for two cases distinguished previously can be used to define the velocity (3.97) completely as a function of x, i.e. to obtain the expressions for phase trajectories. The kinetic energy E_k is described by the second term in Eqs. (3.61) and (3.62) for Case I and by the second term in Eqs. (3.89)–(3.93) for Case II.

3.3.2.1 Case I: Odd-Power Restoring Force

For the systems with an odd-power restoring force and during the motion to the left side, the use of Eqs. (3.61) and (3.97) yields

$$\dot{x}_{\mathrm{I}_{i\leftarrow}} = -\frac{1}{\sqrt{\mu}}\left(\frac{1}{2\mu}\right)^{\frac{\alpha}{2}} e^{\mu x}\sqrt{\Gamma\left[\alpha+1, 2\mu x\right] - \Gamma[\alpha+1, 2\mu x_{i-1}^{+}]}, x_i^{-} \leq x \leq x_{i-1}^{+}. \tag{3.98}$$

During the motion to the right, based on Eqs. (3.62) and (3.97), the part of the phase trajectory is defined by

$$\dot{x}_{\mathrm{I}_{i+1\rightarrow}} = +\frac{1}{\sqrt{\mu}}\left(\frac{1}{2\mu}\right)^{\frac{\alpha}{2}} e^{-\mu x}\sqrt{\Gamma\left[\alpha+1, -2\mu x\right] - \Gamma\left[\alpha+1, -2\mu x_i^{-}\right]}, x_i^{-} \leq x \leq x_{i+1}^{+}, \tag{3.99}$$

$$i = 2j - 1, \quad j \in \mathbb{N}. \tag{3.100}$$

3.3.2.2 Case II: Even-Power Restoring Force

If the restoring force is even, one can combine Eqs. (3.89)–(3.93) with Eq. (3.97) to derive the following expressions for the parts of the phase trajectories, from one relative maximum to the next one:

$$\dot{x}_{\mathrm{II}_{i\leftarrow+}} = -\frac{1}{\sqrt{\mu}}\left(\frac{1}{2\mu}\right)^{\frac{\alpha}{2}} e^{\mu x}\sqrt{\Gamma\left[\alpha+1, 2\mu x\right] - \Gamma[\alpha+1, 2\mu x_{i-1}^{+}]}, 0 \leq x \leq x_{i-1}^{+}, \tag{3.101}$$

$$\dot{x}_{\mathrm{II}_{i\leftarrow-}} = -\frac{1}{\sqrt{\mu}}\left(\frac{1}{2\mu}\right)^{\frac{\alpha}{2}} e^{\mu x}\sqrt{2\Gamma\left[\alpha+1\right] - \Gamma\left[\alpha+1, 2\mu x\right] - \Gamma[\alpha+1, 2\mu x_{i-1}^{+}]},$$

$$x_i^{-} \leq x \leq 0, \tag{3.102}$$

$$\dot{x}_{\mathrm{II}_{i+1\rightarrow-}} = \frac{1}{\sqrt{\mu}}\left(\frac{1}{2\mu}\right)^{\frac{\alpha}{2}} e^{-\mu x}\sqrt{\Gamma\left[\alpha+1, -2\mu x\right] - \Gamma\left[\alpha+1, -2\mu x_i^{-}\right]}, x_i^{-} \leq x \leq 0, \tag{3.103}$$

$$\dot{x}_{\mathrm{II}_{i+1}\to+} = \frac{1}{\sqrt{\mu}}\left(\frac{1}{2\mu}\right)^{\frac{\alpha}{2}} e^{-\mu x}\sqrt{2\Gamma\left[\alpha+1\right]-\Gamma\left[\alpha+1,-2\mu x\right]-\Gamma\left[\alpha+1,-2\mu x_i^-\right]},$$
$$0 \le x \le x_{i+1}^+. \tag{3.104}$$

3.3.2.3 Examples 3.1 and 3.2

The expressions for the phase trajectories derived above are plotted for Examples 3.1 and 3.2. They are shown in Fig. 3.4a and b, for the characteristic values of x_{i-1}^+ and x_i^- given in Tables 3.1 and 3.2 (Fig. 3.4c contains one more example, which will be discussed below). Different types of lines are used to emphasize the use of different expressions per intervals. It should be pointed out that these analytically obtained results were compared with the phase trajectories obtained by integrating the equation of motion numerically and the complete match was found. However, they are not shown here in order not to make the figure hard to interpret. The phase trajectories spiral in from the initial position and have apparent maxima of the velocity in the fourth quadrant.

3.3.3 *Maximal Velocities*

The maximal velocity can be obtained by differentiating Eqs. (3.98) and (3.101) (or Eqs. (3.99) and (3.103)) with respect to x. As a result of that, one derives

$$\frac{\mathrm{d}\dot{x}}{\mathrm{d}x} = -\mu\dot{x} + \frac{x^\alpha}{\dot{x}}. \tag{3.105}$$

In case of a maximum, the left side of Eq. (3.105) is zero, yielding

$$\left(\dot{x}_*\right)^2 = \frac{x_*^\alpha}{\mu}, \tag{3.106}$$

where the subscript '*' stands for the maximal velocity and the corresponding displacement. If plotted in the $x - \dot{x}$ plane, Eq. (3.106) represents the locus of maximal velocities. The loci corresponding to Examples 3.1 and 3.2 are shown in Fig. 3.4a and b as thick lines. It is interesting to note that in the case of a quadratic oscillator, the locus is a straight line, i.e. there is proportionality between the square of the maximal velocities and the corresponding displacements. This case is plotted in Fig. 3.4c.

Equation (3.106) indicates that the ratio of the square of the maximal velocity and the corresponding displacement raised to the power of the restoring force remains constant during the motion and it is equal to the reciprocal of the damping coefficient. Thus, this ratio represents the so-called 'invariant' of the system with a truly nonlinear restoring force and quadratic damping. This conclusion is in agreement with the physical mechanism of the system described by the equation of motion (3.31). Namely, when the velocity is maximal, the acceleration is zero because the motion considered is rectilinear. This fact yields the conclusion about the equilibrium

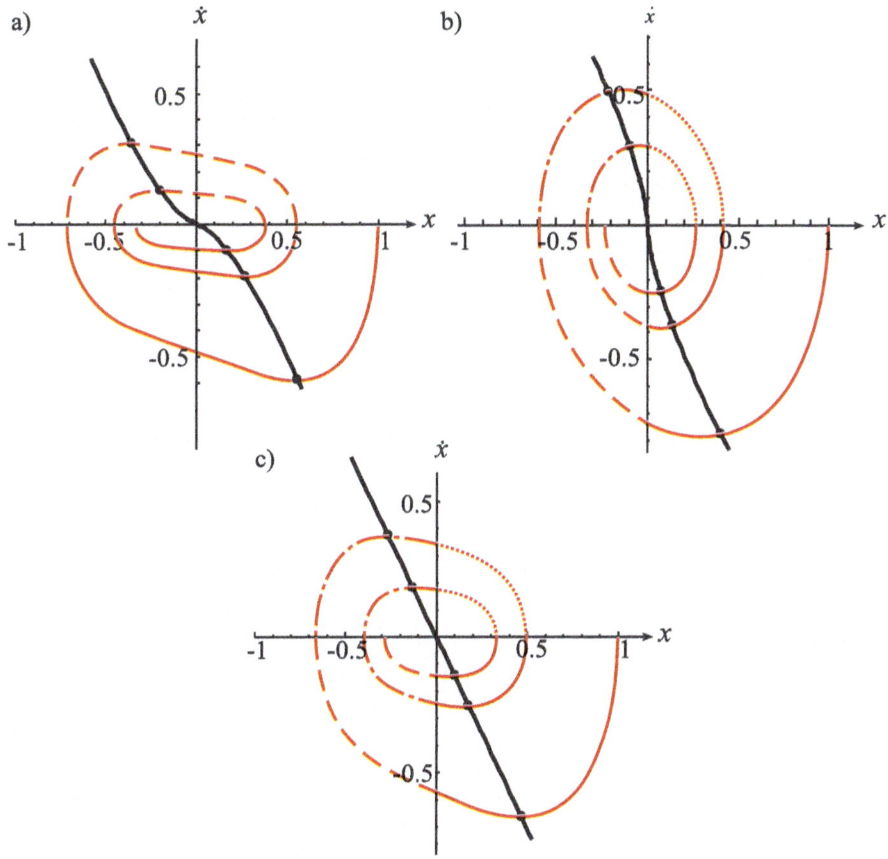

Fig. 3.4 Phase trajectories and loci of maximal velocities (thick solid line) for: **a** pure cubic oscillator $\alpha = 3$; **b** purely nonlinear oscillator $\alpha = 4/3$; **c** pure quadratic oscillator $\alpha = 2$. In all cases, $\mu = 0.5$

between the damping force and the restoring force at the position corresponding to maximal velocities.

3.3.4 Approximate Solutions for Motion

If $\mu << 1$, the oscillator (3.31) can be considered as weakly perturbed purely nonlinear oscillator and its equation of motion can be written in the following form to emphasize this fact:

$$\ddot{x} + \text{sgn}\,(x)\,|x|^{\alpha} = \mu\, f\,(\dot{x})\,, \tag{3.107}$$

$$f(\dot{x}) = -\operatorname{sgn}(\dot{x})\dot{x}^2. \tag{3.108}$$

In order to find the solution for motion in the approximate form, one can use the first-order averaging method [14] for purely nonlinear non-conservative oscillators. It involves the use of a generative solution, which is presented in Chap. 2, Sect. 2.5. Further, the general equations for the response to any $f(x, \dot{x})$ are derived and then the response with the quadratic damping force Eq. (3.107) is solved and presented. Note also that some other approaches for these type of oscillators can be found in [15–18].

3.3.4.1 Conservative Oscillators: Generative Solution

As shown in Chap. 2, Sect. 2.5, purely nonlinear oscillators governed by

$$\ddot{x} + \operatorname{sgn}(x)\,|x|^{\alpha} = 0 \tag{3.109}$$

are characterized with the oscillatory response whose period is exactly defined by

$$T = \sqrt{\frac{8\pi}{\alpha+1}}\frac{\Gamma\left(\frac{1}{\alpha+1}\right)}{\Gamma\left(\frac{\alpha+3}{2(\alpha+1)}\right)}a^{(1-\alpha)/2} = T^* a^{(1-\alpha)/2}, \tag{3.110}$$

where

$$T^* = \sqrt{\frac{8\pi}{\alpha+1}}\frac{\Gamma\left(\frac{1}{\alpha+1}\right)}{\Gamma\left(\frac{\alpha+3}{2(\alpha+1)}\right)}. \tag{3.111}$$

Note that herein a replaces A used in Sect. 2.5.1, but both of them are positive constants. Given the initial conditions (3.38), one has

$$T \equiv T^*. \tag{3.112}$$

The solution for a free response (motion and velocity) can be assumed in the form as the solution of motion of the linear oscillator (see Chap. 2, Sect. 2.2):

$$x = a\cos\psi, \tag{3.113}$$

$$\dot{x} = -a\omega\sin\psi, \tag{3.114}$$

with the complete phase ψ being

$$\psi = \omega(a)\,t + \varphi, \tag{3.115}$$

where φ stands for the initial phase. The angular frequency is noted as being the function of the amplitude a, and this dependence can be derived from the period (3.110)–(3.112) as follows:

$$\omega(a) = \frac{2\pi}{T} = \frac{2\pi}{T^*}\ |a|^{\frac{\alpha-1}{2}}, \tag{3.116}$$

or as

$$\omega(a) = W\ |a|^{\frac{\alpha-1}{2}}, \quad W = \sqrt{\frac{\alpha+1}{2}}\frac{\sqrt{\pi}\Gamma\left(\frac{\alpha+3}{2(\alpha+1)}\right)}{\Gamma\left(\frac{1}{\alpha+1}\right)}. \tag{3.117}$$

3.3.4.2 Non-conservative Oscillator: Approximate Solutions Via Trigonometric Functions

Let us now consider Eq. (3.107) as the perturbation of Eq. (3.109) with $\mu << 1$. Thus, the approximate solution for the response can be assumed in the same form as Eq. (3.113), but with the amplitude and complete phase changing in time

$$x = a(t)\cos\psi(t), \tag{3.118}$$

$$\psi(t) = \int \omega(a)\,dt + \varphi(t). \tag{3.119}$$

Differentiating Eq. (3.118) with respect to time leads to

$$\dot{x} = \dot{a}\ \cos\psi - a\,\omega\sin\psi - a\,\dot{\varphi}\sin\psi. \tag{3.120}$$

Imposing the requirement that this solution should have the form (3.114):

$$\dot{x} = -a(t)\,\omega(a)\sin\psi(t), \tag{3.121}$$

the following constraint is derived:

$$\dot{a}\ \cos\psi - a\,\dot{\varphi}\sin\psi = 0. \tag{3.122}$$

The next step involves differentiating (3.121) with respect to time and substituting it into Eq. (3.107) to obtain

$$-\dot{a}\omega\sin\psi - a\frac{\mathrm{d}\omega}{\mathrm{d}a}\dot{a}\ \sin\psi - a\,\omega\,\dot{\varphi}\cos\psi = \mu\ f\,(a\cos\psi, -a\omega\sin\psi). \tag{3.123}$$

Combining Eqs. (3.122) and (3.123), one can derive

$$\dot{a}\left(\omega + a\frac{\mathrm{d}\omega}{\mathrm{d}a}\sin^2\psi\right) = -\mu\ f\,(a\cos\psi, -a\omega\sin\psi)\sin\psi, \tag{3.124}$$

$$a\,\omega\,\dot{\varphi} + a\frac{d\omega}{da}\dot{a}\;\sin\psi\cos\psi = -\mu\; f\,(a\cos\psi, -a\omega\sin\psi)\cos\psi. \quad (3.125)$$

Recognizing from Eq. (3.117) that $d\omega\,(a)\,/da = (\alpha - 1)\,\omega a^{-1}/2$, and substituting it into the previous equations and performing their averaging over 2π give

$$\frac{da}{dt} = -\frac{2\mu}{\pi\,(\alpha+3)\,\omega\,(a)}\int_0^{2\pi} f\,(a\;\cos\psi, -a\omega\;\sin\psi)\;\sin\,d\psi, \quad (3.126)$$

$$\frac{d\psi}{dt} = \omega\,(a) - \frac{\mu}{2\pi a\;\omega\,(a)}\int_0^{2\pi} f\,(a\;\cos\psi, -a\omega\;\sin\psi)\;\cos\,d\psi. \quad (3.127)$$

When the oscillator is linear, i.e. $\alpha = 1$, the well-known equations for the derivatives of the amplitude and phase of the classical Krylov–Bogoliubov method are derived [5]:

$$\frac{da}{dt} = -\frac{\mu}{2\pi}\int_0^{2\pi} f\,(a\;\cos\psi, -a\omega\;\sin\psi)\;\sin\psi d\psi, \quad (3.128)$$

$$\frac{d\varphi}{dt} = -\frac{\mu}{2\pi a}\cos\psi d\psi. \quad (3.129)$$

3.3.4.3 Application to the System with Quadratic Damping

For the non-conservative term (3.108), Eqs. (3.126) and (3.127) yield

$$a = [1 + \frac{4\sqrt{\frac{2}{\pi}}\,(\alpha+1)^{3/2}\,\Gamma\left(\frac{\alpha+3}{2(\alpha+1)}\right)\mu t}{3\,(\alpha+3)\,\Gamma\left(\frac{1}{\alpha+1}\right)}]^{-\frac{2}{\alpha+1}}, \quad (3.130)$$

$$\psi = -\frac{3\pi\,(\alpha+3)}{16\mu} + \frac{(1 + \frac{4\sqrt{\frac{2}{\pi}}(\alpha+1)^{3/2}\Gamma\left(\frac{\alpha+3}{2(\alpha+1)}\right)\mu t}{3(\alpha+3)\Gamma\left(\frac{1}{\alpha+1}\right)})^{\frac{\alpha-1}{\alpha+1}}\left[3\pi\,(\alpha+3) + \frac{4\sqrt{2\pi}(\alpha+1)^{3/2}\Gamma\left(\frac{\alpha+3}{2(\alpha+1)}\right)\mu t}{\Gamma\left(\frac{1}{\alpha+1}\right)}\right]}{16\mu}. \quad (3.131)$$

It should be noted that for a linear oscillator, the amplitude decay (3.130) and the complete phase become

$$a = \frac{1}{1+\frac{4\mu}{3\pi}t}, \quad (3.132)$$

$$\psi = t, \quad (3.133)$$

implying that damping affects the change of the amplitude, but not the complete phase at this level of approximations. The solutions for the amplitude and complete phase are in complete agreement with the solution found in [3].

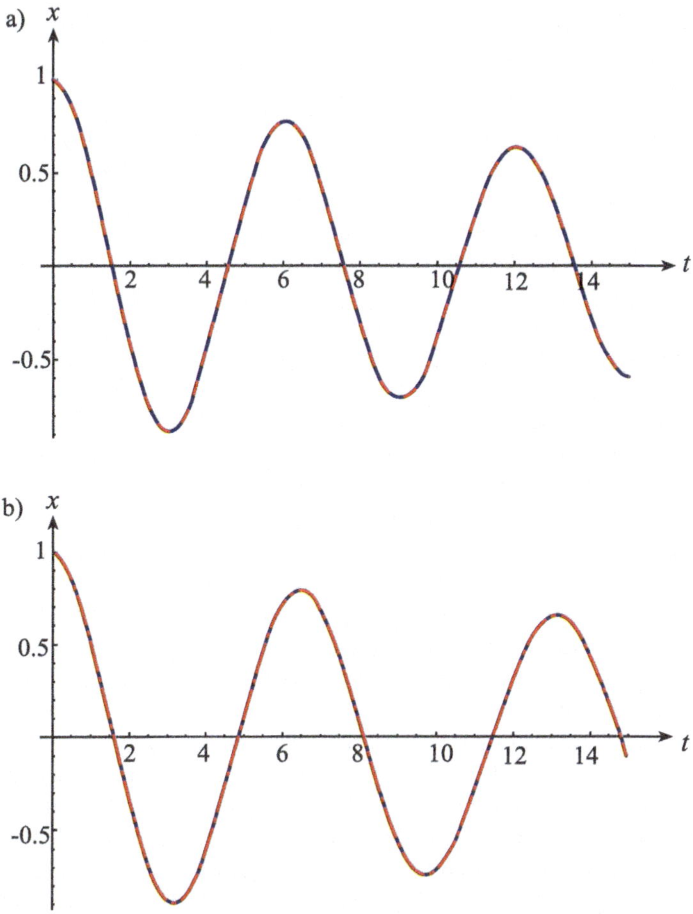

Fig. 3.5 Comparison of the time response obtained numerically (black solid line) and **a** approximate solution (red dashed line) for α =9/11; **b** approximate solution (red dashed-dotted line) for $\alpha =$ 11/9. In both cases, $\mu = 0.1$

For the confirmation of the accuracy of the results derived for $\alpha \neq 1$, the time response obtained analytically (3.130), (3.131) is compared with the numerical results and shown in Fig. 3.5 for two different cases: the under-linear case $\alpha = 9/11$ and the over-linear case $\alpha = 11/9$. In both cases, the coefficient μ is kept constant and equal to 0.1. It is seen that the approximate solutions agree well with the numerical results.

3.4 Purely Nonlinear Oscillators with Fractional Damping: Approximate Solutions via Trigonometric Functions

Fractional calculus has recently attracted great interest among many researchers in science and engineering due to its wide and disparate applications that include viscoelasticity and rheology, biology and bioengineering, electrical engineering, earthquake engineering, signal and image processing, physics, control theory, etc. [19]. Progress has also been made in the theory of oscillations, where fractional derivatives have been included in dynamic models of some generic oscillators: linear [20–24], Duffing [23–25], van der Pol oscillators [23, 26], as well as the systems modelled by the Mathieu equation [27, 28]. In only few of them, characteristic properties with respect to those typical for the linearly damped systems are discussed and the differences are highlighted. Burov and Barkai [20] explored a linear oscillator with fractional damping, distinguishing over-damped motion, under-damped motion and the motion with a few cycles that decays without zero crossings. Naber [21] considered a linear oscillator with the damping generalized to the Caputo fractional derivative and obtained its solution for motion by means of the Laplace transform. He distinguished several different cases when the increase in the order of fractional derivative increases the effects of damping. In [27], the authors studied the Mathieu equation with the damping modelled by the Riemann–Liouville fractional from the viewpoint of the influence of the fractional derivative damping on the location and shape of transition curves.

In this section, free oscillators with the restoring term having a multi-term or single-term power form and with the damping term containing a fractional derivative are investigated, as done in [29]. An approximate method similar to the one presented in Sect. 3.3.4.2 is adjusted and the first-order differential equations for the amplitude and the phase are derived. Physical insights into vibration properties of several exemplary oscillators are given and some differences between them and the ones typical for linearly viscously damped systems are discussed.

The systems considered are modelled by the following non-dimensional equation:

$$\ddot{x} + G(x) = \varepsilon f(x, \dot{x}, D^{Q}x), \tag{3.134}$$

$$G(x) = \sum_{\alpha} c_{\alpha}^{2} \operatorname{sgn}(x) |x|^{\alpha}, \tag{3.135}$$

with the initial conditions

$$x(0) = 0, \ \dot{x}(0) = v_0. \tag{3.136}$$

The restoring force given by Eq. (3.135) has a general form of a sum of power-form terms with positive real exponentsα; c_{α} and v_0 are real positive constants, and there is a restriction that the parameter ε appearing in Eq. (3.134) is small: $\varepsilon << 1$ and f stands for any autonomous force that depends on the displacement x, velocity and/or a fractional derivative term $D^{Q}x$. This term is defined by means of the fractional derivative of order Q ($0< Q<1$), with the Riemann–Liouville formulation of the fractional derivative for $x(0) = 0$ [30]:

$$D^Q x = \frac{1}{\Gamma(1-Q)} \int_0^t \frac{\dot{x}(t-u)}{u^Q} du, \tag{3.137}$$

where Γ is the Euler gamma function.

3.4.1 Approximate Solutions via Trigonometric Functions

In order to find the response of this type of oscillators, the solution for motion and velocity is assumed in the form:

$$x = a(t) \sin \psi(t), \tag{3.138}$$

$$\dot{x} = a(t)\omega \cos \psi(t), \tag{3.139}$$

where

$$\psi = \int_0^t \omega(a)\mathrm{d}t + \theta(t), \tag{3.140}$$

with a and θ representing the amplitude and phase that need to be determined. In addition, the frequency is also an unknown function of the amplitude and this relationship will also be determined.

On the basis of Eq. (3.138), the following expression is derived:

$$\dot{x} = \dot{a} \sin \psi + a\omega \cos \psi + a\dot{\theta} \cos \psi, \tag{3.141}$$

which, after comparing it with Eq. (3.139), imposes the following constraint:

$$\dot{a} \sin \psi + a\dot{\theta} \cos \psi = 0. \tag{3.142}$$

Finding the second derivative of Eq. (3.139) and substituting it together with the assumed solution, Eq. (3.138) into Eq. (3.134), one derives

$$\begin{aligned} &\dot{a}\omega \cos \psi + a\frac{d\omega}{da}\dot{a} \cos \psi - a\omega\dot{\theta} \sin \psi - a\omega^2 \sin \psi \\ &\quad + \sum_\alpha c_\alpha^2 \mathrm{sgn}(a \sin \psi)|a \sin \psi|^\alpha = \varepsilon f, \end{aligned} \tag{3.143}$$

where $f = f(a \sin \psi, a\omega \cos \psi, D^Q(a \sin \psi))$.

The last term on the left-hand side of Eq. (3.143) can be approximated by using the Fourier series expansion [16], i.e.

$$\text{sgn}(\sin\psi)|\sin\psi|^{\alpha} \approx A_{1_\alpha}\sin\psi, \tag{3.144}$$

where A_{1_α} is

$$A_{1_\alpha} = \frac{2\Gamma(1+\frac{\alpha}{2})}{\sqrt{\pi}\Gamma(\frac{3+\alpha}{2})}. \tag{3.145}$$

This approximation can be used now to balance the last term on the left side of Eq. (3.143) with the term in front of it, leading to the following expression for the frequency:

$$\omega^2(a) = \sum_{\alpha} c_{\alpha}^2 A_{1_\alpha}|a|^{\alpha-1}. \tag{3.146}$$

Further, the second term on the left-hand side of Eq. (3.143) contains the derivative of the frequency with respect to the amplitude, which can be calculated now from Eq. (3.146)

$$\frac{\mathrm{d}\omega}{\mathrm{d}a} = \frac{1}{2\omega a}\left[\sum_{\alpha} c_{\alpha}^2 A_{1_\alpha}(\alpha-1)|a|^{\alpha-1}\right]. \tag{3.147}$$

Combining Eqs. (3.143) and (3.147) with the constraint (3.142), one obtains

$$\dot{a} = \frac{\varepsilon}{\pi}\frac{2\omega}{4\omega^2 + (\sum_{\alpha} c_{\alpha}^2 A_{1_\alpha}(\alpha-1)|a|^{\alpha-1})}\int_0^{2\pi} f\cos\psi \mathrm{d}\psi, \tag{3.148}$$

$$a\dot{\theta} = \frac{\varepsilon}{2\pi\omega}\int_0^{2\pi} f\sin\psi \mathrm{d}\psi. \tag{3.149}$$

There is only one term in Eqs. (3.148) and (3.149) that needs additional consideration: the fractional damping term $D^Q(a\sin\psi)$ that appears in f. To adjust it for the technique presented, the procedure developed in [24] is used. First, the time t and the variable u from Eq. (3.137) are rescaled: $d\tau = \omega dt$, $s = \omega u$. This leads to

$$D^Q x = \frac{1}{\Gamma(1-Q)}\int_0^{\tau}\frac{\omega x'(\tau-s)}{\frac{s^Q}{\omega^Q}}\frac{ds}{\omega} = \frac{a\omega^Q}{\Gamma(1-Q)}\int_0^{\tau}\frac{x'(\tau-s)}{s^Q}ds, \tag{3.150}$$

with $x'(\tau-s) = a\cos(\tau+\theta-s)$.

This can be further rewritten as

$$D^Q x = \frac{a\omega^Q}{\Gamma(1-Q)}[\cos(\tau+\theta)I_c + \sin(\tau+\theta)I_s], \tag{3.151}$$

where I_c and I_s stand for the following integrals:

$$I_c = \int_0^{\tau}\frac{\cos s}{s^Q}ds, \tag{3.152}$$

$$I_s = \int_0^{\tau} \frac{\sin s}{s^Q} ds. \tag{3.153}$$

These integrals cannot be evaluated in closed forms for a general value of τ, but can be determined for the case when $\tau \to \infty$:

$$\int_0^{\infty} \frac{\cos s}{s^Q} ds = \Gamma(1-Q)\sin(Q\frac{\pi}{2}) \ , \tag{3.154}$$

$$\int_0^{\infty} \frac{\sin s}{s^Q} ds = \Gamma(1-Q)\cos(Q\frac{\pi}{2}) \ . \tag{3.155}$$

Hence, I_c and I_s in Eq. (3.151) are approximated by the expressions from Eq. (3.154) [24], giving

$$D^Q x = a\omega^Q[\sin(Q\frac{\pi}{2})\cos(\tau+\theta) + \cos(Q\frac{\pi}{2})\sin(\tau+\theta)]. \tag{3.156}$$

This expression can be rewritten in terms of the complete phase defined by Eq. (3.140)

$$D^Q(a\sin\psi) = a\omega^Q[\sin(Q\frac{\pi}{2})\cos\psi + \cos(Q\frac{\pi}{2})\sin\psi], \tag{3.157}$$

which is now written in a suitable form for the application of averaging.

The expression for the damping force of the fractional-order Q defined by Eq. (3.157) can be rewritten by using Eqs. (3.138), (3.139) as

$$\varepsilon D^Q = \varepsilon[\omega(a(t))]^Q \cos(Q\frac{\pi}{2})x + \varepsilon[\omega(a(t))]^{Q-1}\sin(Q\frac{\pi}{2})\dot{x}, \tag{3.158}$$

which gives the insight into the physical nature of this force. In the special cases when $Q = 0$, it corresponds to a linear restoring force $\varepsilon D^0 \equiv \varepsilon x$. When $Q = 1$, it corresponds to a linear viscous force $\varepsilon D^1 \equiv \varepsilon\dot{x}$. For $0 < Q < 1$, it is the function of the displacement and velocity with rheonomic coefficients.

It is important to note that Eq. (3.158) has the same form as the model of the damping force in a viscously damper consisting of a piston moving in a highly viscous fluid considered in [31], which is equal to $C_0(\omega^Q\cos(Q\pi/2)x + \omega^{Q-1}\sin(Q\pi/2)\dot{x})$. This model, derived from a certain force-displacement relationship and then re-expressed by employing the Fourier transform, was found to agree well with experimental results. Unlike the results shown in [31] obtained by fitting experimental results, Eq. (3.158) gives the coefficients in front of x and $\dot{x}$ in the analytical form as depending on the frequency, whose change with time is defined by Eqs. (3.146) and (3.148).

Thus, by using Eq. (3.157), the function f can be expressed as

$$f = f(a\sin\psi, a\omega\cos\psi, a\omega^Q[\sin(Q\frac{\pi}{2})\cos\psi + \cos(Q\frac{\pi}{2})\sin\psi]), \tag{3.159}$$

and by solving the integrals in Eqs. (3.148) and (3.149), the solution for the amplitude and phase can be obtained.

To find the solutions of Eqs. (3.148) and (3.149), the initial values for the amplitude and phase need to be expressed in terms of the initial conditions Eq. (3.136). Combining the condition Eq. (3.136) with the solution Eq. (3.138), one concludes that $\psi(0) = \psi_0 = 0$. Equation (3.138) is satisfied for

$$\dot{x}(0) = v_0 = a_0\omega(a_0)\cos\psi_0 = a_0\sqrt{\sum_\alpha c_\alpha^2 A_{1_\alpha}|a_0|^{\alpha-1}}, \tag{3.160}$$

where a_0 is a constant that needs to be calculated.

In the case of a single-term nonlinearity, i.e. when $G(x) = c_\alpha \mathrm{sgn}(x)|x|^\alpha$, the frequency Eq. (3.146) becomes

$$\omega(a) = c|a|^{\frac{\alpha-1}{2}},\ c = c_\alpha\sqrt{A_{1_\alpha}}. \tag{3.161}$$

For $\alpha = 1$, i.e. when the oscillator is linear, one has $A_{1_1} = 1$ and $\omega(a) = c_1$, which is the well-known result. If α increases from zero, the values of A_{1_α} and the constant c decrease. When compared with the exact frequency, this approximately obtained frequency showed the percentage error smaller than 1% for the power α between 0.4 and 1.8, approximately, while for α from 0 to 0.4 and 1.8 to 3.1, this error is less than 5%. This approximation imposes some limitations: the results are seen to be reliable for the power of the restoring force relatively close to unity. The values of α in the examples considered below will be chosen as such.

Combining Eqs. (3.148) and (3.149) with Eq. (3.161), the first-order differential equations for the amplitude and phase for single-term nonlinearity are derived

$$\frac{\mathrm{d}a}{\mathrm{d}t} = \frac{2\varepsilon}{\pi(\alpha+3)c|a|^{\frac{\alpha-1}{2}}}\int_0^{2\pi} f\cos\psi\,\mathrm{d}\psi, \tag{3.162}$$

$$a\frac{\mathrm{d}\theta}{\mathrm{d}t} = \frac{\varepsilon}{2\pi c|a|^{\frac{\alpha-1}{2}}}\int_0^{2\pi} f\sin\psi\,\mathrm{d}\psi. \tag{3.163}$$

In this case, based on Eq. (3.160), one can determine the sought value of the constant a_0:

$$a_0 = \left(\frac{v_0}{c}\right)^{\frac{2}{\alpha+1}}. \tag{3.164}$$

3.4.1.1 Example 3.3 Generalized van der Pol Oscillator with Fractional Derivative Damping

This example is concerned with the oscillator Eq. (3.134), where

$$f = (1-|x|^\beta)D^Q x, \tag{3.165}$$

with β being an arbitrary positive real constant. When $\alpha = 1$, $\beta = 2$ and $Q = 1$, the model corresponds to a classical van der Pol oscillator.

By substituting Eq. (3.165) into the form Eq. (3.159), and then into Eqs. (3.148) and (3.149), one obtains

$$\dot{a} = \frac{\varepsilon}{\sqrt{\pi}} \frac{2a\omega^Q}{4\omega^2 + (\sum_\alpha c_\alpha^2 A_{1_\alpha}(\alpha - 1)|a|^{\alpha-1})} (\sqrt{\pi} - |a|^\beta \frac{\Gamma(\frac{\beta+1}{2})}{\Gamma(\frac{\beta+4}{2})}) \sin(Q\frac{\pi}{2}), \tag{3.166}$$

$$\dot{\theta} = \frac{\varepsilon}{2\sqrt{\pi}} \omega^{Q-1} (\sqrt{\pi} - 2|a|^\beta \frac{\Gamma(\frac{\beta+3}{2})}{\Gamma(\frac{\beta+4}{2})}) \cos(Q\frac{\pi}{2}). \tag{3.167}$$

Equation (3.166) gives the following steady-state amplitude a_s:

$$|a_s| = \left[\sqrt{\pi} \frac{\Gamma(\frac{\beta+4}{2})}{\Gamma(\frac{\beta+1}{2})} \right]^{\frac{1}{\beta}}. \tag{3.168}$$

This expression implies that the fractional order and the powers of the restoring force do not influence the amplitude of the limit cycle at this level of approximation. According to Eq. (3.166), these parameters have the effect on the rate of how this limit cycle is achieved. The result given by Eq. (3.168) coincides with the amplitude of a limit cycle derived in [32] for $Q = 1$, where its change with β is shown. The difference between the limit cycle of the oscillator with linear viscous damping and the one considered herein is the steady-state value of the phase, which is zero for $Q = 1$ [32], and otherwise

$$\dot{\theta}_s = -\frac{\varepsilon\beta}{2} \omega^{Q-1} \cos(Q\frac{\pi}{2}). \tag{3.169}$$

It should be noted that for $\alpha = 1$, $\beta = 2$, Eqs. (3.166) and (3.167) have the form derived in [26] by the method of multiple expansion, the accuracy of which is confirmed numerically therein. When $\alpha = 1$, $\beta = 2$ and $Q = 1/2$, they are equal to those derived by [24]. The approximate analytical steady-state amplitude Eq. (3.168) is compared with the numerical solution of the equation of motion Eqs. (3.134), (3.165) and plotted in Fig. 3.6 for different multi-term and single-term restoring forces. In general, the agreement between the solutions is good. It should be noted that numerical simulations in this study were carried out by calculating fractional-order derivatives as shown in [33], and $Q = 1/2$ was chosen for comparisons in this work.

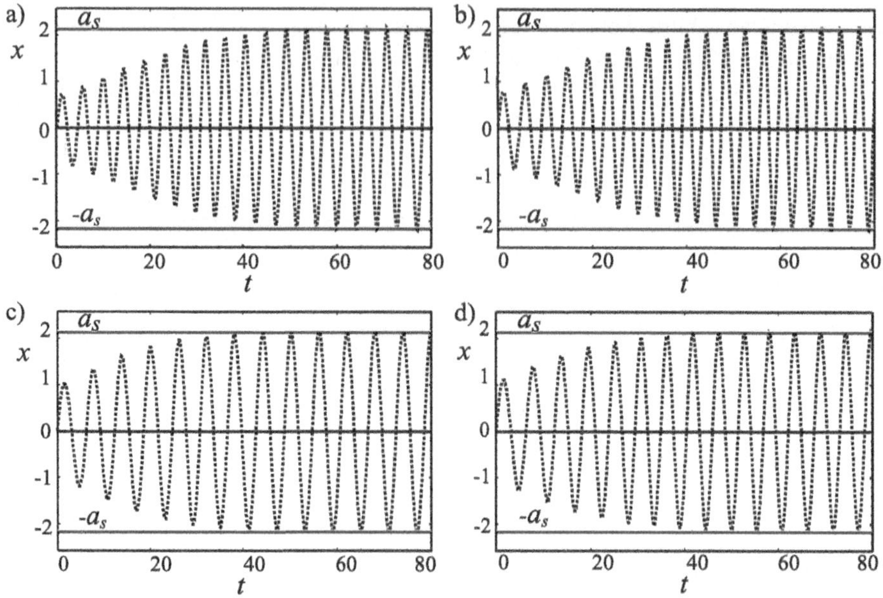

Fig. 3.6 Time response calculated numerically by carrying out direct integration of the equation of motion Eq. (3.134) (dotted line) and the steady-state amplitude Eq. (3.168) (solid line) for ε =0.2, $v_0 = 1$, $\beta = 3/2$, $Q = 1/2$ and: **a** $G(x) = x + \text{sgn}(x)\,|x|^{0.8}$; **b** $G(x) = x + \text{sgn}(x)\,|x|^{1.2}$; **c** $G(x) = \text{sgn}(x)\,|x|^{0.8}$; **d** $G(x) = \text{sgn}(x)\,|x|^{1.2}$

3.4.1.2 Example 3.4 Purely Nonlinear Oscillators with a Fractional Derivative Damping Term

Let us consider the oscillator governed by Eq. (3.134) with a pure single-term nonlinearity $G(x) = c_\alpha \text{sgn}(x)|x|^\alpha$ and with

$$f = -D^Q x. \tag{3.170}$$

Equations (3.162) and (3.163) are now

$$\dot{a} = -\frac{2\varepsilon \sin(Q\frac{\pi}{2})}{(\alpha+3)c^{1-Q}} a^{\frac{2-(\alpha-1)(1-Q)}{2}}, \tag{3.171}$$

$$\dot{\psi} = ca^{\frac{\alpha-1}{2}} + \frac{\varepsilon \cos(Q\frac{\pi}{2})}{2c^{1-Q}} a^{-\frac{(\alpha-1)(1-Q)}{2}}. \tag{3.172}$$

It should be noted that the absolute value has been omitted in Eqs. (3.171), (3.172), as it has been assumed that a >0. The results obtained will then correspond to those related to the upper envelope of the response.

The solutions of Eqs. (3.171), (3.172) are

$$a = [E - Dt]^{\frac{2}{(\alpha-1)(1-Q)}}, \tag{3.173}$$

$$\psi = -\frac{c(1-Q)D}{2-Q}[E - Dt]^{\frac{2-Q}{1-Q}} + \frac{c(1-Q)ED}{2-Q} - \frac{\varepsilon D \cos(Q\frac{\pi}{2})}{2c^{1-Q}} \ln[1 - \frac{D}{E}t], \tag{3.174}$$

where

$$E = a_0^{\frac{(\alpha-1)(1-Q)}{2}}, \tag{3.175}$$

$$D = \frac{\varepsilon(\alpha-1)(1-Q)\sin(Q\frac{\pi}{2})}{(\alpha+3)c^{1-Q}}. \tag{3.176}$$

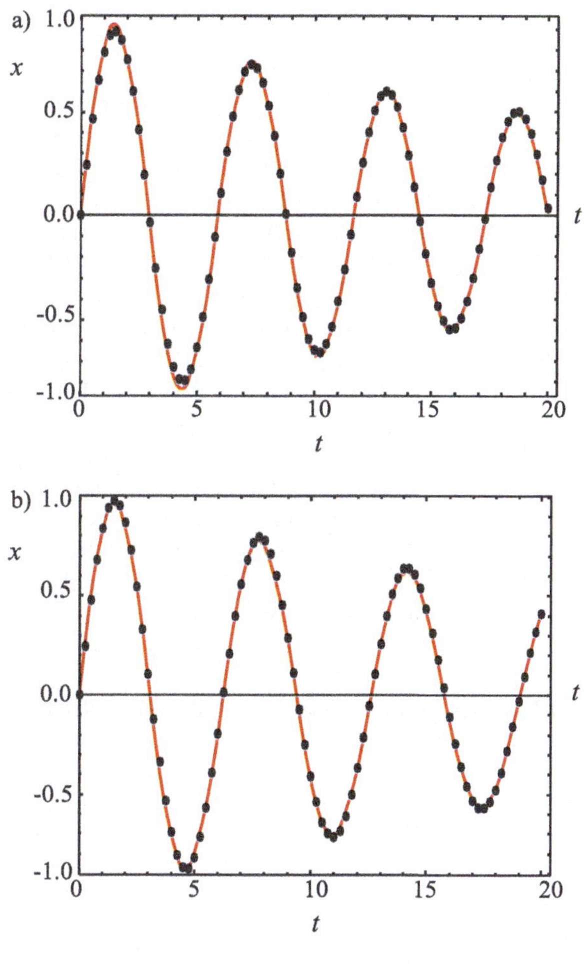

Fig. 3.7 Comparison of the numerical solution (dots) and the analytical solution defined by Eqs. (3.138), (3.173)–(3.176) (solid line) for the time response of the oscillator (3.134), (3.135), (3.170) for $\varepsilon = 0.1$, $c_\alpha = 1$, $v_0 = 1$, $Q = 1/2$: **a** $\alpha = 0.8$; **b** $\alpha = 1.2$

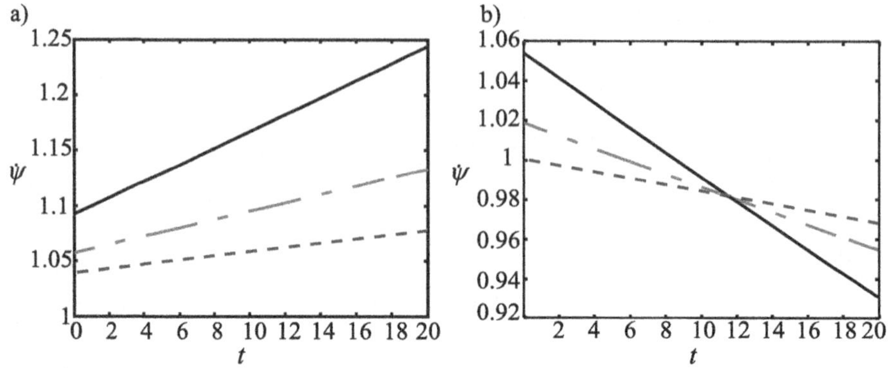

Fig. 3.8 Damped frequency (3.172) versus time for $c_\alpha = 1$, $v_0 = 1$, $Q = 1/2$ and different values of the small parameter $\varepsilon = 0.05$ (dotted line), $\varepsilon = 0.1$ (dashed-dotted line), $\varepsilon = 0.2$ (solid line) for: **a** $\alpha = 0.8$; **b** $\alpha = 1.2$

Due to the limited domain of the logarithmic function, the solution obtained is valid until the moment t when the argument of the logarithmic function becomes equal to zero. The solution defined by Eqs. (3.138), (3.173)–(3.176) is compared with the numerical solution for the under-linear restoring force ($\alpha = 0.8$) and the over-linear one ($\alpha = 1.2$). These cases are shown, respectively, in Fig. 3.7a and b. As seen from this figure, good accuracy of the approximate solution is achieved.

By analysing the amplitude decay of the nonlinear system Eq. (3.171) and the related damped frequency Eq. (3.172), one can distinguish two different cases of the system behaviour depending on the value of the power α: $\alpha < 1$ and $\alpha > 1$. The change of the damped frequency Eq. (3.172) in time is shown in Fig. 3.8a for $\alpha < 1$ ($\alpha = 0.8$) and in Fig. 3.8b for $\alpha > 1$ ($\alpha = 1.2$) for different values of the small parameterα and for the fixed values of the rest of them: $c_\alpha = 1$, $v_0 = 1$, $Q = 1/2$. For the case $\alpha < 1$, the damped frequency increases with time, being higher for higher values of ε. Consequently, the period will have a decreasing trend as time passes. For the case $\alpha > 1$, the damped frequency decreases with time, while the period increases. The rate of its change is higher if ε is higher. The difference in the change of the periods of vibrations for $\alpha = 0.8$ and $\alpha = 1.2$ can also be observed in Fig. 3.7, if one compares them directly on the time response.

3.4.1.3 Example 3.5 Linear Oscillators with a Fractional Derivative Damping Term

A linear oscillator $\alpha = 1$ with the fractional derivative damping force Eq. (3.170) is considered. Equations (3.171) and (3.172) now become

$$\dot{a} = -\frac{\varepsilon \sin(Q\frac{\pi}{2})}{2c_1^{1-Q}} a, \tag{3.177}$$

$$\dot{\psi} = c_1 + \frac{\varepsilon \cos(Q\frac{\pi}{2})}{2c_1^{1-Q}}. \tag{3.178}$$

Their integration yields

$$a = \frac{v_0}{c_1} e^{-\frac{\varepsilon}{2c_1^{1-Q}} \sin(Q\frac{\pi}{2})t}, \tag{3.179}$$

$$\psi = [c_1 + \frac{\varepsilon \cos(Q\frac{\pi}{2})}{2c_1^{1-Q}}]t. \tag{3.180}$$

The approximate solution for motion is given by

$$x = \frac{v_0}{c_1} e^{-\frac{\varepsilon}{2c_1^{1-Q}} \sin(Q\frac{\pi}{2})t} \sin[c_1 + \frac{\varepsilon}{2c_1^{1-Q}} \cos(Q\frac{\pi}{2})]t. \tag{3.181}$$

The approximate analytical solution given by Eq. (3.181) is compared with the numerical solution of the equation of motion, defined by Eqs. (3.134), (3.170) to check its accuracy and they are both plotted in Fig. 3.9. It is seen that the match between the solutions is good.

Frequency of damped oscillations The frequency of linear oscillations with the fractional derivative damping is defined by Eq. (3.178). Investigating $d\dot{\psi}/dQ$, it is found that the frequency has an extremum for

$$-2\cos(Q\frac{\pi}{2}) \log c_1 + \pi \sin(Q\frac{\pi}{2}) = 0. \tag{3.182}$$

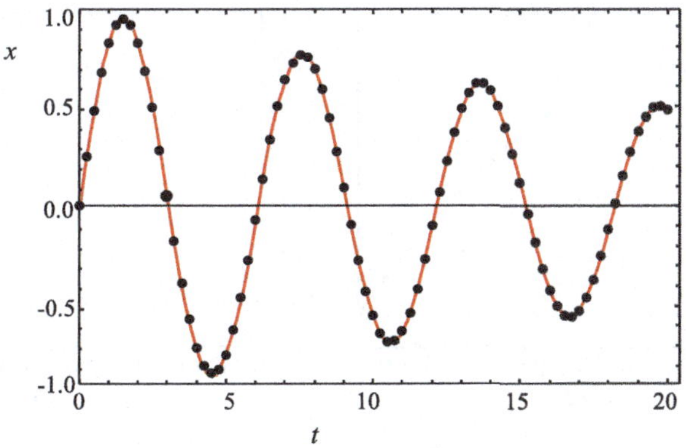

Fig. 3.9 Comparison of the numerical solution (dots) and the analytical solution defined by Eq. (3.181) (solid line) for the time response of the oscillator Eqs. (3.134), (3.135), (3.170) for $\alpha = 1$, $\varepsilon = 0.1$, $c_1 = 1$, $v_0 = 1$, $Q = 1/2$

This is illustrated in Fig. 3.10a: for $c_1 > 1$, as Q increases from zero, the damped frequency increases until it reaches its maximum Q^* and then decreases, which agrees with the results presented in [21]. If $c_1 \leq 1$, the damped frequency monotonically decreases with the increase of Q (Fig. 3.10b). For the sake of comparison, the damped frequency of the corresponding viscously damped linear oscillator (VDL)

$$\dot{\psi}_{VDL} = \frac{2\pi}{\sqrt{c_1^2 - (\frac{\varepsilon}{2})^2}} \tag{3.183}$$

is also shown.

By using Eq. (3.182), the fractional order corresponding to the maximum of the damped frequency for $c_1 > 1$ is obtained

$$Q = \frac{2}{\pi}\arctan(\frac{2\log c_1}{\pi}). \tag{3.184}$$

As c_1 becomes higher, the location of the maximum moves towards higher values of Q, slowly approaching unity as $c_1 \to \infty$.

Logarithmic decrement By using Eq. (3.178), the quasi-period can be obtained as follows:

$$T = \frac{2\pi}{\dot{\psi}} = \frac{2\pi}{c_1 + \frac{\varepsilon\cos(Q\frac{\pi}{2})}{2c_1^{1-Q}}}. \tag{3.185}$$

The solution for the amplitude given by Eq. (3.179) indicates that the amplitude decreases in time following the exponential law.

Analogously to the case of linear viscous damping, the logarithmic decrement can be found. This decrement is defined as the ratio of the successive amplitudes,

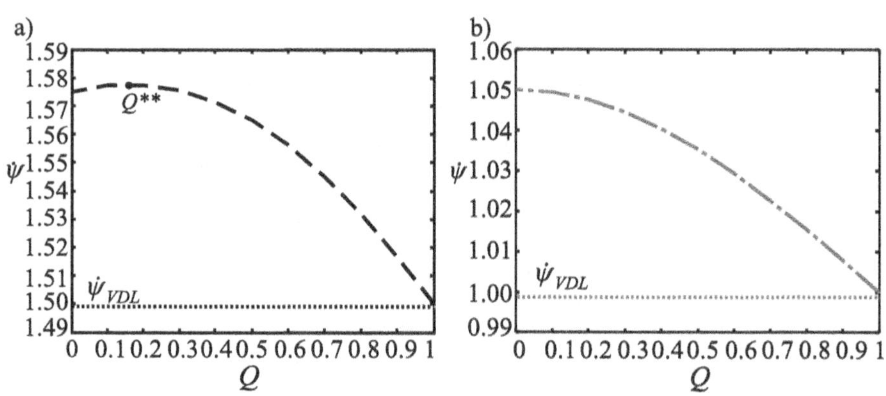

Fig. 3.10 Influence of the fractional order Q on the damped frequency given by Eq. (3.178) for $\varepsilon = 0.1$: **a** $c_1 = 1.5$, **b** $c_1 = 1$. Horizontal lines depict the damping frequency of the corresponding viscously damped linear oscillator

$\Delta = \log(\frac{a(t)}{a(t+T)})$ [34], and can be expressed here as

$$\Delta = \frac{2\varepsilon\pi \sin(Q\frac{\pi}{2})}{2c_1^{2-Q} + \varepsilon \cos(Q\frac{\pi}{2})}. \tag{3.186}$$

When $Q = 1$, this expression becomes $\Delta(Q = 1) = \frac{\varepsilon\pi}{c_1}$, which is the well-known result obtained as the approximation of the decrement for linear viscous damping $\Delta_L = \frac{\varepsilon\pi}{\sqrt{c_1^2-(\frac{\varepsilon}{2})^2}}$ when $(\frac{\varepsilon}{2})^2 << c_1^2$ [34].

Equation (3.186) is used further to analyse the change of the logarithmic decrement with the fractional order Q for fixed values of the damping coefficient ε and the coefficient c_1. The value Q^{**} corresponding to the maximal decrement $\Delta_{\max}$ is found to be defined by the following implicit equation:

$$4c_1^{2-Q} \sin(Q\frac{\pi}{2}) \log c_1 + \pi[\epsilon + 2c_1^{2-Q} \cos(Q\frac{\pi}{2})] = 0. \tag{3.187}$$

This equation indicates that the maximum exists only for $c_1 < 1$, which is illustrated in Fig. 3.11 by plotting the logarithmic decrement for various values of c_1. When $c_1 \geq 1$, this quantity monotonically increases with Q, reaching the value Δ_L at $Q = 1$. It is also seen that when $c_1 < 1$, there is a range of Q for which the logarithmic decrement is higher than the one for linear viscous damping Δ_L. It yields the conclusion that in this range, the ratio of the successive amplitude and the previous one for the oscillator with fractional damping is smaller than this ratio for linear viscous damping.

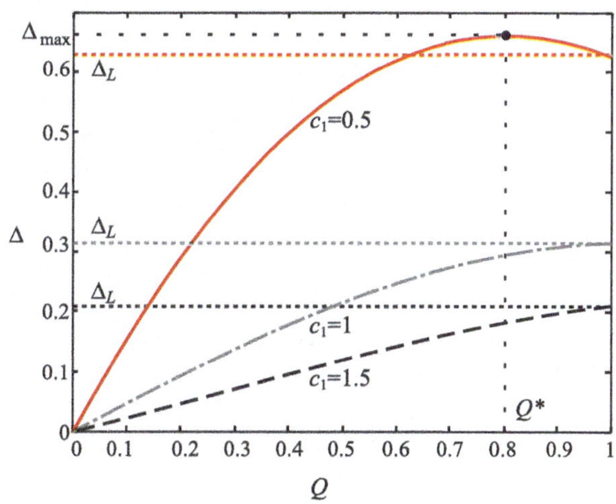

Fig. 3.11 Influence of the fractional order Q on the logarithmic decrement given by Eq. (3.186) for $\varepsilon = 0.1$ and $c_1 = 0.5$ (solid line), $c_1 = 1$ (dashed-dotted line) and $c_1 = 1.5$ (dashed line); the horizontal dotted lines depict the corresponding solutions for the logarithmic decrement of a viscously damped system Δ_L

3.5 Non-conservative Purely Nonlinear Oscillators: Approximate Solutions via Elliptic Functions

The oscillators considered in this section have the equation of motion corresponding to Eq. (3.1) with $f\left(x, \dot{x}, D^{Q}x\right)$ replaced by $\varepsilon f\left(x, \dot{x}\right)$, and with $c_1 = c_3 = 0, c_\alpha = 1$:

$$\ddot{x} + \varepsilon f\left(x, \dot{x}\right) + \mathrm{sgn}\left(x\right)\left|x\right|^{\alpha} = 0, \tag{3.188}$$

where $\varepsilon << 1$ is a small constant parameter.

Among the contributions concerned with the non-conservative oscillators modelled by Eq. (3.188), two main groups of studies can be distinguished. The first group of papers include those concerned with the oscillators with a purely nonlinear restoring force with the motion described by trigonometric functions. In [35], the van der Pol equation with $\alpha = 1/3$ was studied by using phase-space techniques and the Lienard–Levinson–Smith theorem. Waluya and van Horssen [36] used the perturbation method based on integrating factors to derive approximate first integrals for the van der Pol oscillator in which $\alpha = \left(2m + 1\right)/\left(2n + 1\right)$, and $m, n \in N$. In [37], analytical approximations to the period of a generalized van der Pol equation were obtained by using various asymptotic methods. In [14], the Krylov–Bogoliubov method was extended to find the first approximation of motion the frequency of which was calculated from the exact period of the corresponding conservative oscillator. According to the author, the method gives good approximations for $1 \leq \alpha \leq 3$. The second group of papers comprises studies of the oscillators with a polynomial (integer power) restoring force with the motion described by elliptic functions. Quadratic oscillator has been investigated by applying the elliptic perturbation method [38], the elliptic multiple scales method [39] and the elliptic Lindstedt–Poincaré method [40]. Following this work, Chen et al. [41] proposed a perturbation method for determining the homoclinic solution of non-conservative quadratic oscillators with a generative solution that has the form of a hyperbolic function. Cubic oscillators have been studied by using the elliptic multiple scales method [39], the elliptic harmonic balance method [42], the elliptic Krylov–Bogoliubov method [6, 42, 43] and the elliptic Lindstedt–Poincaré method [44].

The aim of this section is to present a fairly accurate approximate method for obtaining free responses based on Jacobi elliptic functions for non-conservative cases without any limitations regarding the value of the power of nonlinearity. Their application is, thus, wide and includes not only some generic nonlinear oscillators, such as the antisymmetric, pure quadratic or pure cubic oscillators, but also those with a non-integer power restoring force. As expressed in the form of the Jacobi cn elliptic function, the solution for motion will include higher harmonics and has excellent accuracy with respect to the exact numerical solution.

3.5.1 Conservative Oscillator: Generative Solution

The first step is to obtain a generative solution for the unperturbed case, i.e. Eq. (3.188) with $\varepsilon = 0$:

$$\ddot{x} + \text{sgn}\,(x)\,|x|^{\alpha} = 0, \tag{3.189}$$

with the initial conditions defined by Eq. (3.38).

As shown in Chap. 2, Sect. 2.5.1, a purely nonlinear oscillator with the amplitude a has the response whose period is exactly defined by

$$T = \sqrt{\frac{8\pi}{\alpha+1}}\frac{\Gamma\left(\frac{1}{\alpha+1}\right)}{\Gamma\left(\frac{\alpha+3}{2(\alpha+1)}\right)}a^{(1-\alpha)/2} = T^{*}a^{(1-\alpha)/2}, \tag{3.190}$$

with T^{*} being

$$T^{*} = \sqrt{\frac{8\pi}{\alpha+1}}\frac{\Gamma\left(\frac{1}{\alpha+1}\right)}{\Gamma\left(\frac{\alpha+3}{2(\alpha+1)}\right)} = 4\sqrt{\frac{\alpha+1}{2}}\frac{\sqrt{\pi}\Gamma\left(1+\frac{1}{\alpha+1}\right)}{\Gamma\left(\frac{1}{2}+\frac{1}{\alpha+1}\right)}. \tag{3.191}$$

A truly nonlinear character of the oscillators under consideration, i.e. the fact that the power α can take any non-negative real value, is the motivation to assume the approximate solution for motion and velocity by using the cn Jacobi elliptic function as utilized in Chap. 2, Sect. 2.5.2.4:

$$x\,(t) = a\text{cn}\,(\omega\,(\alpha)\,t\,|m\,(\alpha))\,, \tag{3.192}$$

$$\dot{x}\,(t) = -\omega\,a\,\text{sn}\,(\omega\,(\alpha)\,t\,|m\,(\alpha))\;\text{dn}\,(\omega\,(\alpha)\,t\,|m\,(\alpha))\,. \tag{3.193}$$

The frequencies ω of the elliptic function and the elliptic parameter m are unknown functions of the power α and need to be specified.

Since the period of the cn Jacobi elliptic function is $4K\,(m)$, where $K\,(m)$ is the complete elliptic integral of the first kind (see Chap. 2, Sect. 2.3.1.2), the frequency $\omega\,(\alpha)$ can be related to this period by

$$\omega\,(\alpha) = \frac{4K\,(m\,(\alpha))}{T} \equiv \frac{4K\,(m\,(\alpha))}{T^{*}}a^{(1-\alpha)/2}. \tag{3.194}$$

By choosing the ratio in Eq. (3.194) to be equal to unity and using the expression T^{*} based on Eq. (3.191), the following implicit equation defining m is derived:

$$K\,(m\,(\alpha)) = \sqrt{\frac{\alpha+1}{2}}\frac{\sqrt{\pi}\Gamma\left(1+\frac{1}{1+\alpha}\right)}{\Gamma\left(\frac{1}{2}+\frac{1}{1+\alpha}\right)}. \tag{3.195}$$

Note that now the frequency becomes

$$\omega = a^{(1-\alpha)/2}. \tag{3.196}$$

This approximate solution constructed for the conservative case can be used further as a generating solution for non-conservative oscillators.

3.5.2 Non-conservative Oscillator: Approximate Solutions via Elliptic functions

It is assumed now that the amplitude and complete phase are not constant anymore, but change with time, so that the trial solution has the form:

$$x(t) = a(t)\ \mathrm{cn}\left(\psi(t)\,|m\right), \tag{3.197}$$

with the unknown complete phase $\psi(t)$ being defined by

$$\psi(t) = \int_0^t \omega(a)\mathrm{d}t + \varphi(t), \tag{3.198}$$

where the phase $\varphi(t)$ also needs to be determined.

Further, the frequency–amplitude relationship $\omega(a)$ can be found from the analogy with this relationship corresponding to the conservative case (3.196):

$$\omega(a) = a(t)^{\frac{\alpha-1}{2}}, \tag{3.199}$$

but with the amplitude $a = a(t)$ varying in time.

The next assumption is that the time derivative of the solution from Eq. (3.197) has the same form as the velocity given by Eq. (3.193), i.e.

$$\dot{x} = -\omega\, a\ \mathrm{sn}\ \mathrm{dn}, \tag{3.200}$$

while Eq. (3.197) gives

$$\dot{x} = -\omega\, a\ \mathrm{sn}\ \mathrm{dn} + \dot{a}\ \mathrm{cn} - a\dot{\varphi}\ \mathrm{sn}\ \mathrm{dn}, \tag{3.201}$$

where the following simplifications of the notation have been introduced:

$$\mathrm{sn} \equiv \mathrm{sn}\left(\psi(t)\,|m\right), \quad \mathrm{cn} \equiv \mathrm{cn}\left(\psi(t)\,|m\right), \quad \mathrm{dn} \equiv \mathrm{dn}\left(\psi(t)\,|m\right). \tag{3.202}$$

Imposing the requirement that Eq. (3.200) has the form (3.201), the following constraint is derived:

$$\dot{a}\ \mathrm{cn} - a\dot{\varphi}\ \mathrm{sn}\ \mathrm{dn} = 0. \tag{3.203}$$

In addition, the trial solution should satisfy the equation of conservative motion as the generating one, for which $\ddot{x} = -\omega^2\, a\mathrm{cn}\ \mathrm{dn}^2 + \omega^2 am\ \mathrm{sn}^2\ \mathrm{cn}$, which reads as

$$-\omega^2\, a\mathrm{cn}\ \mathrm{dn}^2 + \omega^2 am\ \mathrm{sn}^2\ \mathrm{cn} + a^\alpha \mathrm{sgn}\,(\mathrm{cn})\,|\mathrm{cn}|^\alpha = 0. \tag{3.204}$$

Then, differentiating Eq. (3.200) and substituting it into the equation of motion of non-conservative systems (3.188) give

$$\begin{aligned}-\dot{a}\omega\ \mathrm{sn}\ \mathrm{dn} - a\dot{\omega}\ \mathrm{sn}\ \mathrm{dn} - a\omega\dot{\varphi}\ \mathrm{cn}\ \left(1-2m\ \mathrm{sn}^2\right) - \omega^2\, a\mathrm{cn}\ \mathrm{dn}^2 + \\ \omega^2 am\ \mathrm{sn}^2\ \mathrm{cn} + a^\alpha \mathrm{sgn}\,(\mathrm{cn})\,|\mathrm{cn}|^\alpha = -\varepsilon\ f\,(a\ \mathrm{cn}, -a\omega\ \mathrm{sn}\ \mathrm{dn})\,.\end{aligned} \tag{3.205}$$

Taking into account Eq. (3.204), one can transform Eq. (3.205) into

$$\dot{a}\omega\ \mathrm{sn}\ \mathrm{dn} + a\dot{\omega}\mathrm{sn}\ \mathrm{dn} + a\omega\dot{\varphi}\mathrm{cn}\ \left(1-2m\ \mathrm{sn}^2\right) = \varepsilon\ f\,(a\ \mathrm{cn}, -a\omega\ \mathrm{sn}\ \mathrm{dn})\,. \tag{3.206}$$

Combining Eqs. (3.203) and (3.206), one can obtain

$$\dot{a}\omega\left[\mathrm{sn}^2\ \mathrm{dn}^2 + \mathrm{cn}^2\left(1-2m\ \mathrm{sn}^2\right)\right] + a\dot{\omega}\ \mathrm{sn}^2\ \mathrm{dn}^2 = \varepsilon\ f\,(a\ \mathrm{cn}, -a\omega\ \mathrm{sn}\ \mathrm{dn})\ \mathrm{sn}\ \mathrm{dn}, \tag{3.207}$$

$$\dot{\varphi}a\omega\left[\mathrm{sn}^2\ \mathrm{dn}^2 + \mathrm{cn}^2\left(1-2m\ \mathrm{sn}^2\right)\right] + a\dot{\omega}\ \mathrm{sn}\ \mathrm{cn}\ \mathrm{dn} = \varepsilon\ f\,(a\ \mathrm{cn}, -a\omega\ \mathrm{sn}\ \mathrm{dn})\ \mathrm{cn}. \tag{3.208}$$

The time derivative $\dot{\omega}$ is obtained from Eq. (3.199):

$$\dot{\omega} = \omega\frac{\alpha-1}{2a}\dot{a}. \tag{3.209}$$

Introducing Eq. (3.209) into Eqs. (3.207) and (3.208) gives

$$\dot{a}\omega\left[\mathrm{sn}^2\ \mathrm{dn}^2\frac{\alpha+1}{2} + \mathrm{cn}^2\left(1-2m\ \mathrm{sn}^2\right)\right] = \varepsilon\ f\,(a\ \mathrm{cn}, -a\omega\ \mathrm{sn}\ \mathrm{dn})\ \mathrm{sn}\ \mathrm{dn}, \tag{3.210}$$

$$\dot{\varphi}a\omega\left[\mathrm{sn}^2\ \mathrm{dn}^2 + \mathrm{cn}^2\left(1-2m\ \mathrm{sn}^2\right)\right] + \omega\frac{\alpha-1}{2}\dot{a}\mathrm{sn}\ \mathrm{cn}\ \mathrm{dn} = \varepsilon\ f\,(a\ \mathrm{cn}, -a\omega\ \mathrm{sn}\ \mathrm{dn})\ \mathrm{cn}. \tag{3.211}$$

One can proceed by observing that Eqs. (3.210) and (3.211) are periodic in ψ with the period $4K\,(m)$. Therefore, after averaging over this period, one can derive

$$\frac{da}{dt} = \frac{1}{Q_f}\frac{2}{\alpha+3}\frac{1}{\omega\,(a)}\frac{\varepsilon}{4K\,(m)}\int_0^{4K(m)} f\,(a\ \mathrm{cn}, -a\omega\ \mathrm{sn}\ \mathrm{dn})\ \mathrm{sn}\ \mathrm{dn}\ \mathrm{d}\psi, \tag{3.212}$$

$$\frac{\mathrm{d}\varphi}{\mathrm{d}t} = \frac{1}{2Q_f}\frac{1}{a\,\omega\,(a)}\frac{\varepsilon}{4K\,(m)}\int_0^{4K(m)} f\,(a\ \mathrm{cn}, -a\omega\ \mathrm{sn}\ \mathrm{dn})\ \mathrm{cn}\ \mathrm{d}\psi, \tag{3.213}$$

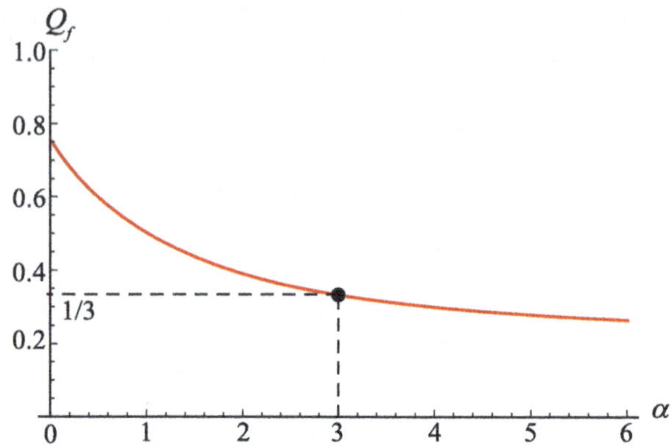

Fig. 3.12 Correction factor Q_f defined by Eqs. (3.214) and (3.195) versus the power α

where Q_f is named a 'correction factor' and is given by

$$Q_f = \frac{(2m - 1) E(m) + (1 - m) K(m)}{3mK(m)}, \tag{3.214}$$

with $E(m)$ being equal to the complete elliptic integral.

By solving Eqs. (3.212) and (3.213), with the parameter m calculated from Eq. (3.195), the solutions for the amplitude and phase of the oscillator in Eq. (3.188) can be obtained.

Some remarks can be made here regarding the forms of Eqs. (3.212) and (3.213) for some special values of the power α. In all of them, the correction factor appears, which represents a constant defined by Eq. (3.214) for the parameter m specified by Eq. (3.195). Thus, this correction factor depends on the power α, as illustrated in Fig. 3.12: it decreases as α increases. In the case of pure cubic oscillators, i.e. when $\alpha = 3$, one can calculate $Q_f = 1/3$, which gives the following equations:

$$\frac{\mathrm{d}a}{\mathrm{d}t} = \frac{\varepsilon}{\omega(a)} \frac{1}{4K(1/2)} \int_0^{4K(1/2)} f(a\ \mathrm{cn}, -a\omega\ \mathrm{sn}\ \mathrm{dn})\ \mathrm{sn}\ \mathrm{dn}\ \mathrm{d}\psi, \tag{3.215}$$

$$\frac{\mathrm{d}\varphi}{\mathrm{d}t} = \frac{3}{2} \frac{\varepsilon}{a\omega(a)} \frac{1}{4K(1/2)} \int_0^{4K(1/2)} f(a\ \mathrm{cn}, -a\omega\ \mathrm{sn}\ \mathrm{dn})\ \mathrm{cn}\ \mathrm{d}\psi, \tag{3.216}$$

with $\omega(a) = a(t)$. Equation (3.215) coincides with the form derived in [6, 43].

When the oscillator is linear, i.e. $\alpha = 1$, the parameter m tends to zero, the elliptic function turns into a cosine function and the correction factor Q_f defined by Eq. (3.214) tends to $1/2$. Then, the well-known equations for the derivatives of the amplitude and phase of the classical Krylov–Bogoliubov method are derived [5]:

$$\frac{\mathrm{d}a}{\mathrm{d}t} = \frac{\varepsilon}{2\pi} \int_0^{2\pi} f(a \cos\psi, -a\omega \sin\psi) \sin\psi \mathrm{d}\psi, \tag{3.217}$$

$$\frac{d\varphi}{dt} = \frac{\varepsilon}{2\pi a}\int_0^{2\pi} f\,(a\,\cos\psi, -a\omega\,\sin\psi)\cos\psi d\psi. \tag{3.218}$$

These equations are also in agreement with Eqs. (3.128), (3.129), where one needs to recognize that ε corresponds to $-\mu$.

The above special cases and their examination allow a fuller understanding of the generality of the derived equations for the amplitude and phase. Several examples will illustrate their use and advantages.

3.5.2.1 Example 3.6 Linear Damping

For the case of linear damping, one has $f\,(x, \dot{x}) = \dot{x}$, so that Eqs. (3.212), (3.213) for the initial conditions $x(0) = 1, \dot{x}(0) = 0$ yield

$$a\,(t) = \exp\left(-\frac{2\varepsilon}{\alpha+3}t\right), \tag{3.219}$$

$$\varphi = 0, \tag{3.220}$$

and the solution for motion is

$$x\,(t) = \exp\left(-\frac{2\varepsilon}{\alpha+3}t\right)\,\text{cn}\left[\frac{\alpha+3}{\varepsilon\,(\alpha-1)}\left[1-\exp\left(-\frac{\alpha-1}{\alpha+3}\varepsilon t\right)\right]\Big|m\right]. \tag{3.221}$$

Equation (3.219) implies that the amplitude changes exponentially with time, and it captures the influence of α. Figure 3.13a contains the comparison of the numerical solution of the equation for motion as well as the approximate solution (3.221) with the amplitude envelope, Eq. (3.219) plotted for $\alpha = 2$ and $\varepsilon = 0.1$. Figure 3.13b shows comparison for $\alpha = 3$ and $\varepsilon = 0.1$, when the approximate solution has the form

$$x\,(t) = \exp\left(-\frac{\varepsilon}{3}t\right)\text{cn}\left(\frac{3}{\varepsilon}\left[1-\exp\left(-\frac{\varepsilon t}{3}\right)\right]\Big|\frac{1}{2}\right). \tag{3.222}$$

The agreement between the solutions shown in Fig. 3.13b is not surprising. To elaborate on that, one can go back to Sect. 3.2.2 in which conservation law as exact solutions for damped Duffing oscillators is considered. Equation (3.28) with $3\sqrt{\frac{c_1}{2}} = \varepsilon$ becomes

$$\ddot{x} + \varepsilon\dot{x} + \frac{2\varepsilon^2}{9}x + c_3x^3 = 0. \tag{3.223}$$

So, for $\varepsilon << 1$, the linear term is very small, and the equation is almost quasi-linear, so it can be related to the pure cubic case whose solution is given by Eq. (3.222).

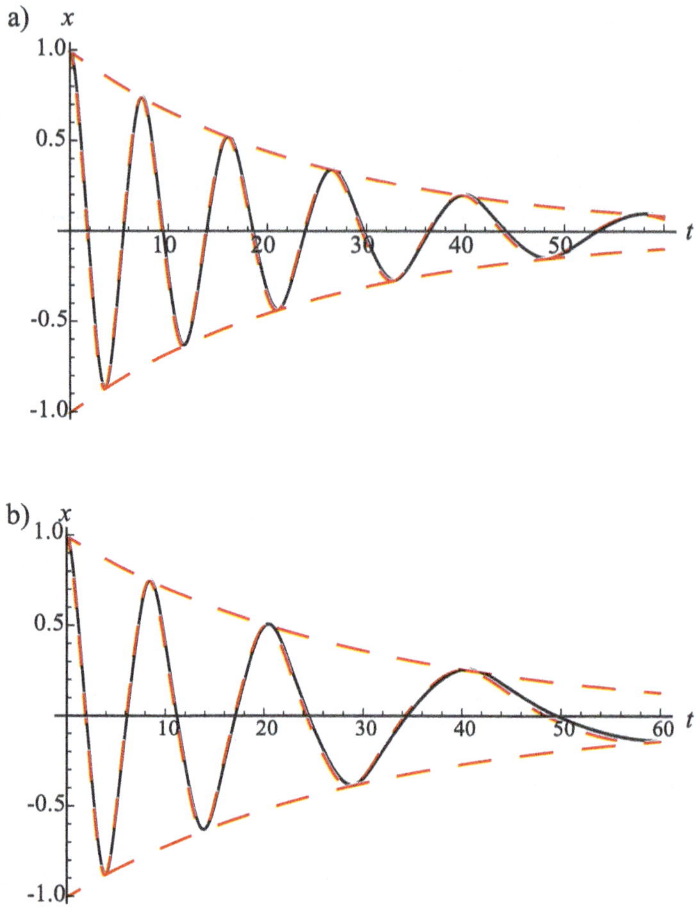

Fig. 3.13 Comparison of the approximate solution for motion (3.221) and the numerical solution of the nonlinear oscillator with linear viscous damping for $\varepsilon = 0.1$ and: **a** $\alpha = 2$; **b** $\alpha = 3$

3.5.2.2 Example 3.7 van der Pol Damping

Here, the case of damping related to the van der Pol oscillator is considered: $f(x, \dot{x}) = -\left(1 - x^2\right)\dot{x}$:

$$\ddot{x} - \varepsilon\left(1 - x^2\right)\dot{x} + \operatorname{sgn}(x)\,|x|^{\alpha} = 0. \tag{3.224}$$

This nonlinear damping mechanism dissipates energy for large displacements and feeds energy for small displacements, as a result of which the system reaches a limit cycle. Of interest here is to obtain analytically the solution for transient motion as well as the amplitude of the limit cycle of the systems with a purely nonlinear restoring force.

Using Eq. (3.212) gives

$$\frac{\mathrm{d}a}{\mathrm{d}t} = \frac{2\varepsilon}{\alpha+3} a\left(1 - Ca^2\right), \tag{3.225}$$

where

$$C = \frac{2\left[1 + m\left(m-1\right)\right] E\left(m\right) - \left(m-2\right)\left(m-1\right) K\left(m\right)}{5m\left[\left(2m-1\right) E\left(m\right) - \left(m-1\right) K\left(m\right)\right]}. \tag{3.226}$$

The steady-state amplitude a_s, i.e. the amplitude of the limit cycle follows from Eq. (3.225) as

$$a_s = C^{-1/2}. \tag{3.227}$$

It is interesting to note that when $\alpha \longrightarrow 1$, one has $m \longrightarrow 0$ and $C \rightarrow \frac{1}{4}$. In this way, the well-known expression for the amplitude equation on the classical van der Pol oscillator with a linear restoring force is obtained [45, 46]:

$$\frac{\mathrm{d}a}{\mathrm{d}t} = \frac{\varepsilon}{2} a\left(1 - \frac{1}{4}a^2\right), \tag{3.228}$$

and

$$a_s(\alpha = 1) = 2. \tag{3.229}$$

The change of the steady-state amplitude a_s with the parameter α, Eqs. (3.227) and (3.226), is plotted in Fig. 3.14 as the solid line, while the red dots depict the numerically obtained amplitudes, calculated from the equation of motion. For $\alpha < 1$, one has $a_s > 2$, and for $\alpha > 1$, it is $a_s < 2$. It should be noted that this dependence is obtained in the first approximation. However, when one wants to determine it by using trigonometric functions, the second approximation is needed. This is because the first approximation of the solution in trigonometric functions does not show the

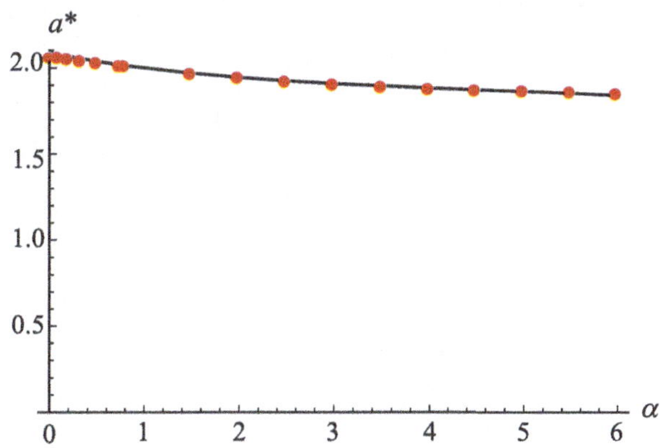

Fig. 3.14 Steady-state amplitude a_s of the oscillator with van der Pol damping obtained approximately Eqs. (3.226), (3.227) (black solid line), amplitudes obtained numerically from the equation of motion (red dots) and the well-known solution for a linear oscillator with van der Pol damping $a_s = 2$ (black dot)

influence of the power of the restoring force on the amplitude of the limit cycle [45], while the approximation obtained here does.

3.5.2.3 Example 3.8 Cubic Oscillator with a Small Linear Restoring Term

Here, the case of a cubic oscillator that is perturbed by a small linear geometric term is considered. This corresponds to $f(x, \dot{x}) = -x$. The initial conditions are taken as $x(0) = 1$, $\dot{x}(0) = 0$. Equations (3.215), (3.216) give

$$a = 1, \tag{3.230}$$

$$\varphi(t) = \frac{3\varepsilon\left[E\left(\frac{1}{2}\right) + \frac{1}{2}K\left(\frac{1}{2}\right)\right]}{K\left(\frac{1}{2}\right)}t \approx 0.68542\varepsilon t. \tag{3.231}$$

The approximate solution is then

$$x(t) = \operatorname{cn}\left((1 + \frac{3\varepsilon\left[E\left(\frac{1}{2}\right) + \frac{1}{2}K\left(\frac{1}{2}\right)\right]}{K\left(\frac{1}{2}\right)})t \,\middle|\, \frac{1}{2}\right). \tag{3.232}$$

The approximate solution given by Eq. (3.232) is plotted together with the numerical solution of the equation for motion in Fig. 3.15 for $\varepsilon = 0.1$. It is seen that they agree well.

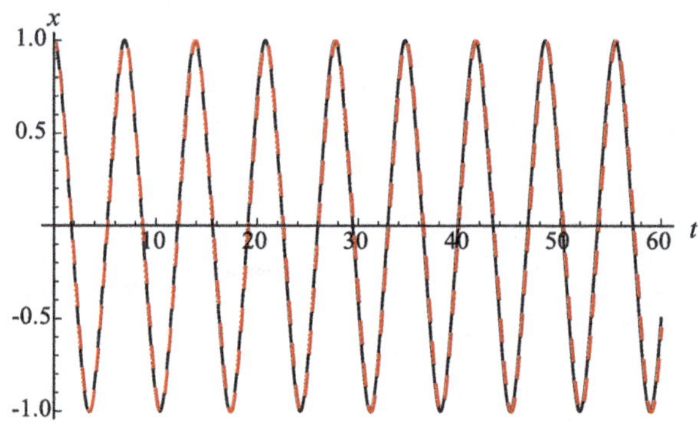

Fig. 3.15 Approximate solution given by Eq. (3.232) for motion (black solid line) and the numerical solution (red dashed line) of the cubic oscillator perturbed by a small linear geometric term

3.6 Non-conservative Oscillators with Constant Restoring Force: Approximate Solutions via Wave Functions

This section regards oscillators with a constant (antisymmetric) restoring force, and it is assumed that they are weakly damped, so that their equation of motion has the form:

$$\ddot{x} + 2\varepsilon \mathrm{sgn}\,(\dot{x})\,|\dot{x}|^n + \mathrm{sgn}\,(x) = 0, \tag{3.233}$$

where $\varepsilon << 1$, and n is a constant that can be equal to 1 or 2, modelling linear viscous or quadratic damping, respectively. The procedure for obtaining solutions for their approximate response, developed in [47], is shown below. It should be noted that the following ending consideration of this chapter can be seen as a continuation of the consideration of ending sections of Chaps. 1 and 2, where also oscillators with a constant (antisymmetric) restoring force are presented; in the former, they are considered in terms of their mechanical and mathematical model, and in the latter, in terms of their exact solutions for motion expressed via Wave functions.

3.6.1 *Conservative Oscillator: Generative Solution*

The first step is to obtain a generative solution for the unperturbed case, i.e. Eq. (3.233) with $\varepsilon = 0$, which corresponds to an antisymmetric/oscillator with a constant restoring force:

$$\ddot{x} + \mathrm{sgn}\,(x) = 0, \tag{3.234}$$

which is considered in detail in Chap. 2, Sect. 2.6 for various cases: general initial conditions, and for the ZID and ZIV oscillator, which correspond, respectively, to zero-valued initial displacement and zero-valued initial velocity.

Its exact solution for motion and velocity can be expressed as (See Chap. 2, Sect. 2.6):

$$x\,(t) = a\,\mathrm{SW}\,(\omega t + \varphi)\left[1 - \mathrm{TW}^2\left(\omega t + \varphi + \frac{1}{4}\right)\right], \tag{3.235}$$

$$\dot{x}\,(t) = 8a\omega\,\mathrm{TW}\left(\omega t + \varphi + \frac{1}{4}\right), \tag{3.236}$$

where SW stands for the square wave function and TW for the triangular wave function, a is the constant amplitude and φ is the constant phase, while

$$\omega = \frac{1}{4\sqrt{2}\sqrt{a}}. \tag{3.237}$$

3.6.2 Non-conservative Oscillator: Approximate Solutions via Wave functions

Based on the generative solution, it is assumed that the amplitude a and phase φ are not constant anymore, but vary in time, so that the approximate (trial) solution has the form

$$x(t) = a(t)\ \mathrm{SW}(\psi(t))\left[1 - \mathrm{TW}^2\left(\psi(t) + \frac{1}{4}\right)\right], \tag{3.238}$$

where the unknown complete phase $\psi(t)$ is

$$\psi(t) = \int_0^t \omega dt + \varphi(t), \tag{3.239}$$

with ω being defined by Eq. (3.237).

The next assumption is that the time derivative of the solution from Eq. (3.238) has the same form as the velocity given by Eq. (3.236), i.e.

$$\dot{x} = 8a(t)\,\omega\,\mathrm{TW}\left(\psi(t) + \frac{1}{4}\right). \tag{3.240}$$

By differentiating Eq. (3.238) and imposing the condition (3.240), the following constraint is obtained:

$$\dot{a}\ \mathrm{SW}(\psi(t))\left[1 - \mathrm{TW}^2\left(\psi(t) + \frac{1}{4}\right)\right] + 8a\dot{\varphi}\ \mathrm{TW}\left(\psi(t) + \frac{1}{4}\right) = 0. \tag{3.241}$$

In addition, the trial solution should satisfy the same condition as the generating one, which reads as

$$\mathrm{sgn}(x) = -\ddot{x} = \mathrm{SW}(\psi(t)). \tag{3.242}$$

Now, differentiating Eq. (3.240) and substituting it into Eq. (3.233) and by using Eq. (3.242), one derives

$$8\dot{a}\omega\ \mathrm{TW}\left(\psi(t) + \frac{1}{4}\right) + 8a\frac{d\omega}{d\tau}\,\dot{a}\,\mathrm{TW}\left(\psi(t) + \frac{1}{4}\right) + 32a\omega\dot{\varphi}\ \mathrm{SW}\left(\psi(t) + \frac{1}{2}\right) = - 2\varepsilon \mathrm{sgn}(\dot{x})\,|\dot{x}|^{\mathrm{n}} \tag{3.243}$$

The time derivative $\frac{\mathrm{d}\omega}{\mathrm{d}t}$ is obtained from Eq. (3.237):

$$\frac{\mathrm{d}\omega}{\mathrm{d}t} = -\frac{\omega}{2a}\frac{\mathrm{d}a}{\mathrm{d}\,t}. \tag{3.244}$$

Substituting Eq. (3.244) into (3.243) and solving it together with Eq. (3.241) with respect to $\dot{a}$ and $\dot{\varphi}$, the following equations are obtained:

$$4\dot{a}\omega = -2\varepsilon \mathrm{sgn}\,(\dot{x})\,|\dot{x}|^n\ \mathrm{TW}\left(\psi\,(t) + \frac{1}{4}\right), \tag{3.245}$$

$$-\,32\dot{\varphi}a\omega = -2\varepsilon \mathrm{sgn}\,(\dot{x})\,|\dot{x}|^n\ \mathrm{SW}\,(\psi\,(t))\left[1 - \mathrm{TW}^2\left(\psi\,(t) + \frac{1}{4}\right)\right]. \tag{3.246}$$

One can proceed by averaging Eqs. (3.245) and (3.246) over the period T by following the rule $\overline{\langle\cdots\rangle} = \frac{1}{T}\int_0^T \langle\cdots\rangle\, \mathrm{d}\psi$. This results in

$$\frac{\mathrm{d}a}{\mathrm{d}t} = -\frac{1}{2\omega}\frac{\varepsilon}{T}\int_0^T \mathrm{sgn}\,(\dot{x})\,|\dot{x}|^n\ \mathrm{TW}\left(\psi + \frac{1}{4}\right)\mathrm{d}\psi, \tag{3.247}$$

$$\frac{\mathrm{d}\varphi}{\mathrm{d}t} = \frac{1}{16a\omega}\frac{\varepsilon}{T}\int_0^T \mathrm{sgn}\,(\dot{x})\,|\dot{x}|^n\ \mathrm{SW}\,(\psi)\left[1 - \mathrm{TW}^2\left(\psi + \frac{1}{4}\right)\right]\mathrm{d}\psi. \tag{3.248}$$

In what follows, Eqs. (3.247) and (3.248) are used to find the response of the damped system with linear viscous and quadratic damping separately.

3.6.3 *Example 3.9 Antisymmetric Oscillator with Linear Viscous Damping*

For the case of linear damping, one has $\mathrm{sgn}(\dot{x})\,|\dot{x}| = \dot{x} = 8a\,(t)\,\omega\mathrm{TW}\big(\psi\,(t) + \frac{1}{4}\big)$, so that Eqs. (3.247) and (3.248) become

$$\frac{\mathrm{d}a}{\mathrm{d}t} = -4a\frac{\varepsilon}{T}\int_0^T \mathrm{TW}^2\left(\psi + \frac{1}{4}\right)\mathrm{d}\psi, \tag{3.249}$$

$$\frac{\mathrm{d}\varphi}{\mathrm{d}t} = \frac{1}{2}\frac{\varepsilon}{T}\int_0^T \mathrm{TW}\left(\psi + \frac{1}{4}\right)\mathrm{SW}\,[\psi]\left[1 - \mathrm{TW}^2\left(\psi + \frac{1}{4}\right)\right]\mathrm{d}\,\psi. \tag{3.250}$$

Solving the integrals leads to

$$\frac{\mathrm{d}a}{\mathrm{d}t} = -\frac{4a\varepsilon}{3}, \tag{3.251}$$

$$\frac{\mathrm{d}\varphi}{\mathrm{d}t} = 0. \tag{3.252}$$

The amplitude changes in accordance with

$$a\,(t) = a\,(0)\exp\left(-\frac{4}{3}\varepsilon t\right), \tag{3.253}$$

while the phase is constant $\varphi = \varphi\,(0)$. Equation (3.239) gives the complete phase:

Table 3.3 Approximate expressions for the time response of antisymmetric oscillators with linear viscous damping

Approximate solutions	ZID oscillator	ZIV oscillator
Motion	$\frac{v_0^2}{2}\exp\left(-\frac{4}{3}\varepsilon t\right)\mathrm{SW}$ $(\psi(t))\left[1-\mathrm{TW}^2\left(\psi(t)+\frac{1}{4}\right)\right]$	$A\exp\left(-\frac{4}{3}\varepsilon t\right)\mathrm{SW}(\psi(t))\left[1-\ \mathrm{TW}^2\left(\psi(t)+\frac{1}{4}\right)\right]$
Velocity	$v_0\exp\left(-\frac{2}{3}\varepsilon t\right)\mathrm{TW}\left(\psi(t)+\frac{1}{4}\right)$	$\sqrt{2A}\exp\left(-\frac{2}{3}\varepsilon t\right)\mathrm{TW}\left(\psi(t)+\frac{1}{4}\right)$
$\psi(t)$	$-\frac{1}{4v_0}\frac{3}{2\varepsilon}\left[1-\exp\left(\frac{2}{3}\varepsilon t\right)\right]$	$-\frac{1}{4\sqrt{2A}}\frac{3}{2\varepsilon}\mathrm{SW}(\psi(t))\left[1-\exp\left(\frac{2}{3}\varepsilon t\right)\right]+\frac{1}{4}$

$$\psi(t)=-\frac{1}{4\sqrt{2a(0)}}\frac{3}{2\varepsilon}\left[1-\exp\left(\frac{2\varepsilon}{3}t\right)\right]+\varphi(0). \tag{3.254}$$

The solutions for motion and velocity are

$$x(t)=a(0)\exp\left(-\frac{4}{3}\varepsilon t\right)\mathrm{SW}(\psi(t))\left[1-\mathrm{TW}^2\left(\psi(t)+\frac{1}{4}\right)\right], \tag{3.255}$$

$$\dot{x}=\sqrt{2a(0)\exp\left(-\frac{4}{3}\varepsilon t\right)}\mathrm{TW}\left(\psi(t)+\frac{1}{4}\right). \tag{3.256}$$

Table 3.3 contains the corresponding approximate solutions for the ZID and ZIV oscillator (the initial velocity in the former is labelled by v_0, and the initial displacement in the latter by A). Note that in the former case $\varphi(0)=0$, while in the latter $\varphi(0)=1/4$. These expressions for the ZID and ZIV oscillator are plotted in Figs. 3.16 and 3.17 together with the results of numerical integration of the equation of motion (3.233). It is seen that both solutions for motion and velocity agree well with numerical results.

3.6.4 *Example 3.10 Antisymmetric Oscillator with Quadratic Damping*

The analytical model for the quadratic damping force can be written as shown below:

$$(\dot{x})|\dot{x}|^2=64a^2\omega^2\,\mathrm{sgn}\left(\mathrm{TW}\left(\psi+\frac{1}{4}\right)\right)\mathrm{TW}\left(\psi+\frac{1}{4}\right), \tag{3.257}$$

so that Eqs. (3.247) and (3.248) give

$$\frac{\mathrm{d}a}{\mathrm{d}t}=-4\sqrt{2}a^{3/2}\frac{\varepsilon}{T}\int_0^T\mathrm{sgn}\left(\mathrm{TW}\left(\psi+\frac{1}{4}\right)\right)\mathrm{TW}^3\left(\psi+\frac{1}{4}\right)\mathrm{d}\psi, \tag{3.258}$$

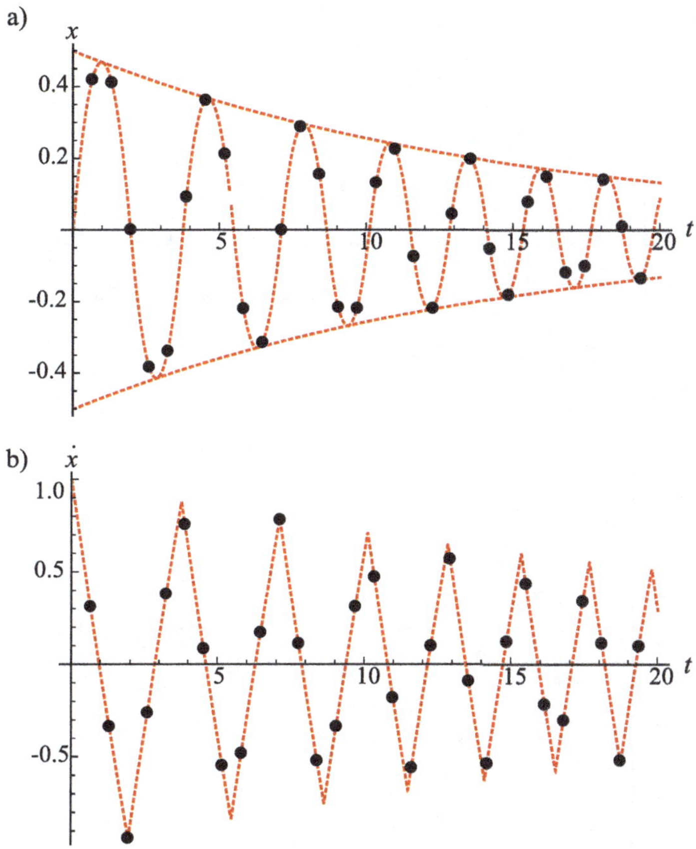

Fig. 3.16 Approximate analytical solutions (red dotted lines) and numerical solutions (black dots) for the antisymmetric ZID oscillator with linear viscous damping, $\varepsilon = 0.05$, $v_0 = 1$: **a** displacement, Eq. (3.255) response; **b** time history of velocity, Eq. (3.256)

$$\frac{\mathrm{d}\varphi}{\mathrm{d}t} = \sqrt{\frac{a}{2}}\frac{\varepsilon}{T}\int_0^T \mathrm{sgn}\left(\mathrm{TW}\left(\psi + \frac{1}{4}\right)\right)\mathrm{SW}(\psi)\left[\mathrm{TW}\left(\psi + \frac{1}{4}\right) - \mathrm{TW}^3\left(\psi + \frac{1}{4}\right)\right]\mathrm{d}\psi. \tag{3.259}$$

The integral in Eq. (3.258) is equal to 1/4, while the one in (3.259) is zero. Thus, Eq. (3.258) simplifies to

$$\frac{\mathrm{d}a}{\mathrm{d}t} = -\sqrt{2}\varepsilon a^{3/2}, \tag{3.260}$$

yielding

$$a(t) = \frac{1}{\left(\frac{1}{\sqrt{a(0)}} + \frac{\sqrt{2}\varepsilon}{2}t\right)^2}, \tag{3.261}$$

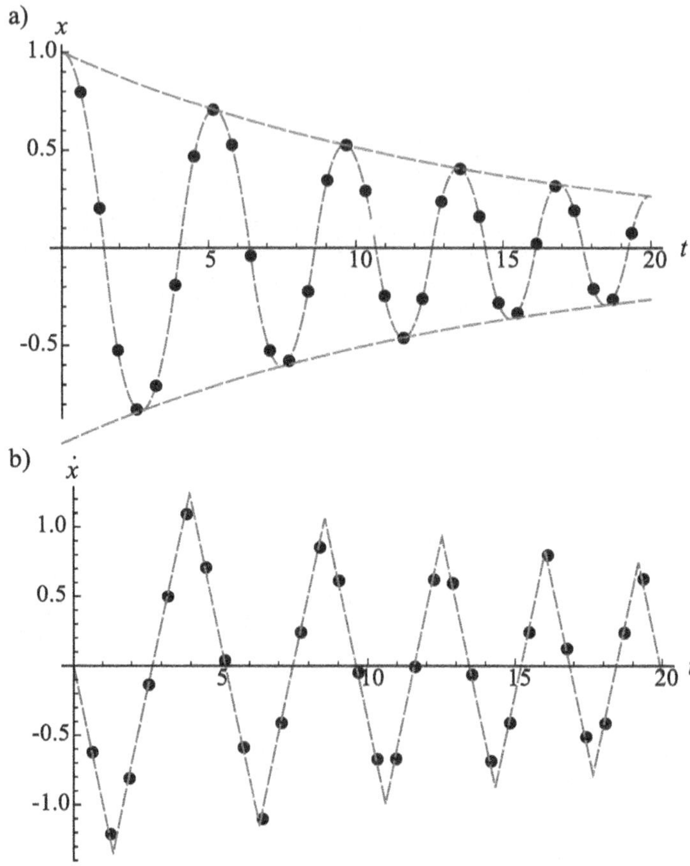

Fig. 3.17 Approximate analytical solutions (red dashed lines) and numerical solutions (black dots) for the antisymmetric ZIV oscillator with linear viscous damping, $\varepsilon = 0.05$, $A = 1$: **a** displacement response; **b** time history of velocity

while $\varphi = \varphi(0)$. The complete phase is

$$\psi(t) = \frac{1}{4\sqrt{2}}\left(\frac{1}{\sqrt{a(0)}}t + \frac{\sqrt{2}\varepsilon}{4}t^2\right) + \varphi(0), \tag{3.262}$$

which gives the following solutions for the motion and velocity:

$$x(t) = \frac{1}{\left(\frac{1}{\sqrt{a(0)}} + \frac{\sqrt{2}\varepsilon}{2}t\right)^2} \mathrm{SW}(\psi(t))\left[1 - \mathrm{TW}^2\left(\psi(t) + \frac{1}{4}\right)\right], \tag{3.263}$$

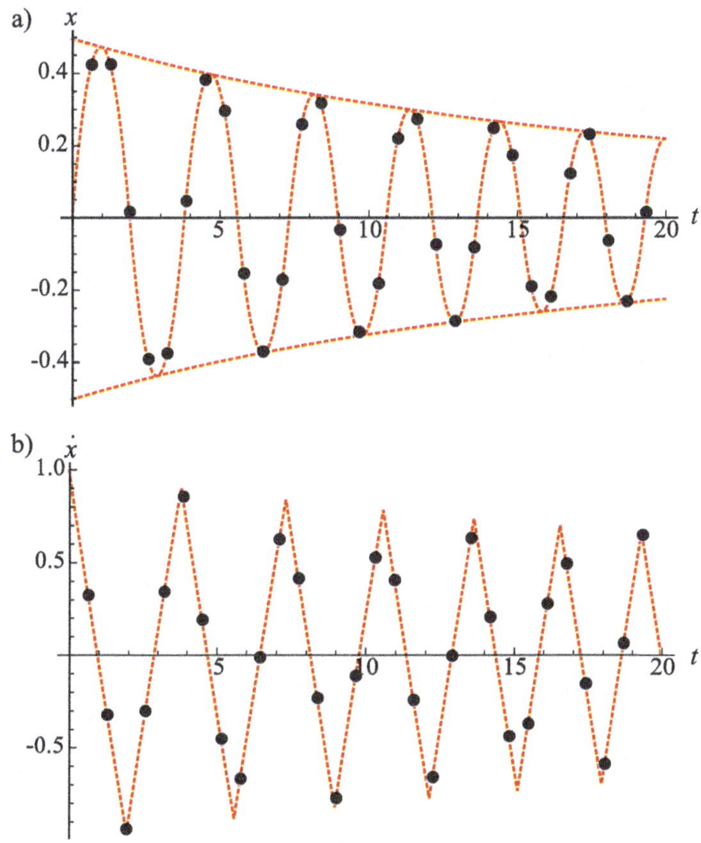

Fig. 3.18 Approximate analytical solutions (red dotted lines) and numerical solutions (black dots) for the antisymmetric ZID oscillator with quadratic damping, $\varepsilon = 0.05$, $\upsilon_0 = 1$: **a** displacement response; **b** time history of velocity

$$\dot{x} = \frac{\sqrt{2}}{\frac{1}{\sqrt{a(0)}} + \frac{\sqrt{2}\varepsilon}{2}t} \mathrm{TW}\left(\psi(t) + \frac{1}{4}\right). \tag{3.264}$$

Table 3.4 contains the corresponding approximate solutions for the ZID and ZIV oscillator. The initial phase and amplitude are the same as in the previous example. Figures 3.18 and 3.19 show plots of these solutions as well as the results of numerical integration of the equation of motion (3.233). A good match between them is seen.

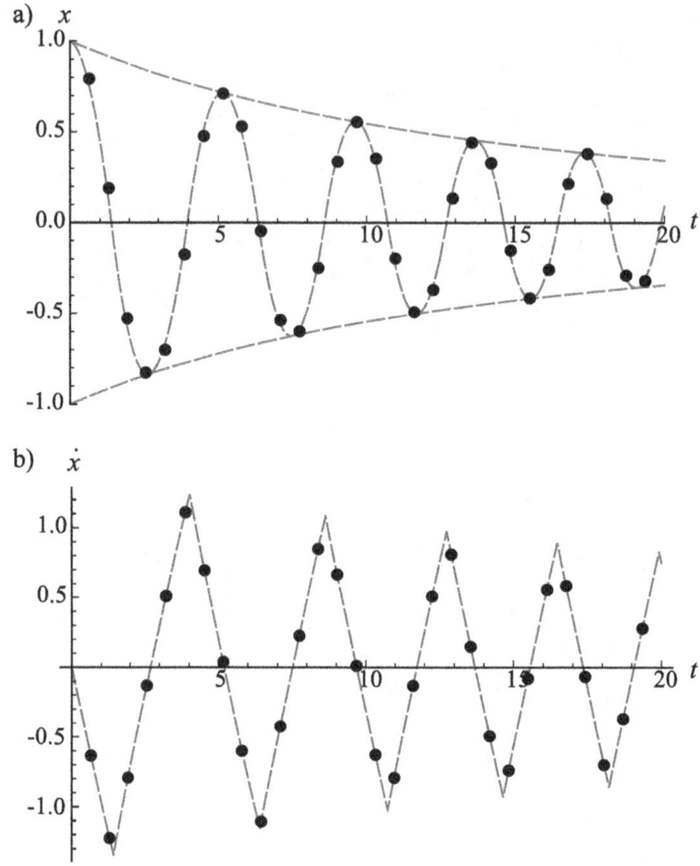

Fig. 3.19 Approximate analytical solutions (red dashed lines) and numerical solutions (black dots) for the antisymmetric ZIV oscillator with linear quadratic damping, $\varepsilon = 0.05$, $A = 1$: **a** displacement response; **b** time history of velocity

Table 3.4 Approximate expressions for the time response of antisymmetric oscillators with quadratic damping

Approximate solutions	ZID oscillator	ZIV oscillator
Motion	$\frac{1}{(\frac{\sqrt{2}}{v_0}+\frac{\sqrt{2}\varepsilon}{2}t)^2}\mathrm{SW}(\psi(t))\left[1-\mathrm{TW}^2\left(\psi(t)+\frac{1}{4}\right)\right]$	$\frac{1}{(\frac{1}{\sqrt{A}}+\frac{\sqrt{2}\varepsilon}{2}t)^2}\mathrm{SW}(\psi(t))\left[1-\mathrm{TW}^2\left(\psi(t)+\frac{1}{4}\right)\right]$
Velocity	$\frac{\sqrt{2}}{\frac{\sqrt{2}}{v_0}+\frac{\sqrt{2}\varepsilon}{2}t}\mathrm{TW}\left(\psi(t)+\frac{1}{4}\right)$	$\frac{\sqrt{2}}{\frac{1}{\sqrt{A}}+\frac{\sqrt{2}\varepsilon}{2}t}\mathrm{TW}\left(\psi(t)+\frac{1}{4}\right)$
$\psi(t)$	$\frac{1}{4\sqrt{2}}(\frac{\sqrt{2}}{v_0}t+\frac{\sqrt{2}\varepsilon}{4}t^2)$	$\frac{1}{4\sqrt{2}}(\frac{1}{\sqrt{A}}t+\frac{\sqrt{2}\varepsilon}{4}t^2)$

References

1. Vujanovic BD, Jones SE (1989) Variational methods in nonconservative phenomena. Academic, Boston
2. Vujanovic BD, Atanackovic TM (2004) An introduction to modern variational techniques in mechanics and Engineering. Birkhauser, Boston
3. Nayfeh AH, Mook DT (1979) Nonlinear oscillations. Wiley, New York
4. Ruzicka JE, Derby TE (1971) Influence of damping in vibration isolation. T.S.a.V.I, Centre, US Department of Defense, Washington DC
5. Bogoliubov NN, Mitropolski YA (1974) Asymptotic methods in the theory of non-linear vibrations. Nauka, Moscow (In Russian)
6. Yuste SB (1992) Quasi-pure-cubic oscillators studied using a Krylov-Bogoliubov method. J Sound Vibr 158:267–275
7. Dasarathy BV (1970) Analysis of a class of nonlinear systems. J Sound Vibr 11:139–144
8. Jordan DW, Smith P (1999) Nonlinear ordinary differential equations: an introduction to dynamical systems. Oxford University Press, Oxford
9. Andronov AA, Vitt SE (1981) Hajkinm theory of vibrations. Moskva: Nauka (In Russian)
10. Klotter K (1955) Free oscillations of systems having quadratic damping and arbitrary restoring force. J Appl Mech-Trans ASME 22:493–499
11. Kovacic I, Rakaric Z (2011) Study of oscillators with a non-negative real-power restoring force and quadratic damping. Nonlinear Dyn 64:293–304
12. Chaudhry MA, Zubair MS (2002) A class of incomplete gamma functions with applications. Chapman & Hall/CRC, Boca Raton
13. functions.wolfram.com/GammaBetaErf/Gamma2/03/01/01/0002/. Accessed 9 June 2010
14. Cveticanin L (2009) Oscillator with fraction order restoring force. J Sound Vibr 320:1064–1077
15. Cveticanin L, Kovacic I, Rakaric Z (2010) Asymptotic methods for vibrations of the pure non-integer order oscillator. Comput Math Appl 60:2616–2628
16. Kovacic I (2011) The method of multiple scales for forced oscillators with some real-power nonlinearities in the stiffness and damping force. Chaos Solitons Fractals 44:891–901
17. Rakaric Z, Kovacic I (2013) An elliptic averaging method for harmonically excited oscillators with a purely nonlinear non-negative real-power restoring force. Commun Nonlinear Sci Numer Simul 18:1888–1901
18. Kovacic I, Rakaric Z, Cveticanin L (2010) A non-simultaneous variational approach for the oscillators with fractional-order power nonlinearities. Appl Math Comput 217:3944–3954
19. Sabatier J, Agrawal PP, Tenreiro Machado JA (2007) Advances in fractional calculus, theoretical developments and applications in physics and engineering. Springer, Dordrecht (2007)
20. Burov S, Barkai E (2008) Fractional Langevin equation: overdamped, underdamped, and critical behaviours. Phys Rev 78:031112 (18pp)
21. Naber M (2010) Linear fractionally damped oscillator. Int J Differ Equ 197020 (12 pp)
22. Rudinger F (2006) Tuned mass damper with fractional derivative damping. Eng Struct 28:1774–1779
23. Wahi P, Chatterjee A (2003) Averaging for oscillations with light fractional order damping. In: Proceedings of design engineering technical conferences and computers and information in engineering conference, Chicago, USA, 2–6 September, 2003, paper VIB 48386 (7pp)
24. Wahi P, Chatterjee A (2004) Averaging oscillations with small fractional damping and delayed terms. Nonlinear Dyn 38:3–22
25. Chen LC, Zhu WQ (2009) Stochastic averaging of strongly nonlinear oscillators with small fractional derivative damping under combined harmonic and white noise excitations. Nonlinear Dyn 56:231–241
26. Xie F, Lin X (2009) Asymptotic solution of the van der Pol oscillator with small fractional damping. Phys Scrip T136:014033 (4pp)
27. Rand RH, Sah SM, Suchorsky MK (2010) Fractional Mathieu equation. Commun Nonlinear Sci Numer Simul 11:3254–3262

28. Kovacic I, Rand R, Sah MS (2018) Mathieu's equation and its generalizations: overview of stability charts and their features. Appl Mech Rev 70:020802 (22 pages)
29. Kovacic I, Zukovic M (2012) Oscillators with a power-form restoring force and fractional derivative damping: application of averaging. Mech Res Commun 41:37–43
30. Ross B (1975) A brief history and exposition of the fundamental theory of fractional calculus. In: Fractional calculus and its applications. Lecture notes in mathematics, vol 457. Springer, Berlin, pp 1–36
31. Makris N, Constantinou MC (1991) Fractional-derivative Maxwell model for viscous dampers. J Struct Eng 117:2708–2724
32. Kovacic I (2011) A limit cycle and relaxation oscillations in a generalized van der Pol oscillator. Commun Nonlinear Sci Numer Simul 16:1640–1649
33. Podlubny I (1999) Fractional differential equations. Academic, San Diego
34. Rao SS (2004) Mechanical vibrations. Pearson Prentice Hall, New Jersey
35. Mickens RE (2002) Analysis of nonlinear oscillators having non-polynomial elastic terms. J Sound Vibr 255:789–792
36. Waluya SB, van Horssen WT (2003) On the periodic solutions of a generalized nonlinear Van der Pol oscillator. J Sound Vibr 268:209–215
37. Andrianov IV, van Horssen WT (2006) Analytical approximations of the period of a generalized nonlinear van der Pol oscillator. J Sound Vibr 295:1099–1104
38. Chen SH, Yang XM, Cheung YK (1998) Periodic solutions of strongly quadratic nonlinear oscillators by the elliptic perturbation method. J Sound Vibr 212:771–780
39. Lakrad F, Belhaq M (2002) Periodic solutions of strongly nonlinear oscillators by the multiple-scales method. J Sound Vibr 258:677–700
40. Chen SH, Yang XM, Cheung YK (1999) Periodic solutions of strongly quadratic nonlinear oscillators by the elliptic Lindstedt-Poincare method. J Sound Vibr 227:1109–1118
41. Chen SH, Chen YY, Sze KY (2009) A hyperbolic perturbation method for determining homoclinic solution of certain strongly nonlinear autonomous oscillators. J Sound Vibr 322:381–392
42. Yuste SB, Bejarano JD (1986) Construction of approximate analytical solutions to a new class of nonlinear oscillator equations. J Sound Vibr 11:347–350
43. Yuste SB, Bejarano JD (1990) Improvement of a Krilov-Bogoliubov method that uses Jacobi elliptic functions. J Sound Vibr 139:151–163
44. Yang CH, Zhu SM, Chen SH (2004) A modified elliptic Lindstedt-Poincare method for certain strongly nonlinear oscillators. J Sound Vibr 273:921–932
45. Mickens RE (1981) An introduction to nonlinear oscillations. Cambridge University Press, Cambridge
46. Hagedorn P (1981) Nonlinear oscillations. Clarendon Press, Oxford
47. Kovacic I (2016) On the response of antisymmetric constant force oscillators: exact and approximate solutions. Commun Nonlinear Sci Numer Simul 32:305–316

Chapter 4
Forced Oscillators

4.1 Introduction

This section is mainly concerned with nonautonomous oscillators governed by

$$\ddot{x} + G(x) = F(x, \dot{x}, \Omega t), \tag{4.1}$$

where the nonlinear restoring term(s) are labelled by $G(x)$, while the term F can be separated into

$$F(x, \dot{x}, \Omega t) = F_1(x, \dot{x}) + F_2(\Omega t), \tag{4.2}$$

where the first term is autonomous and comprises, for example, nonconservative forces, while the second term is non-autonomous and stems from the external periodic forcing of the frequency Ω.

Techniques for obtaining both the exact solutions and the approximate solutions for the response of these nonlinear oscillators are presented. The former techniques involve a tuning approach, in which the external excitation is designed to achieve a desired form of the response, both regarding the amplitude and the frequency. The idea is based on Hsu's approach of the utilization of exact closed-form solutions for free conservative oscillations and a tuned external excitation proportional to the displacement [1]. The exact closed-form solution for the response of a harmonically forced linear oscillator out of resonance is well known and contains only one harmonic, which is for the convenience of the reader presented in Sect. 4.2. However, the regular response of harmonically excited nonlinear oscillators is, in general, characterized by a multi-frequency response. Nevertheless, it has been shown by the author of this book and her coauthors how a variety of nonlinear oscillators can be excited with a tuned excitation to achieve also a single-frequency response [2–7]. Some of these results are presented in Sect. 4.3 for Duffing oscillators [6] and in Sect. 4.7 for Purely Nonlinear Oscillators (PNOs) [4]. Tuning is also considered in Sect. 4.4 from the viewpoint of adjusting the nonlinearity in a restoring force [5] to achieve

I. Kovacic, *Nonlinear Oscillations*,
https://doi.org/10.1007/978-3-030-53172-0_4

a single-frequency response. In addition, the technique for obtaining approximate solutions of harmonically excited PNOs in terms of elliptic functions [8] is shown in Sect. 4.8, and it represents the extension of the averaging method presented in Chap. 3, Sect. 3.5.

4.2 Forced Response of Duffing-Type Oscillators: Exact Solutions

4.2.1 Motivation for the Methodology

Before focusing on nonlinear oscillators, let us explain the motivation and methodology on the example of a Simple Harmonic Oscillator (SHO):

$$\ddot{x} + c_1\, x = 0, \tag{4.3}$$

where $c_1 > 0$. Its free response that corresponds to the following initial conditions:

$$x\,(0) = A, \quad \dot{x}\,(0) = 0, \tag{4.4}$$

has the form (see Chap. 2, Sect. 2.2):

$$x = A\cos\left(\sqrt{c_1}\, t\right). \tag{4.5}$$

Although the initial amplitude A and the coefficient c_1 in front of the linear term (the square of the natural frequency ω, where $\omega = \sqrt{c_1}$) in Eq. (4.3) can be made equal to unity by an appropriate normalization, they are left in this form for the sake of clarity and further generalizations.

If one takes the external excitation F as proportional to the displacement

$$F = Bx, \tag{4.6}$$

where B stands for the coefficient of proportionality, the equation of motion is

$$\ddot{x} + (c_1 - B)\; x = 0. \tag{4.7}$$

The solution satisfying $c_1 > B$ is easily obtained based on Eqs. (4.3) and (4.5) in the form:

$$x = A\cos\left(\Omega\, t\right), \tag{4.8}$$

$$\Omega = \sqrt{c_1 \; - B}. \tag{4.9}$$

Equations (4.8) and (4.6) can now be used to transform Eq. (4.7) into the equation of motion of the forced SHO

$$\ddot{x} + c_1 x = BA\cos(\Omega t), \tag{4.10}$$

whose response outside a resonance region has the form given by Eq. (4.8). Note that since there is no damping in the system and the excitation acts from $t = 0$, the resulting response does not have a transient part and a steady-state part, but it goes straight to the latter [9]. It should be emphasized that there are three parameters involved in the solution (A, B and Ω) and only one relationship between two of them, Eq. (4.9). This implies that A can be arbitrarily chosen, while one a priory decides whether to define B or Ω. The other parameter should then be calculated based on Eq. (4.9).

Thus, the external force is assumed here to be proportional to the displacement to yield the forced response in a straightforward way, but it can actually be interpreted as being proportional to the restoring force. This interpretation enables one to apply the same methodology to nonlinear oscillators, which will be demonstrated subsequently.

4.2.2 *Duffing Oscillators*

Let us consider now a classical conservative Duffing Oscillator (DO)

$$\ddot{x} + c_1 x + c_3 x^3 = 0. \tag{4.11}$$

It is assumed here that $c_1 > 0$ and $c_3 > 0$, i.e. the oscillator is of a hardening type—Hardening Duffing Oscillator (HDO). As shown in Chap. 2, Sect. 2.3.2 its free response can be expressed in closed-form in terms of the Jacobi cn elliptic function, whose frequency ω and positive elliptic parameter m are (see Chap. 2, Sect. 2.3.2 or [2, 10])

$$x = A\mathrm{cn}(\omega t\,|m) = A\mathrm{cn}(\omega t, k), \tag{4.12}$$

$$\omega = \sqrt{c_1 + c_3 A^2}, \tag{4.13}$$

$$m = k^2 = \frac{c_3 A^2}{2\left(c_1 + c_3 A^2\right)}. \tag{4.14}$$

Note that the previous equations can be expressed in terms of the elliptic modulus k, whose square corresponds to the elliptic parameter $m = k^2$ (see Chap. 2, Sect. 2.3).

This solution holds for the initial conditions (4.4). The insight into the content of this response with respect to its harmonics is given in Appendix C. This closed-form response includes odd multiplications of the first harmonic (third, fifth, etc.), which is one of the basic differences of this response with respect to the previously discussed SHO. In addition, both the frequency of the elliptic function (4.13) ω and its elliptic parameter m (4.14) are amplitude-dependent. Note that for $c_3 = 0$, Eq. (4.14) gives $m = 0$, so that the cn function turns into the cosine function [2] and the solution of motion for the SHO is obtained, Eq. (4.5).

Let us now focus on externally excited DOs, assuming that this excitation has the form of the Duffing restoring force

$$\ddot{x} + c_1 x + c_3 x^3 = F, \tag{4.15}$$

$$F = Bx + D x^3, \tag{4.16}$$

where B and D are constants. The following equation of motion is obtained:

$$\ddot{x} + (c_1 - B)\, x + (c_3 - D)\, x^3 = 0. \tag{4.17}$$

Assuming that $c_1 > B$ and $c_3 > D$, comparing Eq. (4.17) to (4.11) and using (4.12)–(4.14), the exact solution for the resulting response x_r is found to be

$$x_r = A\mathrm{cn}\left(\omega_r t \,|m_r\right), \tag{4.18}$$

$$\omega_r = \sqrt{c_1 - B + (c_3 - D)\, A^2}, \tag{4.19}$$

$$m_r = \frac{(c_3 - D)\, A^2}{2\left(c_1 - B + (c_3 - D)\, A^2\right)}. \tag{4.20}$$

This implies that if one wants to excite the oscillator (4.11) with the force (4.16), this force should vary in time as follows:

$$F = BA\mathrm{cn}\left(\omega_r t \,|m_r\right) + DA^3\mathrm{cn}^3\left(\omega_r t \,|m_r\right). \tag{4.21}$$

For the sake of the reader, the Fourier series for the cn function and its cube are given in Appendix C, and they comprise odd harmonics. This implies that x_r is actually a multi-term harmonic response obtained for a multi-term harmonic excitation. These facts will be used during numerical simulations in some of the following examples.

The same concept can be applied to the damping term when it exists in the original equation of motion

$$\ddot{x} + \delta\dot{x} + c_1 x + c_3 x^3 = F, \tag{4.22}$$

but now the additional term of the external excitation on the right-hand side will be added as $\delta\dot{x}_r\,(t)$ [11]. So, by finding the first time derivative of Eq. (4.18) (see Appendix B for the time derivatives of Jacobi elliptic functions), one can write down the following equation of motion of the externally excited damped Duffing oscillator:

$$\begin{aligned}\ddot{x} + \delta\dot{x} + c_1 x + c_3 x^3 =& BA\mathrm{cn}\left(\omega t \,|m\right) + DA^3\mathrm{cn}^3\left(\omega t \,|m\right) - \\ & \delta\omega\left[BA + 3DA^3\mathrm{cn}^2\left(\omega t \,|m\right)\right]\mathrm{sn}\left(\omega t \,|m\right)\mathrm{dn}\left(\omega t \,|m\right).\end{aligned} \tag{4.23}$$

Its steady-state response will have the form (4.18)–(4.20).

4.2.3 Simplification to the Case of Harmonic Excitation: Related Approximations

It is interesting to point out that this class of exact solutions comprise the well-known approximate solution for externally excited Duffing oscillator governed by Eq. (4.1) with the harmonic excitation

$$F = F_0 \cos \frac{2\pi}{T} t = F_0 \cos \Omega t, \tag{4.24}$$

where T is the forcing period and Ω is the forcing frequency.

Let us go back to the Eq. (4.17), assuming that $c_1 = 1$ and $D = 0$. Equations (4.19) and (4.20) now becomes

$$\omega_r = \sqrt{1 - B + c_3 A^2}, \tag{4.25}$$

$$k = \sqrt{\frac{c_3 A^2}{2\left(1 - B + c_3 A^2\right)}}. \tag{4.26}$$

The amplitude of the response and the constant B can be expressed as follows:

$$A = \sqrt{\frac{2}{c_3}} \omega_r \, k, \tag{4.27}$$

$$B = 1 - \omega_r^2 \left(1 - 2k^2\right). \tag{4.28}$$

Thus, the excitation (4.21) is given by

$$F = A[1 - \omega_r^2 \left(1 - 2k^2\right)] \mathrm{cn} \left(\omega_r t, k\right). \tag{4.29}$$

Comparing now Eq. (4.24) with (4.29), one can recognize that

$$F_0 = A[1 - \omega_r^2 \left(1 - 2k^2\right)], \tag{4.30}$$

while the equality of periods of the cn and cosine functions $4K/\omega_r = 2\pi/\Omega$ yields

$$\Omega = \frac{\pi}{2K} \omega_r. \tag{4.31}$$

Equation (4.27) can further be used together with Eq. (4.31) and the series expansion $K^2 \approx \frac{\pi^2}{4}\left(1 + \frac{k^2}{2}\right)$ to derive

$$c_3 A^2 - 2\Omega^2 \, k^2 = 0. \tag{4.32}$$

In the same way, Eq. (4.30) yields

$$1 - \Omega^2 + \frac{3}{2}k^2\Omega^2 = \frac{F_0}{A}. \tag{4.33}$$

By combining Eqs. (4.32), (4.33), the well-know frequency-amplitude relationship [12, 13] is obtained for harmonically excited Duffing oscillators:

$$\Omega^2 = 1 + \frac{3}{4}c_3A^2 - \frac{F_0}{A}. \tag{4.34}$$

To determine the solution for the steady-state response, one can start from $x = A\ \mathrm{cn}(\omega_r, k)$ and use the Fourier expansion for the cn function (see Appendix C), expanding the first two coefficients into the series with respect to k^2 as follows:

$$x = A\ \mathrm{cn}\,(\omega_r t, k) \approx A\ C_1 \cos\left(\frac{\pi}{2K}\omega_r t\right) + A\ C_3 \cos\left(3\frac{\pi}{2K}\omega_r t\right) + \cdots \tag{4.35}$$

$$\begin{aligned} x &= A\left(1 - \frac{k^2}{16}\right)\cos\Omega t + A\frac{k^2}{16}\cos(3\Omega t) \\ &= \left(A - \frac{1}{32}c_3A^3\right)\cos\Omega t + \frac{1}{32}c_3A^3\cos(3\Omega t). \end{aligned} \tag{4.36}$$

This solution is in agreement with the well-known solution obtained by the classical method of multiple scales [12, 13].

4.3 Tuning the Excitation in Odd-Parity Oscillator to Make It Respond as a Simple Harmonic Oscillator

4.3.1 Tuning the Excitation in a Hardening Duffing Oscillator

It is easy to see from Eq. (4.17) that the form of the external force, i.e. the values of B and D can be used as 'control parameters' to define the type of resulting oscillators. For example, by choosing $D = c_3$, $c_1 > B$, the initially hardening DO turns into the SHO, Eq. (4.7). By using its solution for motion in the form given by Eqs. (4.8), (4.9) and introducing it into Eq. (4.16), and then into Eq. (4.15), the following equation of motion of the externally excited hardening Duffing oscillator that will have the exact solution as a one-harmonic response like the SHO is derived:

$$\ddot{x} + c_1x + c_3\,x^3 = BA\cos\left(\sqrt{c_1 - B}t\right) + c_3\,A^3\cos^3\left(\sqrt{c_1 - B}t\right). \tag{4.37}$$

This can be written down as [6]

$$\ddot{x}+c_1x+c_3x^3=\left(BA+\frac{3}{4}c_3\,A^3\right)\cos\left(\sqrt{c_1-B}t\right)+\frac{1}{4}c_3\,A^3\cos\left(3\sqrt{c_1-B}t\right), \tag{4.38}$$

with the resulting response having the form given by Eqs. (4.8), (4.9).

To illustrate this resulting behaviour, a numerical solution is calculated both for the case of the free oscillator (4.11) and the forced oscillator (4.38). These responses are presented in Fig. 4.1. Numerical solutions of Eq. (4.38) are depicted by the black solid line, the analytical response (4.8), (4.9) as the red dots and these solutions coincide. To illustrate the change of the response, the numerical solution of the original equation of motion of the HDO is plotted as the grey dotted line. Two different values of the amplitude are chosen ($A = 1$ in Fig. 4.1a and $A = 1/2$ in Fig. 4.1b) and it is seen that in the original HDO the period is amplitude-dependent, while in the excited HDO (the resulting SHO), it is amplitude-independent.

If the damping term $\delta\dot{x}$ exists in the original equation of motion, the exact solution for the response can still be obtained as the one for an undamped SHO (a one-harmonic steady-state response), but now the additional term of the external excitation should be added to the right-hand side as $\delta\dot{x}_r(t)$. Consequently, the following equation of motion of the externally excited damped hardening Duffing restoring force:

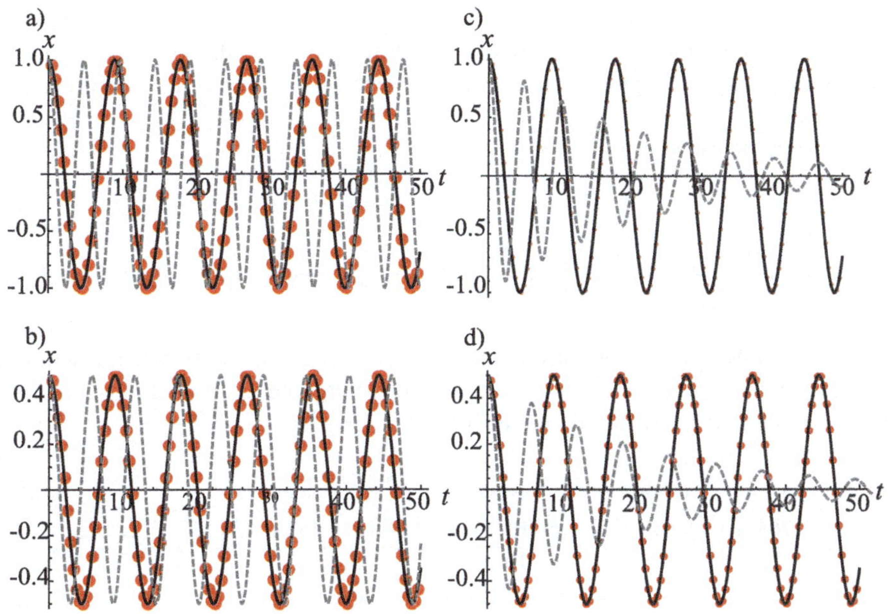

Fig. 4.1 Time response corresponding to $c_1 = 1$, $c_3 = 1$, $B = 0.5$, $D = c_3$, and: **a** $A = 1$, Eq. (4.38), **b** $A = 1/2$, Eq. (4.38), **c** $A = 1$, $\delta = 0.1$, Eq. (4.39), **d** $A = 1/2$, $\delta = 0.1$, Eq. (4.39). Numerical solutions of Eq. (4.38) or (4.39)—black solid line, the analytical response (4.8), (4.9)—red dots, the numerical solution of the original equation of motion of of Eq. (4.38) or (4.39) with the right-hand side being zero—grey dotted line

$$\ddot{x} + \delta\dot{x} + c_1 x + c_3 x^3 = \left(BA + \frac{3}{4}c_3 A^3\right)\cos\left(\sqrt{c_1 - B}t\right) + \frac{1}{4}c_3 A^3 \cos\left(3\sqrt{c_1 - B}t\right) - \delta A\sqrt{c_1 - B}\sin\left(\sqrt{c_1 - B}t\right), \tag{4.39}$$

will have the response defined by Eqs. (4.8), (4.9). The confirmations corresponding to $\delta = 0.1$ are provided in Fig. 4.1c for the same parameters as in Fig. 4.1a, and also in Fig. 4.1d for the same parameters as in Fig. 4.1b.

4.3.2 *Tuning the Excitation in Oscillators with Higher Order Odd-Power form Nonlinearity of the Restoring Force*

The methodology presented previously can be extended in a straightforward manner to achieve a single-frequency response as in the SHO of a fairly wide class of oscillators whose restoring force is multi-term and includes odd-power nonlinearities [5]:

$$\ddot{x} + \sum_{n=0}^{j} c_{2n+1}x^{2n+1} = BA\cos(\sqrt{c_1 - B}t) + \sum_{n=1}^{j} c_{2n+1}A^{2n+1}\cos^{2n+1}(\sqrt{c_1 - B}t). \tag{4.40}$$

This type of oscillators will be referred to as 'Duffing-like oscillators' as they do not contain only the linear-plus-cubic nonlinearity but also higher order odd-powered nonlinearity.

By using the transformation

$$\cos^N(\sqrt{c_1 - B}t) = \frac{2}{2^N}\sum_{i=0}^{\frac{N-1}{2}}\binom{N}{i}\cos[(N - 2i)\sqrt{c_1 - B}t], \quad N = 2n + 1, \tag{4.41}$$

the powers of the cosine function on the right-hand side of Eq. (4.40) can be rewritten in terms of the cosine functions of the multiplied arguments. If an approximation up to $j = 11$ is of interest, one can derive

$$\ddot{x} + c_1 x + c_3 x^3 + c_5 x^5 + c_7 x^7 + c_9 x^9 + c_{11} x^{11} = \left(BA + \frac{768c_3A^3 + 640A^5c_5 + 560A^7c_7 + 504A^9c_9 + 462A^{11}c_{11}}{1024}\right)\cos(\sqrt{c_1 - B}t) + \frac{A^{11}c_{11}}{1024}\cos(11\sqrt{c_1 - B}t), \tag{4.42}$$

with the resulting response having the form given by Eqs. (4.8), (4.9).

Similarly to Eq. (4.39), the existence of the damping term $2\zeta\dot{x}$ yields the following equation of motion for the externally excited damped hardening Duffing-like oscillator:

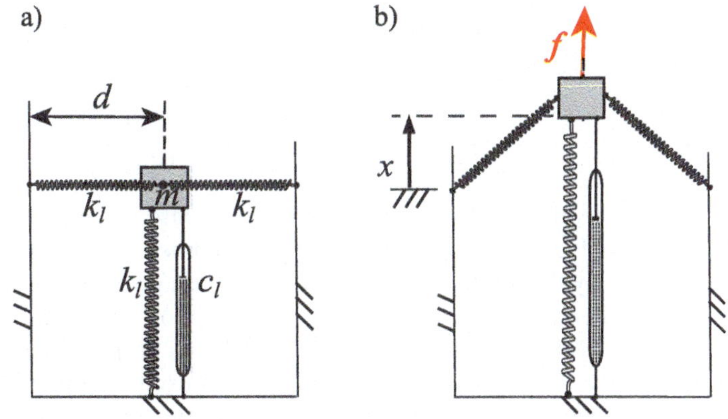

Fig. 4.2 Practical implementation of a geometrically nonlinear oscillator: **a** configuration at the static equilibrium condition; **b** deformed configuration due to the application of a force f

$$\ddot{x} + 2\zeta\dot{x} + c_1 x + c_3 x^3 + c_5 x^5 + c_7 x^7 + c_9 x^9 + c_{11} x^{11} =$$
$$(BA + \frac{768 c_3 A^3 + 640 A^5 c_5 + 560 A^7 c_7 + 504 A^9 c_9 + 462 A^{11} c_{11}}{1024}) \cos(\sqrt{c_1 - B}t) +$$
$$2\zeta A\sqrt{c_1 - B}\sin(\sqrt{c_1 - B}t), \tag{4.43}$$

which will have the response defined by Eqs. (4.8), (4.9).

To validate these theoretical results, numerical experiments are performed on the model of the system presented in Fig. 4.2, where Fig. 4.2a shows the system at rest, and Fig. 4.2b shows it in a deformed configuration under the action of the excitation force f. The system consists of three linear springs geometrically arranged to achieve hardening stiffness nonlinearity. The vertical linear spring is combined with two lateral linear springs, which inclines as the oscillating mass moves, as shown in Fig. 4.2b, generating nonlinear geometric restoring terms. This system behaves as a Duffing oscillator only for small amplitudes of oscillation, i.e. when the displacement x is about 40% of the distance d, indicated in Fig. 4.2a, as shown in [14]. When the amplitude of oscillations increases, the actual spring restoring force is better approximated when more terms (other than the third) are included into the Taylor series expansion.

To better illustrate the effect of Taylor series approximation, the following system parameters are adopted for simulations: stiffness coefficient $k_l = 100$ N/m, distance $d = 0.2$ m, damping coefficient $c_l = 2$ Ns/m and mass $m = 1$ kg.

The spring restoring force of the system in Fig. 4.2 is given in [14], as

$$f = k_l x \left(3 - \frac{2d}{\sqrt{x^2 + d^2}}\right), \tag{4.44}$$

which can be expanded using the Taylor series in the form of the left-hand side of Eq. (4.43), with the stiffness coefficients given by

$$c_1 = k_l, \quad c_3 = \frac{k_l}{d^2}, \quad c_5 = \frac{-3k_l}{4d^4}, \quad c_7 = \frac{5k_l}{8d^6}, \quad c_9 = \frac{-35k_l}{64d^8}, \quad c_{11} = \frac{63k_l}{128d^{10}}. \tag{4.45}$$

Figure 4.3a–d shows the results of simulations in the case where the equivalent spring of the system in Fig. 4.2 is approximated by only a 'linear-plus-cubic' stiffness term and it is excited by the multi-term harmonic excitation given by the right-hand side of Eq. (4.43), where $c_5 = c_7 = c_9 = c_{11} = 0$. A solution is sought in the form of Eqs. (4.8), (4.9), where $B = 80$ N/m and $A = 0.15$ m. In particular, Fig. 4.3a shows the time history of the specially designed external excitation provided to the oscillator, Fig. 4.3b shows the time history of the displacement, starting from rest, i.e. when $x(0) = 0, \dot{x}(0) = 0$. Figure 4.3c contains the Fourier coefficients extracted from the displacement time history at steady state, and Fig. 4.3d presents the phase trajectory at the steady state. It can be seen that the system response is not exactly harmonic, due to the presence of a small third harmonic component, as evident in Fig. 4.3c.

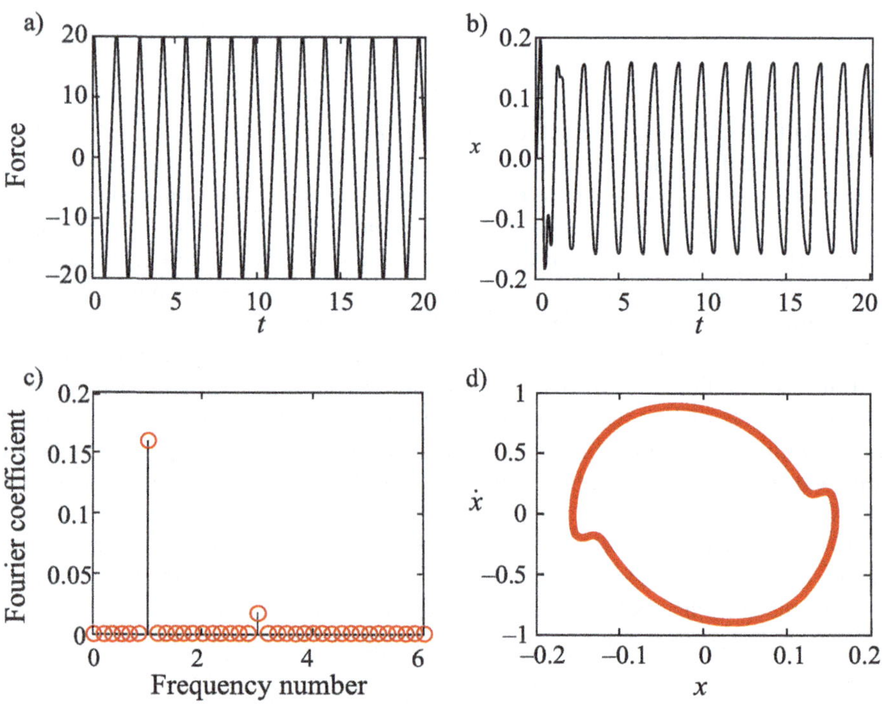

Fig. 4.3 Response of the model in Fig. 4.2 (using numerical simulations) when approximated to a Duffing oscillator with linear-plus-cubic stiffness coefficients: **a** specially designed external excitation; **b** displacement; **c** frequency content of the displacement; **d** phase trajectory

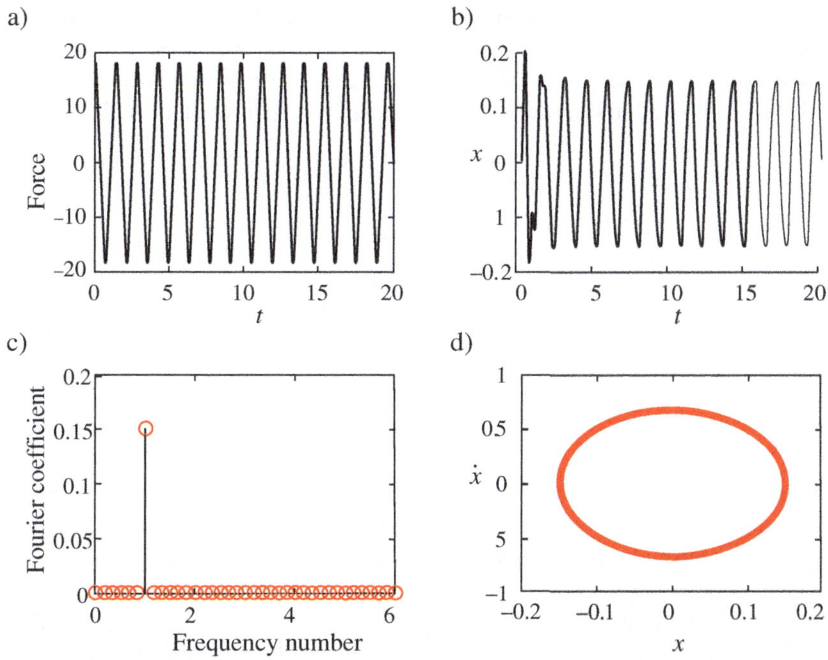

Fig. 4.4 Response of the model in Fig. 4.2 (using numerical simulations) when approximated by an oscillator with stiffness coefficients up to the eleventh-order: **a** specially designed external excitation; **b** displacement; **c** frequency content of the displacement; **d** phase trajectory

However, when all the 11 coefficients from Eq. (4.45) are included, different results are obtained as illustrated in Fig. 4.4a–d. In this case, the system response is clearly (co) sinusoidal, with only one harmonic in its frequency content, as shown in Fig. 4.4c.

4.4 Tuning Nonlinearities in a System Excited by Certain Two-Term Excitation to Produce Single Harmonic Steady-State Response

The previous cases have been concerned with tuning the excitation to get a desirable response for a set of given nonlinearities and stiffness parameters. However, it can be of interest to assume that the excitation is known a priori as multi-term, and to obtain the stiffness and damping parameters that would yield a simple harmonic response. This is the objective of this section, and it is of fundamental nature, since nonlinear systems are known to have a multi-frequency response to one-frequency excitation, where even new frequencies appear in the outcome, unlike in linear systems whose response contains only the frequency(ies) of the external excitation.

To address this issue, Eq. (4.38) with the response required to have the form $x = A\cos\Omega t$, is used first in the undamped case. It is then written as [5]

$$\ddot{x} + c_1 x + c_3 x^3 = F_1\cos(\Omega t) + F_2\cos(3\Omega t), \tag{4.46}$$

where

$$F_1 = BA + \frac{3}{4}c_3 A^3, \tag{4.47}$$

$$F_2 = \frac{1}{4}c_3 A^3, \tag{4.48}$$

$$\Omega = \sqrt{c_1 - B}. \tag{4.49}$$

Assuming that F_1, F_2 and Ω are known a priori, one deals with three Eqs. (4.47)–(4.49) and four unknown parameters, so one parameter can be fixed in advance (for example, the amplitude A or the parameter of proportionality B, under the condition $c_1 > B$).

Introducing the non-dimensional parameters $x^* = \frac{x}{A}$, $t^* = \sqrt{c_1}t$, $\gamma = c_3A^2/c_1$, $F_1^* = F_1/(c_1A)$, $F_2^* = F_2/(c_1A)$, $\Omega^* = \Omega/c_1$, Eq. (4.46) can be presented in a non-dimensional form (with the stars omitted) as

$$\ddot{x} + x + \gamma x^3 = F_1\cos(\Omega t) + F_2\cos(3\Omega t), \tag{4.50}$$

and with the primes denoting the differentiation with respect to new time. Equations (4.47)–(4.49) now give:

$$\gamma = 4F_2, \tag{4.51}$$

$$F_1 = 1 - \Omega^2 + 3F_2. \tag{4.52}$$

Note that since we deal with the HDO, it is required that $\gamma > 0$ and, thus, also $F_2 > 0$. The condition $c_1 > B$ with Eq. (4.49) also imposes $F_1 < 3F_2$.

The next case considered is when the excitations in Eq. (4.46) have a phase shift, which actually implies that they can be presented as including both the cosine and sine terms on the right-hand side. The governing equation can then be assumed in the form

$$\begin{aligned}\ddot{x} + 2\zeta_1\dot{x} + 2\zeta_3\dot{x}^3 + c_1x + c_3x^3 = {} & F_1\cos(\Omega t) + F_2\cos(3\Omega t)\\ & + G_1\sin(\Omega t) + G_2\sin(3\Omega t).\end{aligned} \tag{4.53}$$

The previously presented methodology enables one to determine the additional equations linking ζ_1 and ζ_2 with G_1 and G_2. The response $x = A\cos\Omega t$ will be achieved when

$$-2\zeta_1 A\Omega - \frac{3}{2}\zeta_3 A^3\Omega^3 = G_1, \tag{4.54}$$

$$\frac{1}{2}\zeta_3 A^3\Omega^3 = G_2, \tag{4.55}$$

which yields

$$\zeta_1 = -\frac{G_1 + 3G_2}{2A\Omega}, \tag{4.56}$$

$$\zeta_3 = \frac{2G_2}{A^3\Omega^3}. \tag{4.57}$$

Thus, five Eqs. (4.47)–(4.49), (4.56) and (4.57) with the condition $c_1 > B$ can be used to determine six unknown parameters of the oscillator considered: c_1, c_1, ζ_1, ζ_3, A and B. Given this fact, there is a certain level of flexibility and arbitrariness in this procedure, and one of these six parameters can be fixed a priori. This can be the amplitude A, as it can be made equal to unity after non-dimensionalization.

Numerical experiments are performed on a Duffing-like mechanical oscillator, and they are compared to the classical Duffing oscillator with linear-plus-cubic stiffness terms, to validate the theoretical discussion presented above. The model used in the numerical experiments is the same to that illustrated in Fig. 4.2, where the stiffness k_l, the geometric distance d and the damping c_l are now inversely determined based on the Taylor series expansion as $k_l = c_1$, $d = \sqrt{k/c_3}$ and $c_l = 2\zeta_1$. Two examples are presented below to highlight the fundamental behaviour of the system response.

4.4.1 *Example 4.1*

In the first application example, the multi-term excitation is considered with $F_1 = 1$ N, $F_2 = 1/3$ N, $G_1 = -7/4$ N, $G_2 = 1/4$ N, $\Omega = 1$ rad/s and $A = 1$ m, so that from Eqs. (4.47) to (4.49) and (4.56)–(4.57) one obtains $c_1 = 1$ N/m, $c_3 = 4/3$ N/m^3, $\zeta_1 = 1$ Ns/m, $\zeta_3 = 1$ Ns^3/m^3 and also $d = 0.86603$ m from Eq. (4.45).

The results of the simulation are presented in Fig. 4.5. Figure 4.5a and b shows the time history of the excitation force and its frequency content, respectively. It can be clearly seen that the excitation contains both a first and third harmonic. Figure 4.5c and d shows the time history of the displacement response and its frequency content, respectively, where the two oscillators mentioned above are considered. It can be seen that the response of the classical Duffing oscillator is (co) sinusoidal, as expected by the theory, but the response of the Duffing-like oscillator in Fig. 4.2, is characterized by a higher first harmonic and a tiny third component.

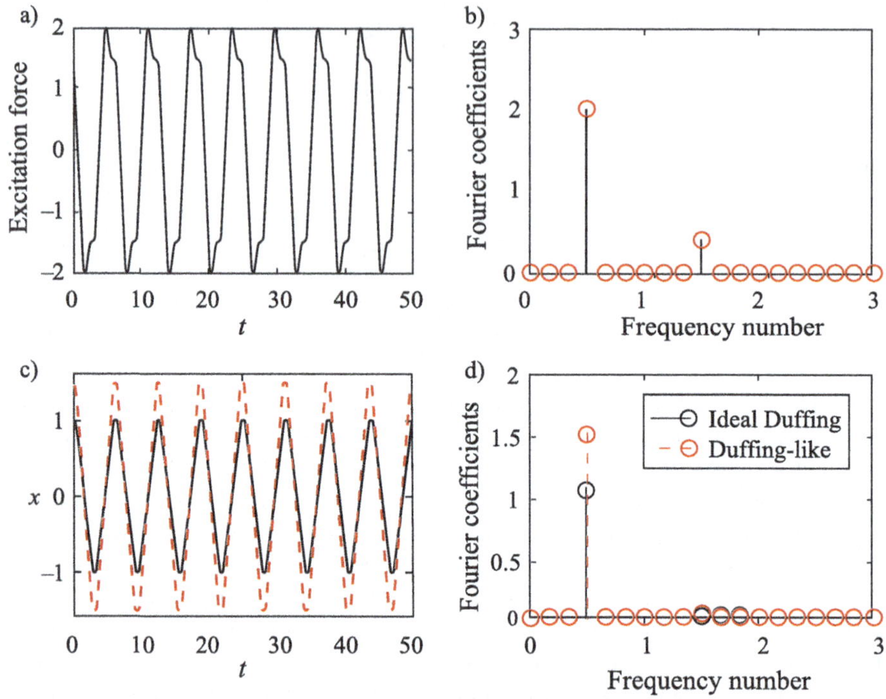

Fig. 4.5 Response of the system with specially designed stiffness in the case when the classical model does not match its physical counterpart in Fig. 4.2: **a** multi-term excitation; **b** corresponding frequency content; **c** system response; **d** corresponding frequency content

The discrepancy in the dynamic behaviour of the Duffing-like system is due to two main effects. The first reason is that the equivalent stiffness of the system in Fig. 4.2 (for the parameters listed above) is softer at higher displacements that the one of the classical Duffing oscillator. This can be seen in Fig. 4.6, which shows the force-deflection curve of the Duffing-like mechanical oscillator compared to that of the classical Duffing oscillator. The second reason for the differences in Fig. 4.5c and d is because of the damping. In fact, the Duffing-like mechanical oscillator in Fig. 4.2 includes a linear viscous damping, while the classical Duffing oscillator involves a damping force with linear and cubic terms.

4.4.2 *Example 4.2*

In order to make the mechanical model in Fig. 4.2 that behaves as the classical model, the following two conditions should be satisfied, $A < 0.4d$ and $\zeta_3 << 1$ Ns^3/m^3. These conditions are achieved in this second application example, where $F_1 = 1$ N, $F_2 = 1/3$ N, $G_1 = -1/2$ N, $G_2 = 1/24$ N, $\Omega = 4$ rad/s and $A = 1$ m, so that

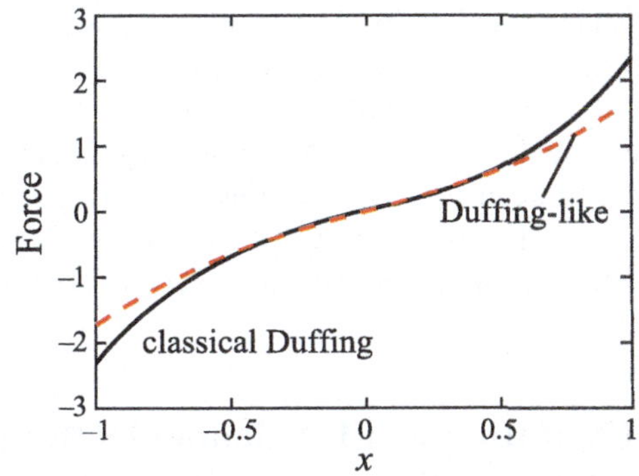

Fig. 4.6 Force-deflection curve of the equivalent spring for the classical Duffing oscillator (solid line) and the Duffing-like mechanical oscillator from Fig. 4.2 (dashed line)

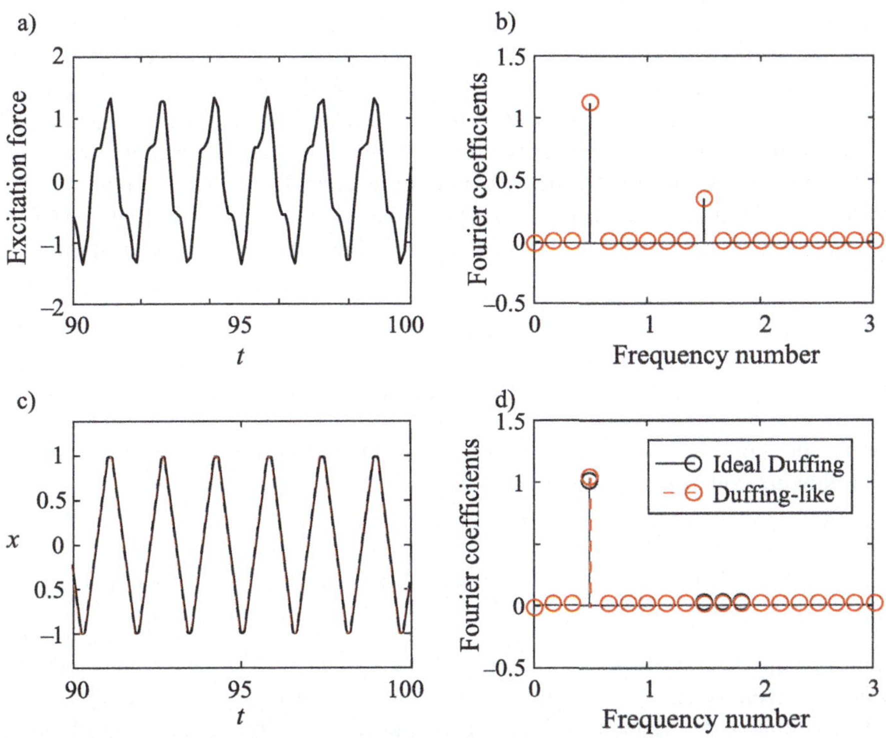

Fig. 4.7 Response of the system with specially designed stiffness in the case when the classical model matches its physical counterpart in Fig. 4.2: **a** multi-term excitation and **b** corresponding frequency content; **c** system response; **d** corresponding frequency content

from Eqs. (4.47)–(4.49) and Eqs. (4.56)–(4.57) one obtains $c_1 = 16$ N/m, $c_3 = 4/3$ N/m^3, $\zeta_1 = 0.09375$ Ns/m, $\zeta_3 = 0.00260417$ Ns^3/m^3 and $d = 3.4641$ m. Results are presented in Fig. 4.7, in a similar way as done in Fig. 4.5. Figure 4.7a and b shows the time history of the excitation force and its frequency content, respectively, while Figs. 4.7c and d show the time history of the displacement response and its frequency content, respectively. It can be seen now that the responses of the two systems are very similar, since the two requirements mentioned above are now satisfied.

4.5 Tuning the Excitation in a Hardening Duffing Oscillator to Make It Respond as Other Types of Free Cubic Oscillators

There are several possibilities for the choice of the external force to change the type of the oscillators governed by Eq. (4.17). The case $D = 0$, when the external force is proportional to the displacement, i.e. to the linear part of the restoring force, is considered in [3]. The general case $B \neq 0$, $D \neq 0$, when the external force is related to the overall restoring force is analysed in [1]. Here, the case that is not covered in these publications, but it is by the author in [6], is considered: $B = c_1$ and $c_3 > D$. With these conditions used, Eq. (4.17) turns into a free Pure Cubic Oscillator (PCO). Note that due to Eq. (4.14), one has now $m_r = 1/2$ and the exact forced response is

$$x_r = A \operatorname{cn}\left(\omega_r t \,\middle|\, \frac{1}{2}\right), \tag{4.58}$$

$$\omega_r = \sqrt{c_3 - D}A. \tag{4.59}$$

By introducing this solution into Eq. (4.16), and then into Eq. (4.15) with $B = c_1$, the following equation of motion is obtained

$$\ddot{x} + c_1 x + c_3 x^3 = c_1 A \operatorname{cn}\left(\sqrt{c_3 - D}At \,\middle|\, \frac{1}{2}\right) + D\, A^3 \operatorname{cn}^3\left(\sqrt{c_3 - D}At \,\middle|\, \frac{1}{2}\right). \tag{4.60}$$

There are three parameters (A, D and ω_r) here and one relationship between them, which is given by Eq. (4.59). Thus, two of them are arbitrary. If the amplitude is to be calculated, the frequency-amplitude relationship (4.59) needs to be used. Note that for $c_3 \approx D$, the value of A can be very large, which might be undesirable (the possibility for large values of A applies also in some cases below and it should be carefully dealt with).

In order to illustrate the results derived, let us make the requirement that the period of the response T is fixed. This period is related to the frequency of the cn function via the relationship $T = 4K(m_r)/\omega_r$, where K is the elliptic integral of the first kind, which gives $\omega_r = 4K(m)/T$. By choosing D as well, the amplitude A can be calculated from Eq. (4.59). The corresponding example with the period of 10s is shown in Fig. 4.8a (the rest of the parameters c_1, c_3, D are given in the figure caption, while the amplitude is calculated to be $A = 2\sqrt{2}K(1/2)/5 \approx 1.04882$). As another example, the additional requirement is introduced: let the period, the parameters c_1 and c_3 stay the same, but the amplitude is reduced for 50% ($A = \sqrt{2}K(1/2)/5 \approx 1.04882/2$). Equation (4.59) leads $D = -1$, and the corresponding responses are presented in Fig. 4.8b. Numerical solutions of the original equation of motion of the HDO are also plotted as the grey dotted line, clearly illustrating the change caused.

To provide additional insights into the harmonic content of the response for the case shown in Fig. 4.8b, the Fourier series expansion for the cn function with the elliptic parameter $m = 1/2$ given in Appendix C is used. It gives

$$\begin{aligned}\frac{x_r}{A} \approx\ & 0.95501\cos(0.62832t) + 0.04305\cos(1.88496t) + \\ & 0.00186\cos(3.14160t) + 0.00008\cos(4.39823t) + \cdots,\end{aligned} \tag{4.61}$$

leading to the conclusion that the first harmonic takes 95.5% of the overall response, the third harmonic 4.3% and the rest of them all together contribute less than 0.2%. The corresponding force from the right-hand side of Eq. (4.60) is

$$F = c_1 A \,\mathrm{cn}\left(\sqrt{c_3 - D}At \left| \frac{1}{2}\right.\right) + D\, A^3 \mathrm{cn}^3\left(\sqrt{c_3 - D}At \left| \frac{1}{2}\right.\right), \tag{4.62}$$

$$\begin{aligned}F_{app} \approx\ & 0.401959\cos(0.62832t) - \\ & 0.0175306\cos(1.88496t) - 0.00481469\cos(3.14160t) + \cdots\end{aligned} \tag{4.63}$$

The expression for the force Eq. (4.63) defines how this tuned excitation can be generated—as a multi-term harmonic excitation. Figure 4.9 is now plotted and contains: (i) the numerical solution of Eq. (4.60) (black thick solid line), (ii) the numerical solution of Eq. (4.60) where the right-hand side is approximated by Eq. (4.63) (grey dotted line), (iii) the numerical solution of Eq. (4.60) where the right-hand side is approximated by first two terms from Eq. (4.63) (black dashed line), (iv) the numerical solution of Eq. (4.60) where the right-hand side is approximated only by the first term from Eq. (4.63) (red solid line). It is seen that the approximations described under (ii) and (iii) are in good agreement with the exact one, while the last one described under (iv) shows some discrepancy around the maximal displacement, as better seen in the enlarger part of the time-response plotted in Fig. 4.9b.

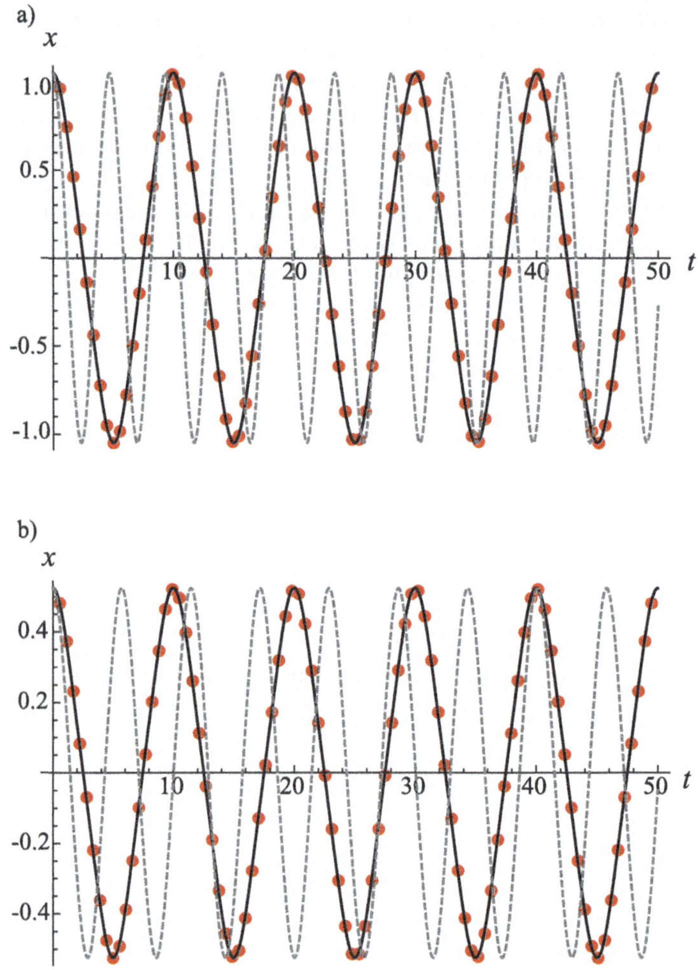

Fig. 4.8 Time response corresponding to $c_1 = 1$, $c_3 = 1$ and: **a** $D = 0.5$, $A = 2\sqrt{2}K\,(1/2)\,/5$, **b** $D = -1$, $A = \sqrt{2}K\,(1/2)/5$. Numerical solutions of Eq. (4.60)—black solid line, the analytical response (4.58), (4.59)—red dots, the numerical solution of the original equation of motion of the HDO (4.11)—grey dotted line

As in the previous sections, the damped case is straightforward. The existence of a damping term $\delta\dot{x}$ on the left-hand side of Eq. (4.60) requires the same type of the term with $\dot{x}$ being the first time derivative of Eq. (4.58) to be added to its right-hand side. The following equation of motion is then obtained:

$$\ddot{x} + \delta\dot{x} + c_1 x + c_3 x^3 = c_1 A \operatorname{cn}\left(\sqrt{c_3 - D}At \,\middle|\, \tfrac{1}{2}\right) + D\, A^3 \operatorname{cn}^3\left(\sqrt{c_3 - D}At \,\middle|\, \tfrac{1}{2}\right)$$
$$-\delta\sqrt{c_3 - D}A \operatorname{sn}\left(\sqrt{c_3 - D}At \,\middle|\, \tfrac{1}{2}\right) \operatorname{dn}\left(\sqrt{c_3 - D}At \,\middle|\, \tfrac{1}{2}\right). \tag{4.64}$$

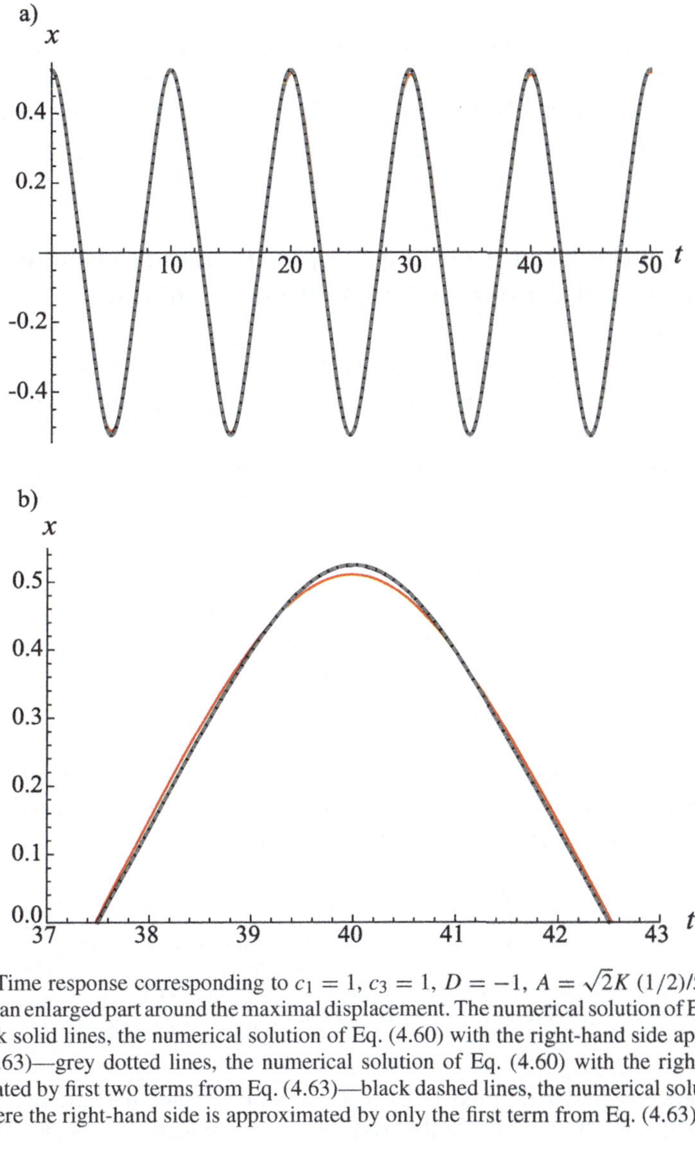

Fig. 4.9 Time response corresponding to $c_1 = 1$, $c_3 = 1$, $D = -1$, $A = \sqrt{2}K\,(1/2)/5$: **a** several periods; **b** an enlarged part around the maximal displacement. The numerical solution of Eq. (4.60)—black thick solid lines, the numerical solution of Eq. (4.60) with the right-hand side approximated by Eq. (4.63)—grey dotted lines, the numerical solution of Eq. (4.60) with the right-hand side approximated by first two terms from Eq. (4.63)—black dashed lines, the numerical solution of Eq. (4.60) where the right-hand side is approximated by only the first term from Eq. (4.63)—red solid lines

Developing the external force from the right-hand side of Eq. (4.64) for the same parameters as previously, the force is calculated to have additional harmonics:

$$F_{app,d} \approx F_{app} - 0.0314672\sin(0.62832t) - 0.00425541\sin(1.88496t) - 0.000306514(3.14160t). \tag{4.65}$$

Similarly to the case shown in Fig. 4.9, three additional terms can be simplified to only two of them to get reasonably good approximations, while keeping only one with the smallest angular frequency would result is slightly reduced accuracy, especially around the maximal displacement.

4.6 Tuning the Excitation in a Simple Harmonic Oscillator to Make It Respond as Free Duffing Oscillators

It is seen from Eq. (4.17) with $c_3 = 0$ that the form of the external force, i.e. the values of the parameters B and D can be adjusted to make the excited SHO respond as free BDOs. For example, by choosing $c_1 < B$, $D < 0$ (and also $-DA^2 > 2|c_1 - B|$ to achieve oscillations around the origin in the BDO), the original SHO can be turned into the Bistable Duffing Oscillators (BDOs) that oscillates around the origin (see Chap. 2, Sect. 2.3.5) with the response

$$x_r = A\mathrm{cn}\left(\omega_r t \,|m_r\right), \tag{4.66}$$

$$\omega_r = \sqrt{c_1 - B - DA^2}, \tag{4.67}$$

$$m_r = \frac{-DA^2}{2\left(c_1 - B - DA^2\right)}. \tag{4.68}$$

According to the notation from Chap. 2, this will corresponds to the BDO1 (full-swing mode). By introducing this solution into Eq. (4.16) with $c_3 = 0$, and then into Eq. (4.15), the required equation of motion is obtained

$$\ddot{x} + c_1 x = BA\mathrm{cn}\left(\sqrt{c_1 - B - DA^2}\;t \left| \frac{-DA^2}{2\left(c_1 - B - DA^2\right)}\right.\right) + DA^3\mathrm{cn}^3\left(\sqrt{c_1 - B - DA^2}t \left| \frac{-DA^2}{2\left(c_1 - B - DA^2\right)}\right.\right). \tag{4.69}$$

Note that now five parameters exist (A, B, D, ω_r and m_r) and there are two relationships between them, Eqs. (4.67) and (4.68). Thus, three parameters can be arbitrarily chosen with the exception of the group B, ω_r and m_r, when DA^2 can be calculated, but then there is no expression to obtain D and A.

Let us illustrate the case when A, ω_r and m_r are fixed: $A = 1, \omega_r = 1$, and $m_r = 3/4$. Then, it follows $D = -3/2$, $B = 3/2$. By using these parameters, Fig. 4.10a is produced in terms of phase trajectories: the numerical solution of Eq. (4.69) is depicted by the black solid line, the analytical response (4.66) by the red dots, and the numerical solution of the original SHO equation of motion (4.3) by the grey dotted line. It is seen that the circular phase trajectory corresponding to the original SHO is changed

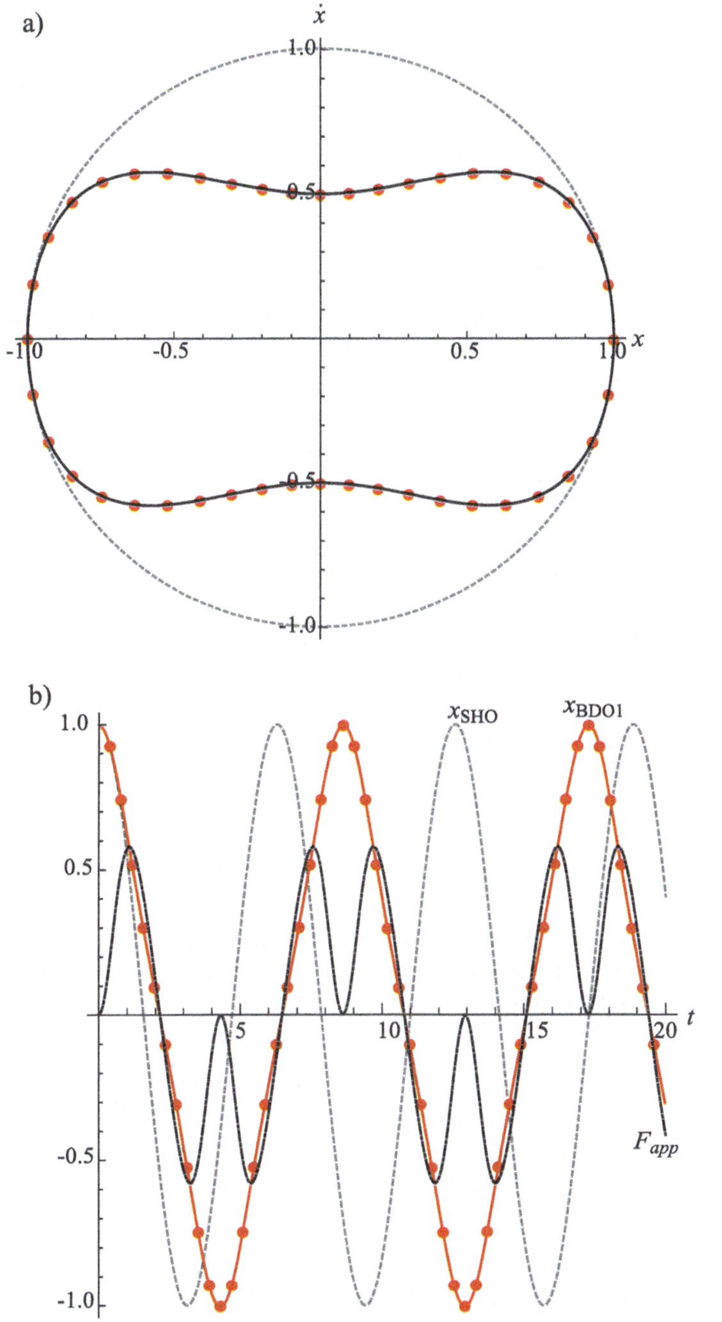

Fig. 4.10 The case corresponding to $c_1 = 1, c_3 = 0, A = 1, D = -3/2, B = 3/2$. **a** Phase trajectory; **b** Time responses and external excitation. Numerical solutions of Eq. (4.69)—black solid line, the analytical response (4.66), (4.67), (4.68)—red dots, the numerical solution of the original equation of motion of the SHO (4.3)—grey dotted line, the external excitation (4.70)—black dashed-dotted line

into the 'deformed' phase trajectory of the BDO1 around the origin. The solutions of Eqs. (4.69) agree completely with the plot of Eq. (4.66). The corresponding force that appears on the right-hand side of Eq. (4.69) can be represented as the following Fourier series expansion:

$$F_{app} \approx 0.42605\cos(0.72840t) - 0.318965\cos(2.18519t) - \\ 0.08896\cos(3.64198t) - 0.01556\cos(5.09877t)\ldots \quad (4.70)$$

Thus, the external excitation needed consists of odd integer multiplications of the first harmonic whose angular frequency is 0.72840. Numerical simulations have shown that this approximation agrees well with the exact form of the excitation shown on the right-hand side of Eq. (4.69), and that further truncations deteriorate this accuracy. Figure 4.10b shows how this force changes in time. For the sake of comparison, the time response of the original SHO and the resulting BDO are also plotted.

It is interesting to make this original SHO that oscillates around a trivial equilibrium oscillate around a non-trivial equilibrium, which is obtained as the behaviour of BDOs in a half-swing mode [12], which is labelled in Chap. 2 as the BDO2 (half-swing mode). These bistable oscillators can be obtained from Eq. (4.17) with $c_3 = 0$ by fixing $c_1 < B$, $D < 0$. By using the exact-closed form for such bistable oscillators in a half-swing mode and the corresponding condition $0 < -DA^2 < 2|c_1 - B|$ (see Chap. 2, Sect. 2.3.5 or [2]), the response is found to be

$$x_r = A\mathrm{dn}(\omega_r t\,|m_r), \quad (4.71)$$

$$\omega_r = \sqrt{\frac{-DA^2}{2}}, \quad (4.72)$$

$$m_r = \frac{2(-DA^2 - |c_1 - B|)}{-DA^2}, \quad (4.73)$$

where dn stands for the Jacobi dn function. By introducing this solution into Eq. (4.16) with $c_3 = 0$, and then into Eq. (4.15), the equation of motion of the externally excited SHO that generates this response is derived

$$\ddot{x} + c_1 x = BA\mathrm{dn}\left(\sqrt{\frac{-DA^2}{2}}t\,\middle|\,\frac{2(-DA^2 - |c_1 - B|)}{-DA^2}\right) + \\ DA^3\mathrm{dn}^3\left(\sqrt{\frac{-DA^2}{2}}t\,\middle|\,\frac{2(-DA^2 - |c_1 - B|)}{-DA^2}\right). \quad (4.74)$$

Let us illustrate again the case when A, ω_r and m_r are fixed: $A = \pm 1$, $\omega_r = 1$, and $m_r = 3/4$. Then, Eqs. (4.72) and (4.73) give $D = -2$, $B = 9/4$. Figure 4.11a contains the phase trajectories obtained for these parameters: the numerical solution

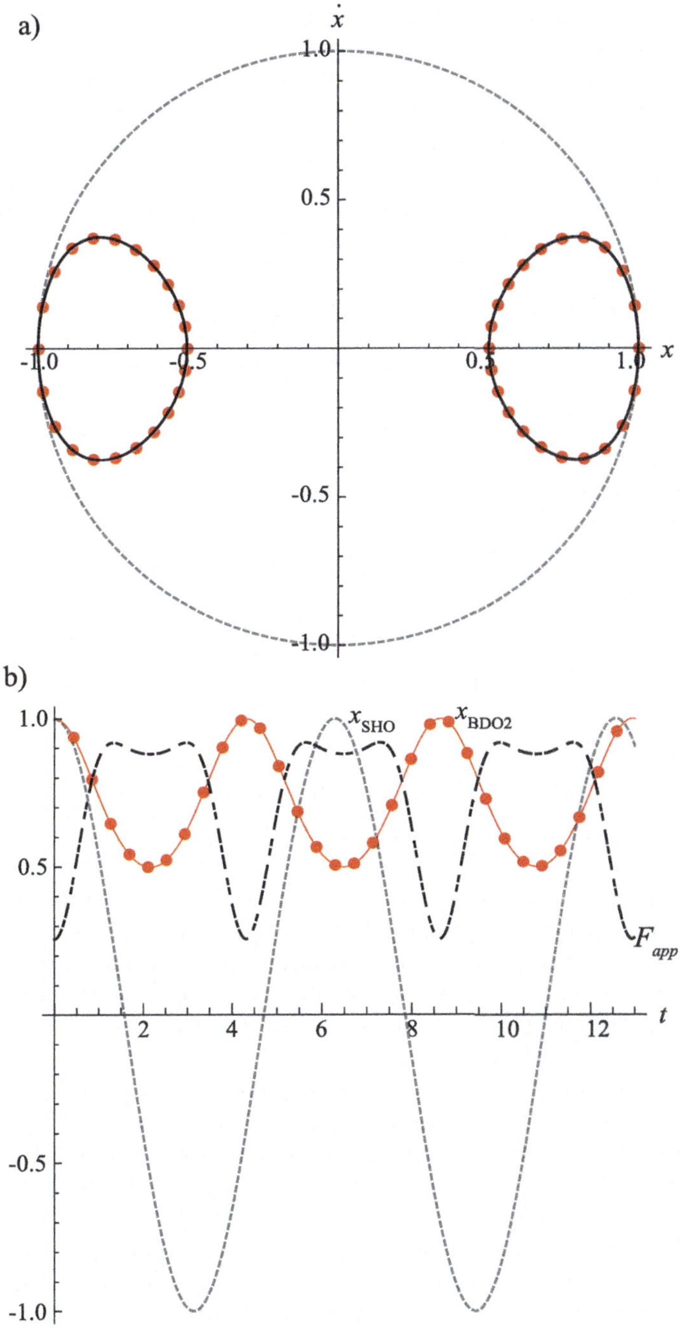

Fig. 4.11 The case corresponding to $c_1 = 1$, $c_3 = 0$, $A = \pm 1$, $D = -2$, $B = 9/4$. **a** Phase trajectory; **b** Time responses and external excitation. Numerical solutions of Eq. (4.74)—black solid line, the analytical response (4.71), (4.72), (4.73)—red dots, the numerical solution of the original equation of motion of the SHO (4.3)—grey dotted line, the external excitation (4.77)—blue dashed-dotted line

of Eq. (4.74) is depicted by the black solid line, the analytical response (4.71) by the red dots, and the numerical solution of the original equation of motion of the SHO (4.3) by the grey dotted line. The circular phase trajectory corresponding to the original SHO is changed into two closed orbits surrounding non-zero equilibria (the left one is obtained for $A = -1$ and the right one for $A = 1$). The solutions of Eqs. (4.74) and (4.71) agree with each other.

The solution (4.71) corresponding to $A = 1$ can be expressed as (see Appendix C for the Fourier series expansion for the Jacobi dn function):

$$\frac{x_r}{A} \approx 0.72840 + 0.24815\cos(1.45679t) + 0.021445\cos(2.91358t) + 0.00184\cos(4.37037t)\ldots. \tag{4.75}$$

It comprises a constant term (an offset) and harmonics whose frequencies are integer multiplications of the first one.

The corresponding force from the right-hand side of Eq. (4.74) is

$$F = BA\mathrm{dn}\left(\sqrt{\frac{-DA^2}{2}}t \,\middle|\, \frac{2\left(-DA^2 - |c_1 - B|\right)}{-DA^2}\right) + DA^3\mathrm{dn}^3\left(\sqrt{\frac{-DA^2}{2}}t \,\middle|\, \frac{2\left(-DA^2 - |c_1 - B|\right)}{-DA^2}\right), \tag{4.76}$$

$$F_{app} \approx 0.72840 - 0.27848\cos(1.45679t) - 0.16060\cos(2.91358t) - 0.03330\cos(4.37037t)\ldots. \tag{4.77}$$

Note that the approximation for the dn function is obtained on the basis of the analytical expressions from Appendix C, while the Fourier coefficients for the cube of the dn function were calculated numerically from the definition of the Fourier coefficient and the option NIntegrate in Wolfram Mathematica. The way this force changes in time in plotted in Fig. 4.11b together with the time response of the original SHO and the resulting BDO.

The expression for the force Eq. (4.77) implies that this force is a multi-term excitation with an offset and certain harmonics, and this is how it can be formed in a lab. To check which level of the approximation of this force is acceptable, Fig. 4.12 is created, comprising: (i) the numerical solution of Eq. (4.74) (black solid line), (ii) the numerical solution of Eq. (4.74) where the right-hand side is approximated by Eq. (4.77) (grey dotted line), (iii) the numerical solution of Eq. (4.74) where the right-hand side is approximated by first three terms from Eq. (4.77) (black dashed line). It is seen that the approximation defined under (ii) is in good agreement with the exact one, while the one described under (iii) shows some discrepancies. Further approximations lead to worse results, but are not shown here in order not to make the figure cluttered.

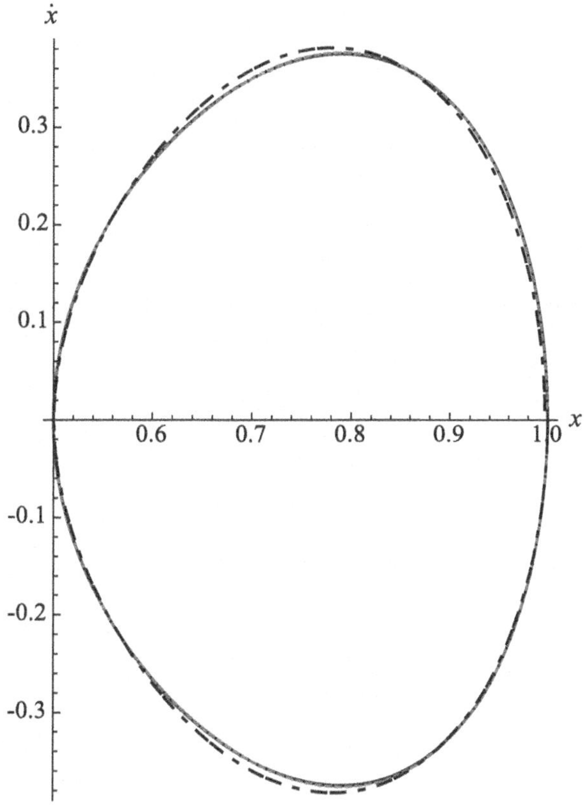

Fig. 4.12 Phase trajectory corresponding to $c_1 = 1$, $c_3 = 0$, $A = 1$, $D = -2$, $B = 9/4$. The numerical solution of Eq. (4.74)—black solid lines, the numerical solution of Eq. (4.74) with the right-hand side approximated by Eq. (4.77)—grey dotted lines, the numerical solution of Eq. (4.74) with the right-hand side approximated by first three terms from Eq. (4.77)—black dashed line

4.7 Forced Response of Purely Nonlinear Oscillators: Exact Solutions via Ateb Functions

This section is concerned with externally excited oscillators governed by Eq. (4.1) whose restoring term $G(x)$ is purely nonlinear:

$$\ddot{x} + c_\alpha \operatorname{sgn}(x) |x|^\alpha = F. \tag{4.78}$$

The main task is to define the form of the external force F so as to obtain the exact solution for the resulting response. To that end, let us first focus on free oscillators presented in Chap. 2, Sect. 2.5. It is shown therein that Purely Nonlinear Oscillators (PNOs) governed by

$$\ddot{x} + c_\alpha \operatorname{sgn}(x) |x|^\alpha = 0, \tag{4.79}$$

where $c_\alpha > 0$ and α is any positive real number has an exact, closed-form solution given in terms of the cosine Ateb (ca) function

$$x = A\,\mathrm{ca}\,(\alpha, 1, \omega_{\mathrm{ca}} t)\,, \tag{4.80}$$

$$\omega_{\mathrm{ca}} = |A|^{(\alpha-1)/2} \sqrt{\frac{c_\alpha (\alpha+1)}{2}}. \tag{4.81}$$

Note that this solution corresponds to the initial conditions: $x(0) = A$, $\dot{x}(0) = 0$.

The solution for the forced response is sought in the Ateb-type form as well

$$x = A\,\mathrm{ca}\,(\alpha, 1, \omega_r t)\,, \tag{4.82}$$

with the amplitude A stands for the steady-state amplitude of forced vibrations now, while the angular frequency is different and it is labelled by ω_r.

The excitation is assumed first in the form related both to the form of the restoring force given by $\mathrm{sgn}(x)\,|x|^\alpha$, as well as to the form of the exact solution for their free response (4.80), but with a new frequency ω_r introduced:

$$F = F_0 \mathrm{sgn}\,(\mathrm{ca}\,(\alpha, 1, \omega t))\,|\,\mathrm{ca}\,(\alpha, 1, \omega t)|^\alpha\,. \tag{4.83}$$

Note that given Eq. (4.82), the excitation (4.83) can be written down as

$$F = \frac{F_0}{\mathrm{sgn}\,(A)\,|A|^\alpha} \mathrm{sgn}\,(x)\,|x|^\alpha\,, \tag{4.84}$$

as a result of which the equation of motion (4.78) turns into

$$\ddot{x} + \left(c_\alpha - \frac{F_0}{\mathrm{sgn}\,(A)\,|A|^\alpha}\right) \mathrm{sgn}\,(x)\,|x|^\alpha = 0. \tag{4.85}$$

By comparing Eq. (4.85) with Eq. (4.79), it is seen that they just have different coefficients in front of the geometric term, i.e. the forcing changes the stiffness coefficient c_α. The solution (4.81) can be used to obtain ω_r:

$$\omega_r = |A|^{(\alpha-1)/2} \sqrt{\left(c_\alpha - \frac{F_0}{\mathrm{sgn}\,(A)\,|A|^\alpha}\right) \frac{(\alpha+1)}{2}}. \tag{4.86}$$

This expression can further be transformed to derive the following amplitude-frequency equation:

$$-\frac{2}{\alpha+1}\omega_r^2 A + c_\alpha \mathrm{sgn}\,(A)\,|A|^\alpha = F_0. \tag{4.87}$$

The solutions of Eq. (4.87) enable one to find the amplitude(s) corresponding to different values of the frequency ω_r and to plot frequency-response (amplitude-frequency) curves. This is dealt with subsequently.

4.7.1 Frequency-Response Curves

Depending whether A is positive or negative, Eq. (4.87) can be, respectively, separated into

$$-\frac{2}{\alpha+1}\omega_{r1}^2 A + c_\alpha A^\alpha = F_0, \tag{4.88}$$

$$\frac{2}{\alpha+1}\omega_{r2}^2 |A| - c_\alpha |A|^\alpha = F_0. \tag{4.89}$$

Instead of expressing A as a function of $\omega_{r1}$and ω_{r2}, it is easier to derive the reverse

$$\omega_{r1} = \sqrt{\frac{\alpha+1}{2}\frac{c_\alpha A^\alpha - F_0}{A}}, \tag{4.90}$$

$$\omega_{r2} = \sqrt{\frac{\alpha+1}{2}\frac{c_\alpha |A|^\alpha + F_0}{|A|}}, \tag{4.91}$$

where ω_{r1} and ω_{r2} correspond to two branches of the amplitude-frequency curves. These branches will be located around the backbone curves, whose expression obtains for $F_0 = 0$, as follows:

$$\omega_{bc} = |A|^{(\alpha-1)/2}\sqrt{c_\alpha \frac{\alpha+1}{2}}. \tag{4.92}$$

This is actually the frequency-amplitude relationship of the unforced conservative case and obviously coincides with Eq. (4.81).

To find the values of the amplitude and frequency at which the number of possible solutions changes (they will be labelled by $\hat{A}$ and $\hat{\omega}_r$, respectively), i.e. where the saddle-node (SD) point occurs, one can differentiate Eq. (4.87) with respect to A to derive

$$\omega_r^2 = |A|^{\alpha-1} c_\alpha \frac{\alpha(\alpha+1)}{2}. \tag{4.93}$$

Substituting this back into Eq. (4.87) and then into Eq. (4.93), one can derive the following expressions for the amplitude $\left|\hat{A}\right|$ and frequency $\hat{\omega}_r$ at which the SD point exists

$$\left|\hat{A}\right| = \left[\frac{F_0}{c_\alpha |\alpha-1|}\right]^{1/\alpha}, \tag{4.94}$$

$$\hat{\omega}_r = \sqrt{\frac{c_\alpha \alpha(\alpha+1)}{2}\left[\frac{F_0}{c_\alpha |\alpha-1|}\right]^{\frac{\alpha-1}{2\alpha}}}. \tag{4.95}$$

Table 4.1 Expressions for the backbone curve ω_{bc}, the frequency $\hat{\omega}_r$ and the amplitude $\left|\hat{A}\right|$ of the SD point for several integer powers of nonlinearity α

α	ω_{bc}	$\hat{\omega}_r$	$\left|\hat{A}\right|$
2	$\sqrt{\lvert A\rvert}\sqrt{\frac{3}{2}c_\alpha}$	$\sqrt{3c_\alpha}\left[\frac{F_0}{c_\alpha}\right]^{\frac{1}{4}}$	$\sqrt{\frac{F_0}{c_\alpha}}$
3	$\lvert A\rvert\sqrt{2c_\alpha}$	$\sqrt{6c_\alpha}\left[\frac{F_0}{2c_\alpha}\right]^{\frac{1}{3}}$	$\left[\frac{F_0}{2c_\alpha}\right]^{1/3}$
4	$\lvert A\rvert^{3/2}\sqrt{\frac{5}{2}c_\alpha}$	$\sqrt{10c_\alpha}\left[\frac{F_0}{3c_\alpha}\right]^{\frac{3}{8}}$	$\left[\frac{F_0}{3c_\alpha}\right]^{1/4}$
5	$\lvert A\rvert^2\sqrt{3c_\alpha}$	$\sqrt{15c_\alpha}\left[\frac{F_0}{4c_\alpha}\right]^{\frac{2}{5}}$	$\left[\frac{F_0}{4c_\alpha}\right]^{1/5}$

Table 4.2 Expressions for the backbone curve ω_{bc}, the frequency $\hat{\omega}_r$ and the amplitude $\left|\hat{A}\right|$ of the SD point for several non-integer powers of nonlinearity α.

α	ω_{bc}	$\hat{\omega}_r$	$\left|\hat{A}\right|$
1/2	$\lvert A\rvert^{-1/4}\sqrt{\frac{3}{4}c_\alpha}$	$\sqrt{\frac{3c_\alpha}{8}}\left[\frac{2F_0}{c_\alpha}\right]^{\frac{-1}{2}}$	$\left[\frac{2F_0}{c_\alpha}\right]^{2}$
2/3	$\lvert A\rvert^{-1/6}\sqrt{\frac{5}{6}c_\alpha}$	$\sqrt{\frac{5c_\alpha}{9}}\left[\frac{3F_0}{c_\alpha}\right]^{\frac{-1}{4}}$	$\left[\frac{3F_0}{c_\alpha}\right]^{3/2}$
3/2	$\lvert A\rvert^{1/4}\sqrt{\frac{5}{4}c_\alpha}$	$\sqrt{\frac{15c_\alpha}{8}}\left[\frac{2F_0}{c_\alpha}\right]^{\frac{1}{6}}$	$\left[\frac{2F_0}{c_\alpha}\right]^{2/3}$
5/3	$\lvert A\rvert^{1/3}\sqrt{\frac{4}{3}c_\alpha}$	$\sqrt{\frac{20c_\alpha}{9}}\left[\frac{3F_0}{2c_\alpha}\right]^{\frac{1}{5}}$	$\left[\frac{3F_0}{2c_\alpha}\right]^{3/5}$

The expressions for the backbone curve (4.92) and the SD point (4.94), (4.95) are used to list their values for several powers of nonlinearity in Table 4.1 for integer powers and in Table 4.2 for non-integer powers. Figure 4.13 contains the graphs of the rescaled SD amplitudes ($|A^*| = \left|\hat{A}\right| / \left[\frac{F_0}{c_\alpha}\right]^{1/\alpha}$) and frequency ($\omega_r^* = \hat{\omega}_r / \left(\sqrt{c_\alpha}\left[\frac{F_0}{c_\alpha}\right]^{\frac{\alpha-1}{2\alpha}}\right)$) versus the power of nonlinearity α. They can be used to estimate where the point SD occurs depending on α. It is labelled that for $\alpha = 1$, one has $\omega_r^* = 1$, so that one can compare other under-linear and over-linear cases with this one.

Equations (4.90) and (4.91) are used to plot the frequency-response branches in Fig. 4.14 for $c_\alpha = 1$ and $F_0 = 1/2$ and four values of α. The backbone curve (4.92) is also presented and the SD point labelled by the asterix. Note that the frequency refers to ω_r. The frequency-response branches corresponding to $\alpha < 1$ are bent to the left-hand side, while for $\alpha > 1$ they are bent to the right-hand side. There are frequency regions in which they have a multi-valued response.

To illustrate the forced response in a time domain, the frequency $\omega_r = 2$ is chosen for $\alpha = 3$ (Fig. 4.14d). Three values of the amplitude are calculated from Eq. (4.87) and two cases are labelled by I and II: $A_{\text{I}} = -0.258652$ and $A_{\text{II}} = 1.52569$, while

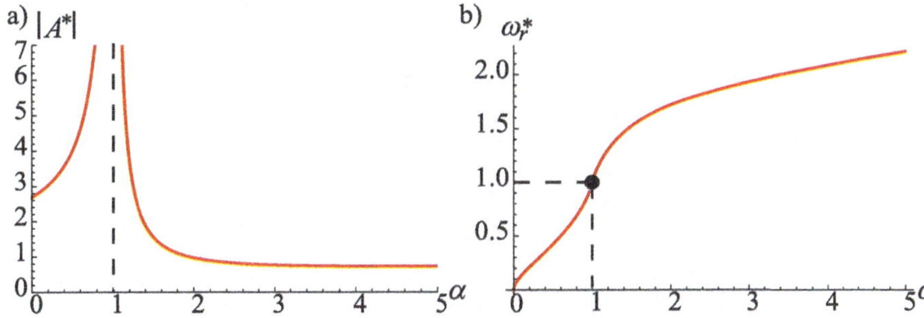

Fig. 4.13 a) Graph of the rescaled SD amplitudes $|A^*|$ versus the power of nonlinearity α; b) Graph of the rescaled SD frequency ω_r^* versus the power of nonlinearity α

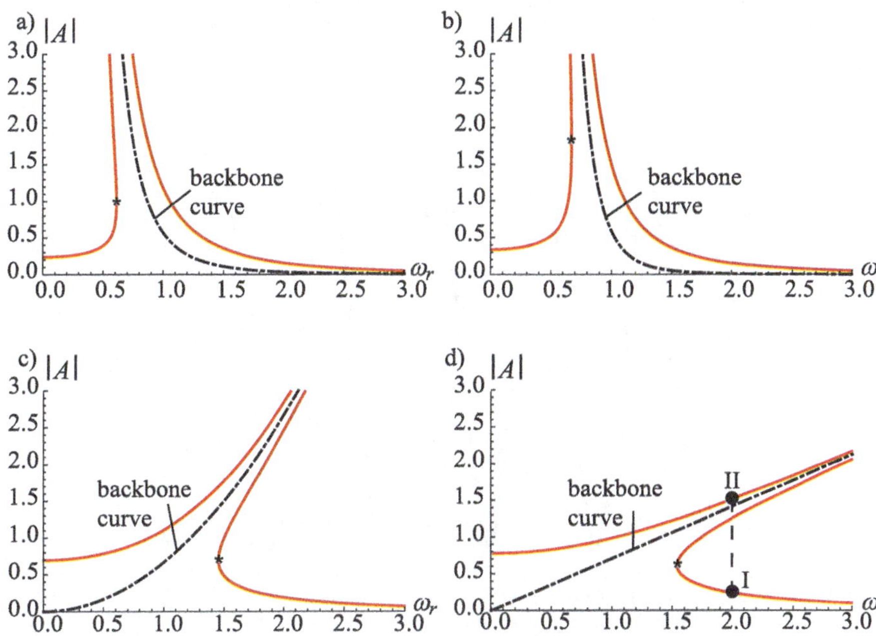

Fig. 4.14 Frequency-response curves (solid line), the SD point (asterix) and the backbone curve (dashed-dotted line) corresponding to $c_\alpha = 1$, $F_0 = 1/2$ and: **a** $\alpha = 1/2$; **b** $\alpha = 2/3$; **c** $\alpha = 2$; **d** $\alpha = 3$

the third one between them is $A = -1.26704$. By using the form (4.82) and adopting the Fourier series from Appendix C, the responses I and II are found to be

$$x_{\mathrm{I}} = -0.25865 \operatorname{ca}(3, 1, 2t) \tag{4.96}$$

$$\approx -0.24701 \cos(1.19814t) - 0.01113 \cos(3.59442t) - 0.00048 \cos(5.9907t) - 0.00002 \cos(8.38698t) \,. \tag{4.97}$$

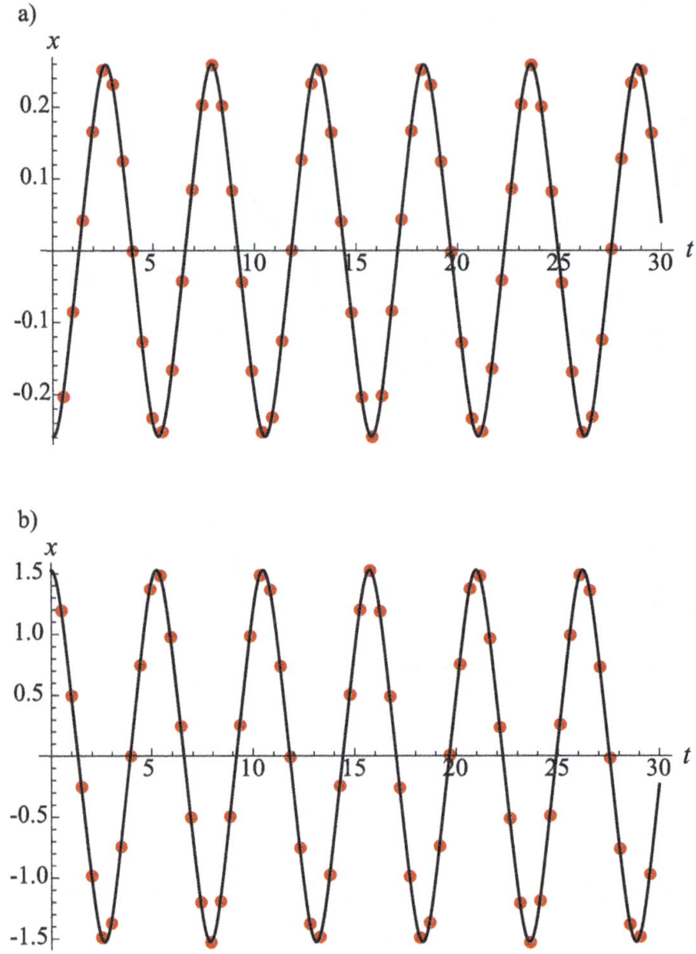

Fig. 4.15 Time response corresponding to $c_\alpha = 1$, $F_0 = 1/2$, $\alpha = 3$, $\omega_r = 2$ and: **a** Case I from Fig. 4.14d; **a** Case II from Fig. 4.14d. The numerical solution of Eqs. (4.78), (4.83)—red dots, Fourier series approximations, Eqs. (4.97), (4.99)—black solid line

and

$$x_{\text{II}} = 1.52569\, \text{ca}\,(3,\, 1,\, 2t) \tag{4.98}$$

$$\approx 1.45704 \cos(1.19814t) + 0.06568 \cos(3.59442t) + 0.00284 \cos(5.9907t) + 0.00012 \cos(8.38698t)\,. \tag{4.99}$$

They are plotted in Fig. 4.15 together with the numerical solutions of Eq. (4.78), (4.83) to demonstrate their mutual agreement. The excitation causing these responses is defined by Eq. (4.83) with $\alpha = 3$.

A remark should be made with respect to the existence of three solutions for the amplitude at the frequency $\omega_r = 2$, as seen in Fig. 4.14d: the solutions corresponding to A_{I} and A_{II} were found to be stable; the solution corresponding to $A = -1.26704$ was found to be unstable, and this is the reason why this solution is not given here in a time domain. The details about the stability check are omitted here, as this was done in a classical way by introducing the perturbation into the equation of motion (4.78), and deriving the linear variational equation [15]. The solution for the forced response was used then in the form of the Fourier series, so that Hill's equation was obtained easily for all the solutions. The corresponding positions in the stability chart with respect to the instability tongues were analysed [15] to detect whether the solutions found are stable or unstable.

4.7.1.1 About Some Simplifications and Approximations

In the linear case, Eqs. (4.78), (4.83) reduce to the harmonically excited linear oscillator

$$\ddot{x} + c_1\, x = F_0 \cos\left(\omega_r t\right), \tag{4.100}$$

where Eq. (4.86) gives

$$\omega_r = \sqrt{c_1 - \frac{F_0}{A}}. \tag{4.101}$$

with the forced response, Eq. (4.82) being

$$x = A\, \mathrm{ca}\left(1, 1, \omega_r t\right) = A\, \cos\left(\omega_r t\right). \tag{4.102}$$

It is easy to verify that Eq. (4.102) satisfies Eq. (4.100).

For the pure cubic case, Eqs. (4.78), (4.83) become

$$\ddot{x} + c_3\, x^3 = F_0 \mathrm{ca}^3\left(3, 1, \omega_r t\right). \tag{4.103}$$

The exact closed-form solution for the forced response is

$$x = A\, \mathrm{ca}\left(3, 1, \omega_r t\right), \tag{4.104}$$

$$\omega_r = |A| \sqrt{2\left(c_3 - \frac{F_0}{A^3}\right)}. \tag{4.105}$$

Given the transformation of the Ateb function into the Jacobi cn function for the case of cubic nonlinearity (see Chap. 2, Sect. 2.5), one can represent the equation of motion (4.103) in the form

$$\ddot{x} + c_3 x^3 = F_0 \mathrm{cn}^3 \left(\omega_{\mathrm{cn}} t \left| \frac{1}{2} \right. \right) \tag{4.106}$$

with the exact solution for the forced response

$$x = A\ \mathrm{cn} \left(\omega_{\mathrm{cn}} t \left| \frac{1}{2} \right. \right), \tag{4.107}$$

$$\omega_{\mathrm{cn}} = |A| \sqrt{\left(c_3 - \frac{F_0}{A^3} \right)}. \tag{4.108}$$

The results given by Eqs. (4.106)–(4.108) can be considered as a special case of Hsu's exact closed-form solution for linear-plus-cubic oscillators with a two-term elliptic forcing (Sect. 3.2 in [1]): they can be obtained from the results derived therein by omitting the linear restoring term and the related excitation term.

Using the Fourier series representation from Appendix C, the external excitation in Eq. (4.106) can be expressed as the following multi-term excitation

$$F_0 \mathrm{cn}^3 \left(\omega_{\mathrm{cn}} t \left| \frac{1}{2} \right. \right) \approx F_1 \cos\left(\tfrac{\pi}{2K} \omega_{\mathrm{cn}} t\right) + F_3 \cos\left(\tfrac{3\pi}{2K} \omega_{\mathrm{cn}} t\right) + F_5 \cos\left(\tfrac{5\pi}{2K} \omega_{\mathrm{cn}} t\right) \ldots, \tag{4.109}$$

$$F_1 = F_0 D_1 \left(\tfrac{\pi}{2K}\right)^2, \quad F_3 = F_0 D_3 \left(\tfrac{3\pi}{2K}\right)^2, \quad F_5 = F_0 D_5 \left(\tfrac{5\pi}{2K}\right)^2, \ldots \tag{4.110}$$

On the other hand, Eq.(4.107) can be expressed as the following Fourier series (see Appendix C):

$$A \mathrm{cn} \left(\omega_{\mathrm{cn}} t \left| \frac{1}{2} \right. \right) \approx A D_1 \cos\left(\frac{\pi}{2K} \omega_{\mathrm{cn}} t\right) + A D_3 \cos\left(3 \frac{\pi}{2K} \omega_{\mathrm{cn}} t\right) + A D_5 \cos\left(5 \frac{\pi}{2K} \omega_{\mathrm{cn}} t\right) \ldots \tag{4.111}$$

This expression, thus, represents the approximate response to the multi-term external excitation (4.109) [13].

4.7.2 On Some Further Choices of the System Parameters

Let us now assume the excitation in the form related directly to the desired forced response, Eq. (4.82) and with the same power-form as the restoring force:

$$F = f_0 \mathrm{sgn}\left(A \mathrm{ca}\left(\alpha, 1, \omega_r t\right)\right) \left|A\ \mathrm{ca}\left(\alpha, 1, \omega_r t\right)\right|^{\alpha}. \tag{4.112}$$

The excitation (4.112) can now be written down as

$$F = f_0 \mathrm{sgn}\left(x\right) |x|^{\alpha}, \tag{4.113}$$

so that the equation of motion (4.78) becomes

$$\ddot{x} + (c_\alpha - f_0)\ \mathrm{sgn}\,(x)\,|x|^\alpha = 0. \tag{4.114}$$

By comparing this equation with Eq. (4.79), it is seen that the excitation changes the stiffness coefficient c_α.

Assuming that $c_\alpha > f_0$, the solution (4.81) can be used to obtain ω_r:

$$\omega_r = |A|^{(\alpha-1)/2}\sqrt{(c_\alpha - f_0)\,\frac{(\alpha+1)}{2}}. \tag{4.115}$$

There are a number of ways in which this relationship can be used depending on the combination of the parameters that is known/given.

A constant positive value of f_0 obviously decreases ω_r, and, consequently increases the resulting period of vibrations for a fixed amplitude A. If the value of f_0 is negative, these characteristics are affected in the opposite way. This implies that the value of f_0 can be used to make the response have a different period although the amplitude and the power of nonlinearity stay the same.

Besides this, one can determine the constant value of f_0 that will result in having a certain desired amplitude at a certain frequency:

$$f_0 = c_\alpha - \frac{2}{\alpha+1}\omega_r^2\,|A|^{1-\alpha}. \tag{4.116}$$

Note that if one defines f_0 as amplitude-dependent $f_0 = F_0/\left(\mathrm{sgn}\,(A)\,|A|^\alpha\right)$, Eq. (4.87) is obtained and the analysis presented previously holds.

This approach also enables one to design the excitation to cause an isochronous response (see Chap. 5 of this book or [16], i.e. the response with an amplitude-independent frequency/period: $\omega_r = C = const$. This is fulfilled for

$$f_0 = c_\alpha - \frac{2}{\alpha+1}C^2\,|A|^{1-\alpha}. \tag{4.117}$$

To illustrate this possibility, let us assume that it is required that the period of the response is 10 s, so that the expression for the period of the ca function gives $\omega_r = C = 2B\,(1/4, 1/2)\,/10$. By using this value together with $c_\alpha = 1$, $\alpha = 3$, as well as three amplitudes $A = 1/4$; $1/2$; 1, the constant f_0 is calculated and then the equation of motion (4.78), (4.112), (4.113) solved numerically. The responses obtained are plotted in Fig. 4.16, confirming that the period is constant as required, while the amplitude varies.

The exact solution can also be derived for a damped case

$$\ddot{x} + \delta\dot{x} + c_\alpha\ \mathrm{sgn}\,(x)\,|x|^\alpha = F, \tag{4.118}$$

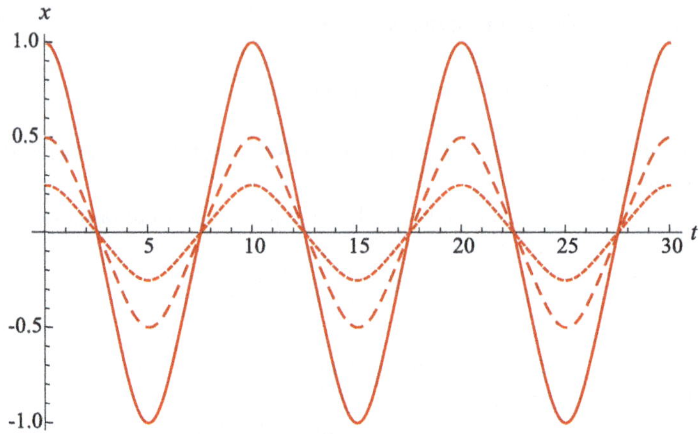

Fig. 4.16 Time response corresponding to $c_\alpha = 1$, $\alpha = 3$, $T = 10$ s and three amplitudes $A = 1/4$; $1/2$; 1 obtained numerically from Eqs. (4.78), (4.112), (4.113) solved numerically

but now the external excitation will be $F = C\mathrm{sgn}(x)\,|x|^\alpha + \delta\dot{x}$, i.e.

$$F = C\,\mathrm{ca}\left(\alpha, 1, |A|^{(\alpha-1)/2}\sqrt{\frac{(c_\alpha - C)(\alpha+1)}{2}}t\right) - \delta C\,|A|^{(\alpha-1)/2}\sqrt{\frac{2(c_\alpha - C)}{\alpha+1}}\,\mathrm{sa}\left(\alpha, 1, |A|^{(\alpha-1)/2}\sqrt{\frac{(c_\alpha - C)(\alpha+1)}{2}}t\right). \quad (4.119)$$

Note that the derivative of the Ateb function used to obtain Eq. (4.119) is given in Chap. 2, Sect. 2.5.

4.7.3 Changing the Period via the Power of Nonlinearity

Another modification that can be introduced is to assume the excitation force in the form

$$F = c_\alpha \mathrm{sgn}(x)\,|x|^\alpha - E\mathrm{sgn}(x)\,|x|^\beta, \quad (4.120)$$

with $E > 0$ (note that for $E < 0$, the resulting response will not be oscillatory) and with β being any positive real number. Now, the equation of motion becomes

$$\ddot{x} + E\ \mathrm{sgn}(x)\,|x|^\beta = 0. \quad (4.121)$$

If $\beta = 1$, this becomes the SHO. Its solution has the form of the cosine function:

$$x_r = A\cos\left(\sqrt{E}t\right), \tag{4.122}$$

which should be substituted into Eqs. (4.120) and (4.78) to derive the equation of motion:

$$\ddot{x} + c_\alpha \operatorname{sgn}(x)|x|^\alpha = c_\alpha \operatorname{sgn}\left(A\cos\left(\sqrt{E}t\right)\right)\left|A\cos\left(\sqrt{E}t\right)\right|^\alpha - EA\cos\left(\sqrt{E}t\right). \tag{4.123}$$

The response will be the same in a damped case with the excitation having the additional term obtained as the first time derivative of (4.122)

$$\ddot{x} + \delta\dot{x} + c_\alpha \operatorname{sgn}(x)|x|^\alpha = c_\alpha \operatorname{sgn}\left(A\cos\left(\sqrt{E}t\right)\right)\left|A\cos\left(\sqrt{E}t\right)\right|^\alpha - EA\cos\left(\sqrt{E}t\right) - \delta A\sqrt{E}\sin\left(\sqrt{E}t\right). \tag{4.124}$$

If $\beta = 3$, Eq. (4.121) becomes the equation of motion of the PCO. The corresponding forced response is given by

$$x_r = A\operatorname{cn}\left(\sqrt{E}At\left|\frac{1}{2}\right.\right). \tag{4.125}$$

After substituting it into Eqs. (4.120) and (4.78), the equation of motion of externally excited PNOs that respond as a free conservative PCO is derived

$$\ddot{x} + c_\alpha \operatorname{sgn}(x)|x|^\alpha = c_\alpha \operatorname{sgn}\left(A\operatorname{cn}\left(\sqrt{E}At\left|\frac{1}{2}\right.\right)\right)\left|A\operatorname{cn}\left(\sqrt{E}At\left|\frac{1}{2}\right.\right)\right|^\alpha - EA^3\operatorname{cn}^3\left(\sqrt{E}At\left|\frac{1}{2}\right.\right). \tag{4.126}$$

The response will be the same for a damped case with the excitation having the additional term obtained as being proportional to the first time derivative of (4.125):

$$\ddot{x} + \delta\dot{x} + c_\alpha \operatorname{sgn}(x)|x|^\alpha = c_\alpha \operatorname{sgn}\left(A\operatorname{cn}\left(\sqrt{E}At\left|\frac{1}{2}\right.\right)\right)\left|A\operatorname{cn}\left(\sqrt{E}At\left|\frac{1}{2}\right.\right)\right|^\alpha - EA^3\operatorname{cn}^3\left(\sqrt{E}At\left|\frac{1}{2}\right.\right) - \tag{4.127}$$

$$\delta A\sqrt{E}\operatorname{sn}\left(\sqrt{E}At\left|\frac{1}{2}\right.\right)\operatorname{dn}\left(\sqrt{E}At\left|\frac{1}{2}\right.\right). \tag{4.128}$$

For any other $\beta \neq \alpha$, the power of nonlinearity is changed from α to β. Equation (4.80) gives the form of the response, and then Eqs. (4.120) and (4.78) yield the equation of motion:

$$x_r = A\,\mathrm{ca}\left(\beta, 1, |A|^{(\beta-1)/2}\sqrt{\tfrac{E(\beta+1)}{2}}t\right), \tag{4.129}$$

$$\begin{aligned} &\ddot{x} + c_\alpha\,\mathrm{sgn}\,(x)\,|x|^\alpha = \\ &c_\alpha \mathrm{sgn}\left(A\,\mathrm{ca}\left(\beta, 1, |A|^{(\beta-1)/2}\sqrt{\tfrac{E(\beta+1)}{2}}t\right)\right)\left|A\,\mathrm{ca}\left(\beta, 1, |A|^{(\beta-1)/2}\sqrt{\tfrac{E(\beta+1)}{2}}t\right)\right|^\alpha \\ &-EA^\beta\,\mathrm{ca}^\beta\left(\beta, 1, |A|^{(\beta-1)/2}\sqrt{\tfrac{E(\beta+1)}{2}}t\right). \end{aligned} \tag{4.130}$$

The response will be the same for a damped case with the excitation having the additional term proportional to the first time derivative of Eq. (4.129):

$$\begin{aligned} \ddot{x} + \delta\dot{x} + c_\alpha\,\mathrm{sgn}\,(x)\,|x|^\alpha &= c_\alpha \mathrm{sgn}\left(A\,\mathrm{ca}\left(\beta, 1, |A|^{(\beta-1)/2}\sqrt{\tfrac{E(\beta+1)}{2}}t\right)\right) \\ &\left|A\,\mathrm{ca}\left(\beta, 1, |A|^{(\beta-1)/2}\sqrt{\tfrac{E(\beta+1)}{2}}t\right)\right|^\alpha - \\ &EA^\beta\,\mathrm{ca}^\beta\left(\beta, 1, |A|^{(\beta-1)/2}\sqrt{\tfrac{E(\beta+1)}{2}}t\right) \\ -\delta\,|A|^{(\beta-1)/2}\sqrt{\tfrac{2E}{\beta+1}}&\mathrm{sa}\left(\beta, 1, |A|^{(\beta-1)/2}\sqrt{\tfrac{E(\beta+1)}{2}}t\right). \end{aligned} \tag{4.131}$$

It is believed that this can be a convenient way for modifying the stiffness characteristic without influencing internal elements, but by using the external excitation. For example, this can be beneficial when the original power of nonlinearity is under-linear ($\alpha < 1$), but it is desirable to make it be over-linear ($\beta > 1$) or vice versa.

Figure 4.17 contains time responses of the original oscillator with the power of nonlinearity $\alpha = 1/2$ as well as time responses for all these three cases: $\beta = 1$ (Fig. 4.17a), $\beta = 3$ (Fig. 4.17b), $\beta = 5 \neq \alpha$ (Fig. 4.17c). It is clearly seen that the change of the power of nonlinearity causes the extension of the period. The shape of vibrations also changes, implying the modification of the harmonic content. For instance, in the example shown in Fig. 4.17c, the response was changed from the PNO with $\alpha = 1/2$ having the Fourier series (see Appendix C):

$$\begin{aligned} x_r \approx\ &1.01511\cos(1.05164t) - 0.01729\cos(3.15491t) + \\ &0.00279\cos(5.25818t) - 0.00085\cos(7.36146t)\ldots \end{aligned} \tag{4.132}$$

to the response of the PNO with $\beta = 5$ having the Fourier series:

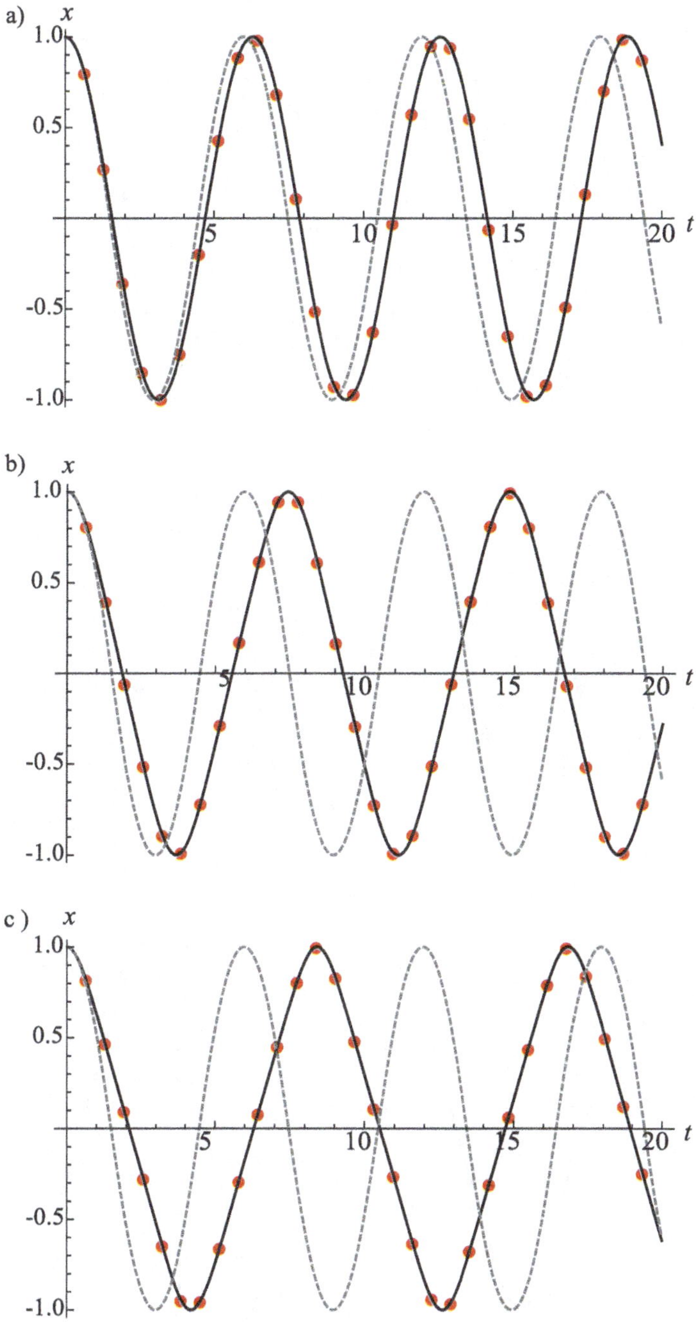

Fig. 4.17 Time responses for purely nonlinear oscillators corresponding to $c_{1/2} = 1$, $\alpha = 1/2$, $E = 1$, $A = 1$ and **a** $\beta = 1$; **b** $\beta = 3$; **b** $\beta = 5$. Numerical solutions of the externally excited original oscillator with the force given by Eq. (4.120)—black solid line, the resulting analytical response—red dots, the numerical solution of the original conservative oscillator—grey dotted line

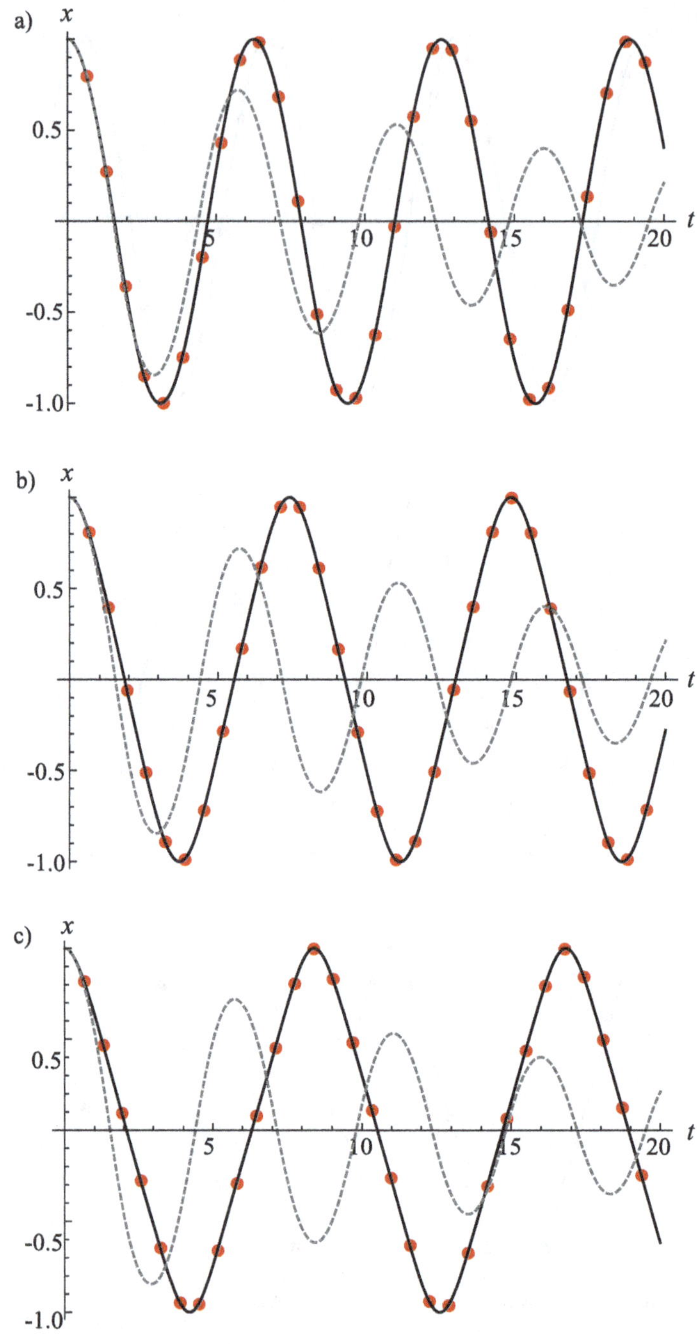

Fig. 4.18 Time responses for purely nonlinear oscillators corresponding to $c_{1/2} = 1$, $\alpha = 1/2$, $E = 1$, $A = 1$, $\delta = 0.1$ and **a** $\beta = 1$; **b** $\beta = 3$; **b** $\beta = 5$. Numerical solutions of the externally excited original oscillator governed by Eq. (4.131)—black solid line, the resulting analytical response—red dots, the numerical solution of the original damped oscillator Eq. (4.131) with the right-hand side being zero—grey dotted line

$$x_r \approx 0.92643\cos(0.74683t) + 0.06339\cos(2.24050t) + \\ 0.00878\cos(3.73417t) + 0.00120\cos(5.22784t)\ldots \quad (4.133)$$

Validations for the damped case (4.131) with $\delta = 0.1$ are provided in Fig. 4.18 for the same sets of parameters used in Fig. 4.17.

4.7.4 Changing the Type of Oscillator

It is also possible to adjust the type of oscillator (4.118) by assuming the external excitation as follows:

$$F = Bx+Dx^3+E\text{sgn}(x)|x|^\alpha + \delta\dot{x}. \quad (4.134)$$

Now, the equation of motion becomes:

$$\ddot{x} + (c_\alpha - E)\ \text{sgn}(x)|x|^\alpha - Bx - Dx^3 = 0. \quad (4.135)$$

By imposing $c_\alpha = E$, the PNO loses its original power-form nonlinearity. Further, by choosing the signs and values of the coefficients B and D, one can end up with the SHO ($B < 0$ and $D = 0$), the SDO ($B < 0$ and $D > 0$), the PCO ($B = 0$ and $D < 0$), the HDO ($B < 0$ and $D < 0$) and the BDO ($B > 0$ and $D < 0$). Their exact solutions for their free response are discussed in detail in Chap. 2.

4.8 Forced Response of Purely Nonlinear Oscillators: Approximate Solutions via Elliptic Functions

This section is concerned with the forced oscillators modelled by Eqs. (4.1), (4.2), with $G(x) = \text{sgn}(x)|x|^\alpha$, $F_1(x, \dot{x}) = -\varepsilon f(x, \dot{x})$ and $F_2(\Omega t) = \varepsilon F\cos(\Omega t)$. The mathematical model actually corresponds to harmonically excited purely nonlinear oscillators (PNOs), governed by the following non-dimensional equation of motion

$$\ddot{x} + \varepsilon f(x, \dot{x}) + \text{sgn}(x)|x|^\alpha = \varepsilon F\cos(\Omega t), \quad (4.136)$$

where ε is a perturbation parameter, $\varepsilon << 1$; $F \equiv \varepsilon F$ and Ω are the magnitude and frequency of harmonic excitation, respectively, with the former being of order ε; α is the power of the restoring force that can be any non-negative real number; f stands here for any autonomous force that depends on the displacement and/or velocity.

These approximate solutions for the steady-state response of such oscillators oscillator have been expressed in terms of harmonic functions [17–19], but the use of elliptic functions have been proven to yield reliable and accurate results [20–32] as these functions contain higher order harmonic components. Elliptic averaging has mainly been used to study free nonlinear oscillations [20–28, 30], while forced nonlinear vibrations have been considered in this way by Roy [29] and Okabe et al. [31, 32]. In Roy's method [29], the equation of motion is transformed into an autonomous system with respect to energy and the so-called resonant phase angle, but the solution for motion was found only for the cases when the exact solution of the unperturbed oscillator is known a priori. These cases include a mathematical pendulum (the restoring force is a sine function of the displacement), a single-well Duffing oscillator with the restoring force consisting of a positive linear and a positive cubic term, and a canonical escape oscillator with (the restoring force has the form of a positive linear-negative quadratic term). In [31, 32], elliptic averaging is proposed for the oscillators the restoring force of which contains a linear and a cubic term of different signs. This section is concerned with the method of elliptic averaging developed in [8] to find the steady-state response of purely nonlinear one-degree-of-freedom harmonically excited oscillators modelled by a general Eq. (4.136).

4.8.1 Conservative Oscillators: Generative Solution

The solution of conservative oscillators needs to be determined first, as it will serve as a generating solution for the motion of the forced oscillators (4.136). According to the assumptions introduced, both the autonomous term $f(x, \dot{x})$ and the harmonic force represent perturbations of the conservative oscillators, whose equation of motion is derived from Eq. (4.136) by putting $\varepsilon = 0$:

$$\ddot{x} + \operatorname{sgn}(x)|x|^{\alpha} = 0. \tag{4.137}$$

Initial conditions are prescribed as $x(0) = a$, $\dot{x}(0) = 0$, where a is a real constant.

As shown in Chap. 2, Sect. 2.5.2.4, the solutions for the displacement and the velocity can be taken in the form of Jacobi elliptic functions:

$$x(t) = a\operatorname{cn}(\psi\,|m), \tag{4.138}$$

$$\dot{x}(t) = -a\omega\operatorname{sn}(\psi\,|m)\operatorname{dn}(\psi\,|m), \tag{4.139}$$

where a is a constant amplitude of motion defined by the initial displacement, ψ is the complete phase and the parameter m is the elliptic parameter. The complete phase is defined by

$$\psi(t) = \omega t + \varphi, \tag{4.140}$$

where ω and φ are the frequency and the phase of the response, respectively.

The next fact to be used in the exact period of PNOs (Chap. 2, Sect. 2.5.1):

$$T = T^* a^{\frac{1-\alpha}{2}}, \tag{4.141}$$

with T^* being

$$T^* = 4\sqrt{\frac{\alpha+1}{2}} \frac{\sqrt{\pi}\Gamma[1+\frac{1}{1+\alpha}]}{\Gamma[\frac{1}{2}+\frac{1}{1+\alpha}]}, \tag{4.142}$$

where Γ represents the Euler gamma function (see Appendix A).

Further, the fact that the period of the Jacobi elliptic function $\mathrm{cn}(\psi\,|m)$ is known is used. It is given by $T = 4K(m)$, where K is the complete elliptic of the first kind (Chap. 2, Sect. 2.3.1). Thus, equating $4K(m)$ with Eq. (4.142), the following expression is derived

$$4K \equiv 4K[m(\alpha)] = 4\sqrt{\frac{\alpha+1}{2}} \frac{\sqrt{\pi}\Gamma\left[1+\frac{1}{1+\alpha}\right]}{\Gamma\left[\frac{1}{2}+\frac{1}{1+\alpha}\right]}. \tag{4.143}$$

This equation defines the parameter m implicitly as a function of the power, which is plotted in Fig. 2.19 in Chap. 2. Some of these values corresponding to different are also calculated numerically and given in Table 2.2 in Chap. 2. It can be seen therein that the values of this parameter are negative for $\alpha < 1$ and positive for $\alpha > 1$. It should be noted that in a special case $\alpha = 1$, i.e. when the oscillator is linear, one has m=0, so that the elliptic cn function turns into the trigonometric cosine function. For a pure cubic oscillator $\alpha = 3$, the parameter m is equal to 0.5.

The frequency of the elliptic function is $\omega = 4K/T$. By using Eqs. (4.141), (4.142) and (4.143), one obtains

$$\omega = a^{\frac{\alpha-1}{2}}. \tag{4.144}$$

This expression implies that there is a specific nonlinear power-form relationship between the amplitude of free oscillations and the frequency of the elliptic function.

Note that the solution defined by Eqs. (4.138)–(4.140) is the solution of the equation of motion (4.137) provided that the following equation is satisfied

$$-a\omega^2 \mathrm{cndn}^2 + a\omega^2 m \mathrm{sn}^2 \mathrm{cn} + a^\alpha \mathrm{sgn}(\mathrm{cn})|\mathrm{cn}|^\alpha, \tag{4.145}$$

where the arguments of the elliptic functions are omitted for brevity. Equation (4.145) is obtained by differentiating Eq. (4.139) once with respect to time and substituting it together with Eq. (4.138) into the equation of motion (4.137). The condition (4.145) represents one of the basic assumptions of the procedure developed, according to which the approximate solution for a free conservative system is such that it satisfies Eq. (4.145). This condition will also play a significant role in the following section, where the approximate solution of a non-conservative forced system is sought.

4.8.2 Forced Oscillators: Approximate Solutions via Elliptic Functions

Using the analogy with the generating solution given by Eqs. (4.138), (4.139), a trial solution of the perturbed oscillators (4.136) can be assumed in the same form, but with the amplitude, the phase and the parameter m varying in time, i.e. as

$$x(t) = a(t)\mathrm{cn}\left(\psi(t)\,|m(t)\right), \tag{4.146}$$

$$\dot{x}(t) = -a(t)\omega\mathrm{sn}\left(\psi(t)\,|m(t)\right)\mathrm{dn}\left(\psi(t)\,|m(t)\right), \tag{4.147}$$

with the complete phase given by Eq. (4.140). Based on Eqs. (4.146), (4.147), the solution for the velocity has the form given by Eq. (4.147) if the following constraint is satisfied

$$\frac{\mathrm{d}a}{\mathrm{d}t}\mathrm{cn} - a\frac{\mathrm{d}\varphi}{\mathrm{d}t}\mathrm{sndn} + a\frac{\mathrm{d}m}{\mathrm{d}t}\frac{\partial\mathrm{cn}}{\partial m} = 0, \tag{4.148}$$

where the arguments of the elliptic functions are again omitted for brevity, but are defined in Eqs. (4.146), (4.147). In order to define $\mathrm{d}m/\mathrm{d}t$, the period $4K(m(t))$ is equated with Eq. (4.141), and the relationship obtained is differentiated with respect to time, yielding

$$\frac{\mathrm{d}m}{\mathrm{d}t} = \frac{\Lambda}{a}\frac{\mathrm{d}a}{\mathrm{d}t}, \tag{4.149}$$

with Λ being

$$\Lambda = \frac{(1-m)(1-\alpha)mK}{E-(1-m)K}, \tag{4.150}$$

where E is the complete elliptic integral (complete elliptic integral of the second kind), $E = E\,(m)$.

Differentiating Eq. (4.147) and substituting it into Eq. (4.136), as well as substituting Eq. (4.149) into Eq. (4.148), one obtains

$$-\frac{\mathrm{d}a}{\mathrm{d}t}\omega\left[\mathrm{sndn} + \Lambda\frac{\partial}{\partial m}(\mathrm{sndn})\right] - a\omega\frac{\mathrm{d}\varphi}{\mathrm{d}t}\mathrm{cn}(1-2m\mathrm{sn}^2) = \\ -\varepsilon f(x,\dot{x}) + F\cos\left(\Omega t\right), \tag{4.151}$$

$$\frac{\mathrm{d}a}{\mathrm{d}t}\left(\mathrm{cn} + \Lambda\frac{\partial\mathrm{cn}}{\partial m}\right) - a\frac{\mathrm{d}\varphi}{\mathrm{d}t}\mathrm{sndn} = 0. \tag{4.152}$$

While deriving Eq. (4.151), the fact that the trial solution should satisfy the same condition as the generating solution given by Eq. (4.145) has been used. In addition, due to the fact that the parameter m and ω are both related to the period, the variation is performed with respect to m only.

Both Eqs. (4.151) and (4.152) contain the derivatives of the elliptic functions with respect to the parameter m. These derivatives can be expressed in the forms given in Appendix B. Using them and solving Eqs. (4.151)–(4.152) with respect to $\frac{da}{dt}$ and $\frac{d\varphi}{dt}$, one follows

$$-\omega D\frac{\mathrm{d}a}{\mathrm{d}t} = -\varepsilon f(a\mathrm{cn}, -a\omega\mathrm{sndn})\mathrm{sndn} + F\cos(\Omega t)\,\mathrm{sndn}, \tag{4.153}$$

$$-a\omega D\frac{\mathrm{d}\varphi}{\mathrm{d}t} = \left[-\varepsilon f(a\mathrm{cn}, -a\omega\mathrm{sndn}) + F\cos(\Omega t)\right]\left[\mathrm{cn} + \Lambda\frac{\partial\mathrm{cn}}{\partial m}\right], \tag{4.154}$$

where

$$D = \mathrm{sn}^2\mathrm{dn}^2 + \mathrm{cn}^2(1-2m\mathrm{sn}^2) + \sigma\left[\left[-\mathrm{sn}^2\mathrm{cn}^2(1-2m\mathrm{sn}^2) + \mathrm{sn}^2\mathrm{dn}^2\mathrm{cn}^2 - \mathrm{sn}^4\mathrm{dn}^2\right]\right], \tag{4.155}$$

and

$$\sigma = \frac{m(1-\alpha)K}{2[E-(1-m)K]}. \tag{4.156}$$

Further, the term on the right-hand side of Eq. (4.154) can be written down as

$$[-\varepsilon f(a\mathrm{cn}, -a\omega\mathrm{sndn}) + F\cos(\Omega t)]\left(\mathrm{cn} + \Lambda\frac{\partial\mathrm{cn}}{\partial m}\right) - \varepsilon f(a\mathrm{cn}, -a\omega\mathrm{sndn})\mathrm{cn} -$$
$$\varepsilon f(a\mathrm{cn}, -a\omega\mathrm{sndn})\sigma_1\left[Z\mathrm{sndn} - m\mathrm{sn}^2\mathrm{cn}\frac{1}{2}[c_1+\sigma_1(z_1-ms_1)]\cos\left(\frac{\pi}{2K}\psi - \Omega t\right)\right], \tag{4.157}$$

where $\sigma_1 = \sigma/m$ and where the periodic Jacobi zeta function has been introduced:

$$Z = E(\theta, m) - \frac{E}{K}F_1(\theta, m), \tag{4.158}$$

where $F_1(\theta,m)$ and $E(\theta,m)$ are the incomplete elliptic integral of the first and second kind, respectively, and $\theta = \mathrm{am}[4K(m),m]$ is the elliptic amplitude function, while $K = K(m)$ has already been defined as the complete elliptic integral of the first kind and E=$E(m)$ as the complete elliptic integral of the second kind. In addition, the elliptic function cn, the products sn dn, Z sn and sn^2 dn have been expanded into the Fourier series, and only the first terms of these expansions are retained (see Table 4.3 and Appendix B). Retaining the first terms only is consistent with the level of approximation of the method as a whole, the idea of which is to take into account higher harmonics through the trail solution given in the form of one elliptic function and keep the rest of the influences in the technique that might add additional terms as simple as possible, as done, for example, in [29].

The last term in Eq. (4.153) can be approximated as follows:

Table 4.3 Fourier series expansions (approximations and the first Fourier coefficients) for certain elliptic functions and their products (note that q is the Nome and K is q is the complete elliptic integral of the first kind)

Function	Approximation	First Fourier coefficient
$\mathrm{cn}(\psi \mid m)$	$\frac{2\pi}{K\sqrt{m}}\sum_{n=0}^{\infty}\frac{q^{n+1/2}}{1+q^{2n+1}}\cos\left[(2n+1)\frac{\pi}{2K}\psi\right]$	$c_1=\frac{2\pi}{K\sqrt{m}}\frac{q^{1/2}}{1+q}$
$\mathrm{sn}(\psi \mid m)\mathrm{dn}(\psi \mid m)$	$a_1\sin(\frac{\pi}{2K}\psi)+\cdots$	$a_1=\frac{\pi^2}{K^2\sqrt{m}}\frac{q^{1/2}}{1+q}$
$\mathrm{Zsn}(\psi \mid m)\mathrm{dn}(\psi \mid m)$	$z_1\cos(\frac{\pi}{2K}\psi)+\cdots$	$z_1=\frac{\pi^3}{2K^3\sqrt{m}}\frac{q^{3/2}}{(1+q)(1-q^2)}$
$\mathrm{sn}^2(\psi \mid m)\mathrm{cn}(\psi \mid m)$	$s_1\cos(\frac{\pi}{2K}\psi)+\cdots$	$s_1=\frac{1}{2m}[-\frac{\pi^3}{2K^3\sqrt{m}}\frac{q^{1/2}}{1+q}+\frac{2\pi}{K\sqrt{m}}\frac{q^{1/2}}{1+q}]$

Table 4.4 Values of some parameters for certain values of the power α

α	m	a_1	d_1	d_5	$\bar{D}$	P
0	-0.506160	1.138383	0.651587	0.15522	0.801715	0.92144
1/4	−0.353627	1.1002	0.607119	0.14637	0.865627	0.907209
1/3	−0.307134	1.088058	0.593378	0.14634	0.884476	0.94263
4/5	−0.081307	1.02479	0.525209	0.130038	0.971148	0.980711
1	0	1	1/2	0	1	1
2	0.3058305	0.893363	0.401005	0.105123	1.09159	1.12228
3	0.5	0.809093	1/3	0.091389	1.125626	1.28507
4	0.629107	0.740723	0.285029	0.081446	1.12668	1.4849

$$F\cos(\Omega t)\,\mathrm{sndn}=\frac{1}{2}Fa_1[\sin(\frac{\pi}{2K}\psi-\Omega t)+\sin(\frac{\pi}{2K}\psi+\Omega t)], \tag{4.159}$$

where a_1 is also defined in Table 4.3 and its values for different α are given in Table 4.4 for m obtained from Eq. (4.143).

The form of the argument of the first sine function in Eq. (4.159) is the motivation for a new variable to be introduced

$$\phi=\frac{\pi}{2K}\psi-\Omega t, \tag{4.160}$$

which represents a difference between the phase of the system response and the phase of excitation (phase shift).

By using Eqs. (4.153)–(4.160), an averaging procedure of terms over a period $4K$ is performed and the general equations for the amplitude and the new phase are obtained

$$\frac{\mathrm{d}a}{\mathrm{d}t} = \frac{\varepsilon}{\omega \bar{D}} \frac{1}{4K} \int_0^{4K} f(a\mathrm{cn}, -a\omega \mathrm{sndn})\mathrm{sndn} d\psi - \frac{F a_1 a^{\frac{1-\alpha}{2}}}{2\bar{D}} \sin\phi, \quad (4.161)$$

$$\frac{\mathrm{d}\phi}{\mathrm{d}t} = \frac{\pi}{2K} a^{\frac{\alpha-1}{2}} - \Omega - \frac{\pi F P}{2K \bar{D} a (\frac{\pi a^{\frac{\alpha-1}{2}}}{2K} + \Omega)} \cos\phi + \frac{\pi}{2K} \frac{\varepsilon}{a\omega \bar{D}} \frac{1}{4K} \int_0^{4K} f(a\mathrm{cn}, -a\omega \mathrm{sndn})\mathrm{cn} d\psi + \frac{\pi}{2K} \frac{\epsilon \sigma_1}{a\omega \bar{D}} \frac{1}{4K} \int_0^{4K} f(a\mathrm{cn}, -a\omega \mathrm{sndn})[Z\mathrm{sndn} - m\mathrm{sn}^2\mathrm{cn}] d\psi, \quad (4.162)$$

where $\bar{D}$ is an averaged value of D given by Eq. (4.155)

$$\bar{D} = 2d_1 + \sigma_1(d_2 + d_3 + d_4), \quad (4.163)$$

and where d_1-d_4 stand for non-zero averaged values of the products existing in Eq. (4.155). They are defined in Table 4.5 and some of their values are included into Table 4.4 as well. Besides this, the following new parameter has been introduced in Eq. (4.162)

$$P = c_1 + \sigma_1(z_1 - m s_1), \quad (4.164)$$

where c_1, z_1 and s_1 are the first Fourier coefficients given in Table 4.3. Note that while deriving Eqs. (4.161) and (4.162), averaging has been performed with respect to the variable ψ and the variable ϕ is held fixed, which is a standard procedure in the method of averaging (see [15, 33] for the related references). In addition, the fact that the frequency of free oscillations modelled by the Jacobi elliptic function is close to the excitation frequency $\omega \approx \Omega$ has been used. This implies that $2\omega \approx \omega + \Omega = \frac{\pi a^{\frac{\alpha-1}{2}}}{2K} + \Omega$. Hence, the response of the system defined in Eq. (4.146) has the form

$$x(t) = a(t)\mathrm{cn}\left[\frac{2K}{\pi}(\Omega t + \phi) \left| m(t) \right.\right], \quad (4.165)$$

where the amplitude a can be obtained from Eq. (4.161), the phase shift ϕ from Eq. (4.162) and the parameter m from Eq. (4.143).

When the oscillator is linear, i.e. $\alpha = 1$, the parameter m tends to zero, the elliptic function turns into a cosine function and Eqs. (4.161) and (4.162) transform into:

$$\frac{\mathrm{d}a}{\mathrm{d}t} = \frac{\varepsilon}{2\pi} \int_0^{2\pi} f(a\cos\psi, -a\omega \sin\psi) \sin\psi d\psi - \frac{F}{2} \sin\phi, \quad (4.166)$$

$$\frac{\mathrm{d}\phi}{\mathrm{d}t} = 1 - \Omega - \frac{F}{a(1+\Omega)} \cos\phi + \frac{\varepsilon}{2\pi a} \int_0^{2\pi} f(a\cos\psi, -a\omega \sin\psi) \cos\psi d\psi. \quad (4.167)$$

These equations have a well-known form obtained by applying the Krylov-Bogoliubov method (see, for example, [34]), while the solution (4.165) becomes

$$x(t) = a(t)\cos(\Omega t + \phi). \tag{4.168}$$

The use of the technique developed is illustrated subsequently by two examples. To illuminate the dynamics of these systems with various integer and non-integer powers of the restoring force, the equations derived for the amplitude (4.161) and phase shift (4.162) are employed to demonstrate some of their potential benefits.

4.8.3 Example 4.3. Forced Oscillators with van der Pol Damping

For the case of a van der Pol damping force defined by $f(x, \dot{x}) = (x^2 - 1)\dot{x}$, the first-order differential equations (4.161) and (4.162) for the amplitude and the phase become

$$\frac{\mathrm{d}a}{\mathrm{d}t} = \frac{\varepsilon a d_1}{\bar{D}} - \frac{\varepsilon a^3 d_5}{\bar{D}} - \frac{1}{2}\frac{F a_1}{\bar{D}} a^{\frac{1-\alpha}{2}} \sin\phi, \tag{4.169}$$

$$\frac{\mathrm{d}\phi}{\mathrm{d}t} = \frac{\pi}{2K} a^{\frac{\alpha-1}{2}} - \Omega - \frac{\pi F P}{2K\bar{D}a\left(\frac{\pi a^{\frac{\alpha-1}{2}}}{2K} + \Omega\right)} \cos\phi, \tag{4.170}$$

where d_5 is defined in Table 4.5, while the values of the parameters $a_1, d_1, d_5, \bar{D}$ and P are given in Table 4.4 for certain values of the power α.

The steady-state motion occurs when $\frac{\mathrm{d}a}{\mathrm{d}t} = 0$ and $\frac{\mathrm{d}\phi}{\mathrm{d}t} = 0$. It corresponds to the solution of

Table 4.5 Averaged values d_i (i=1–5) of some integrands I_i that are the products of certain elliptic functions

i	Integrand I_i	$d_i = \frac{1}{4K}\int_0^{4K} I_i d\psi$
1	$\mathrm{sn}^2(\psi\,\vert m)\mathrm{dn}^2(\psi\,\vert m)$	$\frac{(2m-1)E(\theta,m)-4(m-1)K}{12mK}$
2	$\mathrm{sn}^2(\psi\,\vert m)\mathrm{cn}^2(\psi\,\vert m)\left(1-2m\mathrm{sn}^2(\psi\,\vert m)\right)$	$\frac{-(4m^2+m-6)E+8(m^2+2m-3)K}{60m^2K}$
3	$\mathrm{sn}^2(\psi\,\vert m)\mathrm{cn}^2(\psi\,\vert m)\mathrm{dn}^2(\psi\,\vert m)$	$\frac{2(m^2-m+1)E(\theta,m)-4(m^2-3m+2)K}{60m^2K}$
4	$\mathrm{sn}^4(\psi\,\vert m)\mathrm{dn}^2(\psi\,\vert m)$	$-\frac{(8m^2-3m-2)E(\theta,m)+8(-2m^2+m+1)K}{60m^2K}$
5	$\mathrm{cn}^2(\psi\,\vert m)\mathrm{sn}^2(\psi\,\vert m)\mathrm{dn}^2(\psi\,\vert m)$	$\frac{2(1+(m-1)m)E(\theta,m)-4(2+(m-3)m)K}{60m^2K}$

$$\frac{\varepsilon a d_1}{\bar{D}} - \frac{\varepsilon a^3 d_5}{\bar{D}} = \frac{1}{2}\frac{F a_1}{\bar{D}} a^{\frac{1-\alpha}{2}} \sin\phi, \tag{4.171}$$

$$\frac{\pi}{2K} a^{\frac{\alpha-1}{2}} - \Omega = \frac{\pi F P}{2K\bar{D}a(\frac{\pi a^{\frac{\alpha-1}{2}}}{2K} + \Omega)} \cos\phi. \tag{4.172}$$

Squaring and adding these equations, the amplitude-frequency equation is obtained

$$\frac{4\varepsilon^2 (d_1 - a^2 d_5)^2 a^{\alpha+1}}{a_1^2} + \left[\frac{\Omega^2 a - (\frac{\pi}{2K})^2 a^{\alpha}}{\frac{\pi}{2K}\frac{P}{\bar{D}}}\right]^2 = F^2. \tag{4.173}$$

Dividing Eq. (4.171) with Eq. (4.172), the expression for the steady-state value of ϕ is derived

$$\tan\phi = \frac{\pi P \varepsilon (d_1 - d_5 a^2) a^{\frac{\alpha-1}{2}}}{\bar{D} K a_1 [(\frac{\pi}{2K})^2 a^{\alpha-1} - \Omega^2]}. \tag{4.174}$$

The local stability of the steady-state response can be checked by adding small perturbations δ and η to the steady-state values a_0 and ϕ_0 satisfying Eqs. (4.171) and (4.172), i.e. by substituting $a = a_0 + \delta$ and $\phi = \phi_0 + \eta$ into these equations. By keeping the linear terms in δ and η, one derives

$$\frac{d\delta}{dt} = [\frac{\varepsilon d_1}{\bar{D}} - \frac{3\varepsilon d_5}{\bar{D}} a^2 + \frac{\varepsilon(\alpha-1)}{2\bar{D}}(d_1 - d_5 a^2)]\delta - [\frac{K a_1}{\pi P} a^{\frac{3-\alpha}{2}}((\frac{\pi}{2K})^2 a^{\alpha-1} - \Omega^2)]\eta, \tag{4.175}$$

$$\frac{d\eta}{dt} = [\frac{\pi(\alpha-1)}{4K} a^{\frac{\alpha-3}{2}} + (\frac{(\frac{\pi}{2K} a^{\frac{\alpha+1}{2}} + \alpha\frac{\pi}{2K} a^{\frac{\alpha+1}{2}} + 2a\Omega)((\frac{\pi}{2K})^2 a^{\alpha-1} - \Omega^2)}{2a(\frac{\pi}{2K} a^{\frac{\alpha}{2}} + \sqrt{a}\Omega)^2}]\delta + \frac{\pi P}{K a_1}\frac{(\frac{\varepsilon d_1}{\bar{D}} a - \frac{\varepsilon d_5}{\bar{D}} a^3)}{(\frac{\pi}{2K} a^{\frac{\alpha-1}{2}} + \Omega)} a^{\frac{\alpha-3}{2}} \eta. \tag{4.176}$$

The stability of the steady-state response depends on the eigenvalues of the coefficient matrix on the right side of Eqs. (4.175) and (4.176). The corresponding determinant (det) and the trace (tr) [15] are

$$\det = \frac{a^{-\frac{2+\alpha}{2}}\{a^{2\alpha} a_1^2 \bar{D}^2 \alpha \pi^4 + 4a^{\alpha+1}\pi^2 [2d_1^2(\alpha+1)P^2\varepsilon^2 - 4a^2 d_1 d_5 (3+\alpha) P^2 \varepsilon^2\}}{8a_1^2 \bar{D} P \pi K^2 (a^{\frac{\alpha}{2}}\pi + 2\sqrt{a}\Omega K)} + \frac{a^{-\frac{2+\alpha}{2}}\{2a^4 d_5^2(\alpha+5)P^2\varepsilon^2 - a_1^2 D^2 (1+\alpha)\Omega^2]K^2 + 16a^2 a_1^2 \bar{D}^2 \Omega^4 K^4\}}{8a_1^2 \bar{D} P \pi K^2 (a^{\frac{\alpha}{2}}\pi + 2\sqrt{a}\Omega K)} \tag{4.177}$$

$$\mathrm{tr} = \frac{a^{-\frac{\alpha+1}{2}}(-a_1 d_1 + 5a^2 a_1 d_5 - a_1 d_1 \alpha + a^2 a_1 d_5 \alpha - 4d_1 P + 4a^2 d_5 P)}{2a_1 K(d_1 - 5a^2 d_5 + d_1 \alpha - a^2 d_5 \alpha)}. \quad (4.178)$$

Depending on the sign of the determinant, trace and the expression $S = \mathrm{tr}^2 - 4\det$ conclusions about the stability of steady-state solutions can be determined [15].

Now, the use of the equations derived previously for the study of the entrainment phenomenon in the generalized van der Pol oscillator is illustrated. The case when the power of the restoring force is $\alpha = 4/5$ is considered. In Fig. 4.19, several frequency-response curves defined by Eq. (4.173) are plotted for a fixed value of the ordering parameter $\varepsilon = 0.2$ and for different values of the magnitude of the force F. For higher values of F, such as, for example, F =0.3 shown in Fig. 4.19, the frequency-response curve in continuous, with the apparent maximum. As F decreases, the left and right branch become closer to each other, as plotted for F=0.2. There is a value of the magnitude F when they coalesce, and then separate into two parts as plotted for F=0.1: one running near the frequency axis corresponding to lower amplitudes, labelled by the points L and N, and the other one that is detached and closed, surrounding the point (a^*, Ω^*). This point, marked by a star in Fig. 4.19, characterizes the free limit cycle oscillations. This point lies on the backbone curve $\Omega_{bc} = \frac{\pi a^{\frac{\alpha-1}{2}}}{2K}$, the expression for which was derived by equating the frequency-amplitude relationship (4.144) with the frequency of the cn function in the steady-state response (4.165).

In order to define the parts of these frequency-response curves that are stable and, thus, attainable, the use is made of the expressions for the determinant (4.177) and the trace (4.178) [15]. If the determinant is negative, a saddle occurs, which is unstable. This region is shaded in the Ω-a plane in Fig. 4.19 and the corresponding part of the frequency-response curve is plotted as the dotted line. When the determinant is positive (outside the shaded region), unstable solutions can also occur, which happens for a positive trace. Positive values of the trace are related to the part of the plane below a thick line in Fig. 4.19, corresponding to the zero value of the trace (4.178). There are also two more lines of interest plotted in Fig. 4.19: these are the black solid lines $S = \mathrm{tr}^2 - 4\det = 0$. Steady-state solutions are stable when det>0 and tr<0 [15]. These parts are plotted as the red solid lines. Unstable steady-state solutions that are below the thick line are shown as the dashed lines (note that some of their parts are plotted as the dotted lines as they belong to the shaded region). This implies that along the frequency-response curve A-B-C-D-E-H-J corresponding to F =0.2, the parts A-B and H-J are associated with unstable foci, B-C with unstable nodes, C-D with saddles, D-E with stable nodes and E-H with stable foci. Therefore, only the part D-E-H is realizable. These findings are checked numerically by carrying out direct integration of the equation of motion (4.136). Steady-state numerical results are depicted by dots, confirming very good accuracy of the analytical results.

It can be concluded that the entrainment phenomenon can occur in a very narrow range of the excitation frequency. For F=0.2, this region is between the frequencies corresponding to the points D and H. The entrainment region moves towards higher frequency as the power increases. Starting from $\alpha = 0$, its width decreases until $\alpha = 1$ and then increases again, which yields the conclusion that this region is the narrowest for the linear oscillator [8]. Unlike in the linear systems with van der Pol damping, whose frequency-amplitude plane consists of the curves symmetric with

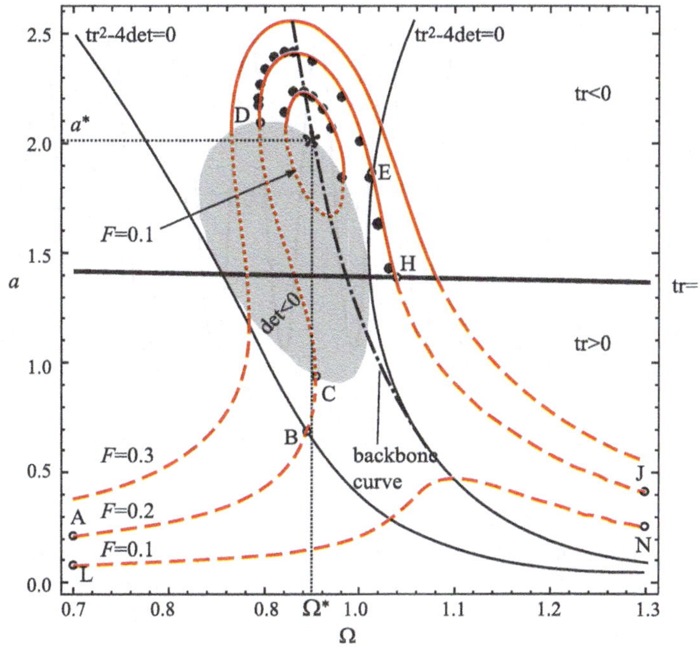

Fig. 4.19 Analytically obtained frequency-response curves defined by Eq. (4.173) of the oscillator governed by Eq. (4.136) for van der Pol damping, when $\alpha = 4/5$, $\varepsilon = 0.2$ and different values of F. Numerical results are depicted by black dots. The lines and zones related to $S = \mathrm{tr}^2 - 4\det$, tr and det are defined by Eqs. (4.177) and (4.178). A backbone curve is plotted as a dashed-dotted line. The mark'*' stands for the characteristics of free limit cycle oscillations

respect to the vertical axis, in the systems with a non-negative power-form restoring force, these curves are bent to the left-hand side for the powers lower the unity and to the right-hand side for the powers higher than unity [35]. Bending to the left-hand side plotted for $\alpha = 4/5$ is clearly seen in Fig. 4.19.

In order to confirm the accuracy of the analytical approximate solution for the steady-state solution given by Eq. (4.165), the case $\alpha = 4/5$, $\varepsilon = 0.2$, F=0.2, shown in Fig. 4.19 is considered, but for $\Omega = 0.95$. The approximate solution obtained is

$$x(t) = 2.394\mathrm{cn}\,[0.98055(-1.915 + 0.95t)\,|-0.081307\,]\,. \tag{4.179}$$

This solution is plotted in Fig. 4.20 together with the numerical solution of the equation of motion (4.136) for $x(0) = 1$, $\dot{x}(0) = 0$. This comparison illustrates excellent accuracy of the solution obtained.

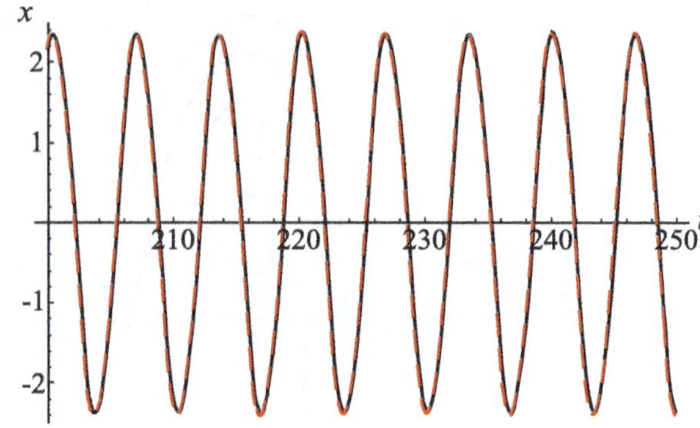

Fig. 4.20 Comparison of the analytical solution for motion Eq. (4.179) (red dashed line) and the corresponding numerical solution (black solid line) of the equation of motion (4.136) for van der Pol damping $\alpha = 4/5$, $F = \varepsilon = 0.2$, $\Omega = 0.95$, $x(0) = 1$ and $\dot{x}(0) = 0$

4.8.4 Example 4.4. Forced Oscillators with Linear Viscous Damping

For the case of a viscous damping force, one has $f(x, \dot{x}) = \dot{x} = -a\omega\text{sndn}$, so that Eqs. (4.161) and (4.162) become

$$\frac{\mathrm{d}a}{\mathrm{d}t} = -\frac{\varepsilon a d_1}{\bar{D}} - \frac{F a_1 a^{\frac{1-\alpha}{2}}}{2\bar{D}} \sin\phi, \tag{4.180}$$

$$\frac{\mathrm{d}\phi}{\mathrm{d}t} = \frac{\pi}{2K} a^{\frac{\alpha-1}{2}} - \Omega - \frac{\pi F P}{2K\bar{D}a(\frac{\pi a^{\frac{\alpha-1}{2}}}{2K} + \Omega)} \cos\phi. \tag{4.181}$$

It should be noted that the fourth term in Eq. (4.162) becomes zero after averaging and the last one is of order $O(\varepsilon^2)$ and, hence, has been neglected.

By equating the left side of Eqs. (4.180) and (4.181) with zero and performing some transformations, the following frequency-amplitude equation of the steady-state response is derived

$$\frac{4\varepsilon^2 d_1^2 a^{\alpha+1}}{a_1^2} + \frac{[\Omega^2 a - (\frac{\pi}{2K})^2 a^\alpha]^2 4K^2\bar{D}^2}{\pi^2 P^2} = F^2, \tag{4.182}$$

as well as the equation for the phase shift

$$\tan\phi = -\frac{\varepsilon d_1 P\pi}{K\bar{D}d_1 a^{\frac{1-\alpha}{2}}[(\frac{\pi}{2K})^2 a^{\alpha-1} - \Omega^2]}. \tag{4.183}$$

It is interesting to note that for $\alpha = 1$ and the fact that $\omega = a^{\frac{\alpha-1}{2}} = 1 \approx \Omega$, Eq. (4.182) transforms into

$$a = \frac{F}{\sqrt{(1-\Omega^2)^2 + \varepsilon^2\Omega^2}}, \tag{4.184}$$

which is the well-known expression of the amplitude of linear viscously damped vibrations [36].

For $F = \varepsilon = 0$, Eq. (4.182) gives the backbone curve

$$\Omega_{bc} = \frac{a^{\frac{\alpha-1}{2}}\sqrt{\frac{\pi}{2}}\Gamma(\frac{1}{2}+\frac{1}{1+\alpha})}{\sqrt{1+\alpha}\Gamma(1+\frac{1}{1+\alpha})}, \tag{4.185}$$

which corresponds to the amplitude-frequency relationship of the conservative oscillator [18], and represents the relationship that can be directly related to Eqs. (4.141) and (4.142). For $\alpha < 1$ the backbone bends towards left and for $\alpha > 1$ towards right. When $\alpha = 1$, Eq. (4.185) gives the resonant frequency of the linear oscillator $\Omega = 1$.

Two branches of the frequency-response curve can be derived from Eq. (4.182)

$$\Omega_{1,2}(a) = [\frac{a^{\alpha-1}\pi^2}{4K^2} \pm \frac{P\pi}{2a^2 a_1 \bar{D}K}\sqrt{a^2 a_1^2 F^2 - 4a^{3+\alpha}\varepsilon^2}]^{\frac{1}{2}}, \tag{4.186}$$

where the sign '+' corresponds to the right branch and the sign '−' to the left one.

In order to determine the number of steady-state solutions, the frequency-amplitude equation (4.182) is considered and the fact that $\Omega = \frac{\pi a^{\frac{\alpha-1}{2}}}{2K}$ is used to express the first term in this equation in terms of the excitation frequency and the amplitude. Equation (4.182) can then be written down as

$$\frac{\pi^2\bar{D}^2}{4K^2P^2}a^{2\alpha} - 2\frac{\bar{D}^2}{P^2}\Omega^2 a^{\alpha+1} + a^2(\frac{16\varepsilon^2 d_1^2 K^2}{a_1^2\pi^2}\Omega^2 + \frac{4K^2\bar{D}^2}{\pi^2 P^2}\Omega^4) - F^2 = 0. \tag{4.187}$$

Following Descartes's Rule of Signs [18], the power-form expression in the amplitude (4.187) has three sign changes and, thus, implies that there can be a maximum number of three positive solutions or one solution.

Local stability of the corresponding steady-state solution a_0 and ϕ_0 is examined by introducing small perturbations δ and η, i.e. by substituting $a = a_0 + \delta$ and $\phi = \phi_0 + \eta$ into Eqs. (4.180) and (4.181). Linearizing these equations, the following equations are derived:

$$\frac{\mathrm{d}\delta}{\mathrm{d}t} = -\frac{\varepsilon d_1(1+\alpha)}{2\bar{D}}\delta - \frac{a_0^{\frac{3-\alpha}{2}} a_1 K[(\frac{\pi}{2K})^2 a^{\alpha-1} - \Omega^2]}{\pi P}\eta, \tag{4.188}$$

$$\frac{\mathrm{d}\eta}{\mathrm{d}t} = [\frac{\pi}{4K}(\alpha-1)a_0^{\frac{\alpha-3}{2}} + \frac{(\frac{\pi}{4K}a_0^{\frac{\alpha+1}{2}} + \alpha\frac{\pi}{4K}a_0^{\frac{1+\alpha}{2}} + 2a_0\Omega)((\frac{\pi}{4K})^2 a^{\alpha-1} - \Omega^2)}{2a_0(\frac{\pi}{4K}a_0^{\frac{\alpha}{2}} + \sqrt{a_0}\Omega)^2}]\delta - \frac{\varepsilon\pi P d_1 a_0^{\frac{\alpha-1}{2}}}{K\bar{D}a_1(\frac{\pi}{4K}a^{\frac{\alpha-1}{2}} + \Omega)}\eta, \tag{4.189}$$

Finding the determinant of this system and equating it with zero, the boundaries of the instability zones are found:

$$\Omega^{s}_{1,2}(a) = \frac{1}{2}\sqrt{\frac{\pi}{2}}\{\frac{1}{a^2 a_1^2 \tilde{D}^2 K^4}[a^{1+\alpha} a_1^2 \tilde{D}^2 (1+\alpha)\pi K^2 \pm \sqrt{a^{2+\alpha} a_1^2 \tilde{D}^2 K^4 [a^{\alpha} a_1^2 \tilde{D}^2 (\alpha-1)^2 \pi^2 - 32 a d_1^2 P^2 (1+\alpha)\varepsilon^2 K]}]\}^{\frac{1}{2}} . \tag{4.190}$$

In Fig. 4.21a, the frequency-response curves is plotted by using Eq. (4.186) for the under-linear case $\alpha = 1/4$ and $F = \varepsilon = 0.1$. The zone of instability defined by Eq. (4.190) is also shown as well as the backbone curve given by Eq. (4.185). Numerical results obtained by integrating the equation of motion (4.136) directly are depicted by the black dots and confirm good accuracy of the analytical results. It should be noted that in this under-linear case, the frequency-response curve is bent to the left, i.e. it is of a softening type as shown in [18] . Besides this, it is interesting that the right branch follows the backbone very closely along a relatively long frequency range from higher frequencies until the jump-down point.

In Fig. 4.21b, the frequency-response curve is plotted for the over-linear case $\alpha = 4$. As it can be seen, the frequency-response curve is bent to the right, which is a behaviour similar to a harmonically excited hardening Duffing oscillator [12]. The agreement between analytical results (red solid line) and numerical results (black dots) is again very good. Figure 4.21b also contains the arrows and black circles indicating the hysteresis and jump phenomena associated with increasing or decreasing forcing frequency Ω. When the frequency is slowly increased, the amplitude a changes following the path 1-2-3-4. However, this path is different when the frequency is slowly decreased, as depicted by 4-5-6-1. This behaviour corresponds to the hysteresis phenomenon. At the frequencies corresponding to Point 2, there is a jump down to Point 3, while at the frequencies corresponding to Point 5, there is a jump up to Point 6.

The peak amplitude can be obtained by finding the intersection of the backbone curve (4.185) and the frequency-response curve (4.182):

$$a_{\max} = 4^{\frac{1}{\alpha+1}} (\frac{a_1 F}{d_1 \varepsilon})^{\frac{2}{\alpha+1}} . \tag{4.191}$$

When $\alpha = 1$, it follows $a_{\max} = F/\varepsilon$, which is the well-known expression for damped forced linear oscillations [36]. Further, for an arbitrary non-negative real α, when $F/\varepsilon \leq 1$, the value of $a_{\max}$ increases as α becomes higher; when $F/\varepsilon > 1$, the peak amplitude $a_{\max}$ becomes smaller as α increases.

The approximate analytical solution for motion defined by Eq. (4.165) and corresponding to $\alpha = 4$, $F = \varepsilon = 0.1$, $\Omega = 0.6$, $x(0) = 0.3$ and $\dot{x}(0) = 0$ is found to be

$$x(t) = 0.8976 \mathrm{cn}\,(1.262(0.6t - 0.628)\,|0.6291) . \tag{4.192}$$

In Fig. 4.22, the steady-state solution (4.192) is shown (dashed line) together with the numerical solution obtained by integrating the differential equation of motion directly (solid line). Excellent accuracy of the analytical solution is confirmed.

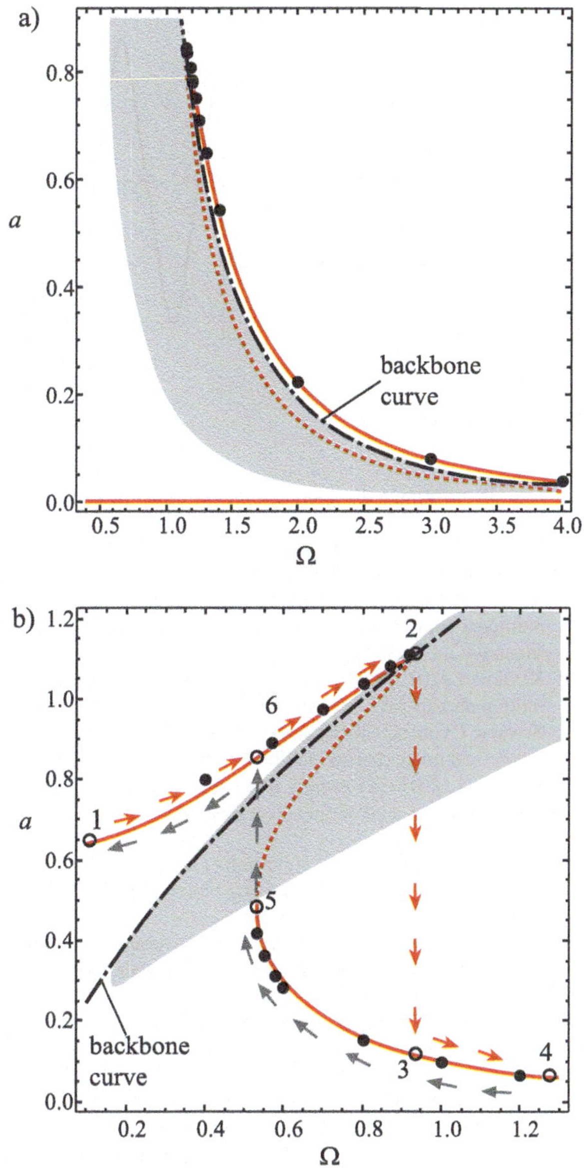

Fig. 4.21 Frequency-response curves of the oscillator governed by Eq. (4.136) for linear viscous damping, $F = \varepsilon = 0.1$ and for: **a** $\alpha = 1/4$; **b** $\alpha = 4$. Analytical results (4.186) are depicted by the solid line (stable) and by the dotted line (unstable); the shaded region (instability zone) is bounded by the curves defined by Eq. (4.190); the backbone curve (4.185) is shown as the dashed-dotted line; numerical results are depicted by the black dots. The arrows and black circles shown in Fig. 4.21b indicate the hysteretic paths 1-2-3-4 and 4-5-6-1

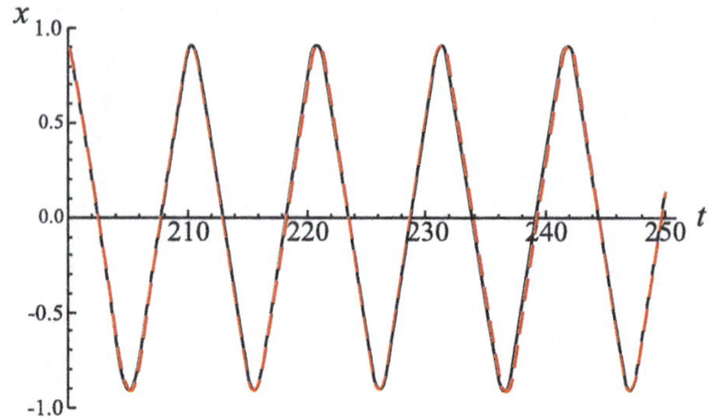

Fig. 4.22 Solutions for motion for linear viscous damping and $\alpha = 4$, $F = \varepsilon = 0.1$, $\Omega = 0.6$, $x(0) = 0.3$, $\dot{x}(0) = 0$: analytically obtained steady-state motion defined by Eq. (4.192) (red dashed line) and the response obtained numerically from the equation of motion (black solid line).

References

1. Hsu CS (1960) On the application of elliptic functions in nonlinear forced oscillations. Q Appl Math 17:393–407
2. Kovacic I, Cveticanin L, Zukovic M, Rakaric Z (2016) Jacobi elliptic functions: a review of nonlinear oscillatory application problems. J Sound Vib 380:1–36
3. Rakaric Z, Kovacic I, Cartmell M (2017) On the design of external excitations in order to make nonlinear oscillators respond as free oscillators of the same or different type. Int J Non-Linear Mech 94C:323–333
4. Kovacic I (2017) On the response of purely nonlinear oscillators: an Ateb-type solution for motion and an Ateb-type external excitation. Int J Nonlinear Mech 92:15–24
5. Kovacic I, Gatti G (2018) Some benefits of using exact solutions of forced nonlinear oscillators: theoretical and experimental investigations. J Sound Vib 436:310–326
6. Kovacic I (2018) Externally excited undamped and damped linear and nonlinear oscillators: exact solutions and tuning to a desired exact form of the response. Int J Nonlinear Mech 102:72–81
7. Kovacic I, Gatti G (2020) Helmholtz Duffing and Helmholtz-Duffing oscillators: exact steady-state solutions. IUTAM Bookseries, vol 37. Springer International Publishing, Berlin, pp 167–177
8. Rakaric Z, Kovacic I (2013) An elliptic averaging method for harmonically excited oscillators with a purely nonlinear non-negative real-power restoring force. Commun Nonlinear Sci Numer Simul 18:1888–1901
9. Balachandran B, Magrab EB (2009) Vibrations. Cengage Learning, Toronto
10. Byrd P, Friedman M (1954) Handbook of elliptic integrals for engineers and scientists. Springer, Berlin
11. Blanchard A, Vakakis A (2018) Exact steady states of the periodically forced and damped Duffing oscillator. J Sound Vib 413:57–65
12. Kovacic I, Brennan MJ (eds) (2011) The duffing equation: nonlinear oscillators and their behaviour. Wiley, Chichester
13. Nayfeh AH, Mook DT (1979) Nonlinear oscillations. Wiley, New York
14. Gatti G, Brennan MJ, Kovacic I (2010) On the interaction of the responses at the resonance frequencies of a nonlinear two degrees-of-freedom system. Phys D 239:591–599
15. Rand RH (2006) Lecture notes on nonlinear vibrations (version 53). http://dspace.library.cornell.edu/handle/1813/28989
16. Kovacic I, Rand R (2013) About a class of nonlinear oscillators with amplitude-independent frequency. Nonlinear Dyn 74:455–465
17. Kovacic I (2010) The method of multiple scales for forced oscillators with some real-power nonlinearities in the stiffness and damping force. Chaos Solitons Fractals 44:891–901

18. Kovacic I (2011) Forced vibrations of oscillators with a purely non-linear power-form restoring force. J Sound Vib 330:4313–4327
19. Kovacic I, Lenci S (2018) Externally excited purely nonlinear oscillators: insights into their response at different excitation frequencies. Nonlinear Dyn 93:119–132
20. Barkham PGD, Soudack AC (1969) An extension to the method of Kryloff and Bogoliuboff. Int J Control 10:377–392
21. Barkham PGD, Soudack AC (1970) Approximate solutions of non-linear nonautonomous second-order differential equations. Int J Control 11:101–114
22. Barkham PGD, Soudack AC (1977) New results on the elliptic function method. Int J Control 26:341–358
23. Christopher PAT (1973) An approximate solution to a strongly non-linear second- order, differential equation. Int J Control 17(3):597–608
24. Cap FF (1974) Averaging method for the solution of non-linear differential equations with periodic non-harmonic solutions. Int J Non-Linear Mech 9:441–450
25. Yuste SB, Bejarano JD (1989) Extension and improvement to Krylov and Bogoliubov methods using elliptic functions. Int J Control 49:1127–1141
26. Yuste SB, Bejarano JD (1990) Improvement of a Krylov-Bogoliubov method that uses Jacobi elliptic functions. J Sound Vib 139:151–163
27. Coppola VT, Rand RH (1990) Averaging using elliptic functions: approximation of limit cycles. Acta Mech 81:125–142
28. Rand RH (1990) Using computer algebra to handle elliptic functions in the method of averaging in symbolic computations and their impact on mechanics. In: Noor AK, Elishakoff I, Hulbert G (Eds) American Society of Mechanical Engineers, PVP-vol 205, pp 311–326, New York
29. Roy RV (1994) Averaging method for strongly non-linear oscillators with periodic excitations. Int J Non-Linear Mech 29:737–753
30. Vakakis AF, Rand RH (2004) Non-linear dynamics of a system of coupled oscillators with essential stiffness non-linearities. Int J Non-Linear Mech 39:1079–1091
31. Okabe T, Kondou T (2009) Improvement to the averaging method using the Jacobian elliptic function. J Sound Vib 320:339–364
32. Okabe T, Kondou T, Ohnishi J (2011) Elliptic averaging methods using the sum of Jacobian elliptic delta and zeta functions as the generating solution. Int J Non-Linear Mech 46:159–169
33. Rakaric Z, Kovacic I (2011) Approximations for motion of the oscillators with a non-negative real-power restoring force. J Sound Vib 330:321–336
34. Mynorsky N (1974) Non-linear oscillations. Robert E. Kreiger Publishing Company, New York
35. Kovacic I (2013) Harmonically excited generalized van der Pol oscillators: entrainment phenomenon. Meccanica 48:2415–2425
36. Thomsen JJ (2003) Vibrations and stability, advanced theory, analysis, and tools, 2nd edn. Springer, Berlin

Chapter 5
Nonlinear Isochronous Oscillators

5.1 Introduction

It is demonstrated in Chap. 2 that a conservative system like the free, undamped Duffing oscillators

$$\ddot{x} + x \pm x^3 = 0, \tag{5.1}$$

is characterized by a relationship between the amplitude of vibration A of a typical periodic motion and its period or frequency ω. When plotted as a curve in the $\omega - A$ plane, this relation presents a backbone curve (see Fig. 2.11 in Chap. 2). The backbone curve for Eq. (5.1) is bent to the right for the upper sign, i.e. for the Hardening Duffing Oscillator (HDO) and to the left for the lower sign, i.e. for the Softening Duffing Oscillator (SDO).

When the Duffing oscillator is driven by a periodic forcing function $F \cos \Omega t$, in the presence of small damping $\epsilon\dot{x}$:

$$\ddot{x} + x + x^3 = F \cos \Omega t - \epsilon\dot{x}, \tag{5.2}$$

the resulting steady-state motion exhibits hysteresis and jump phenomenon in the frequency-amplitude plane (see, for example, [1] or [2], or Chap. 4, Example 4.4). The illustration of the hysteresis and jump phenomena associated with increasing or decreasing forcing frequency Ω are shown in Fig. 5.1: when the frequency is increased, the amplitude A changes following the path 1-2-3-4, as indicated by the red arrows; when it is decreased, a different path is followed 4-5-6-1, as indicated by the grey arrows. At the frequencies corresponding to 2 and 3, the amplitude decreases discontinually (this is a jump-down point), while at the frequencies corresponding to 5 and 6, it increases discontinually (this is a jump-up point). The usual explanation of these phenomena [1, 2] involves relating the backbone curve of the unforced Eq. (5.1) with the response of the forced Eq. (5.2) (see Fig. 5.1). Note that these phenomena are illustrated here for the HDO only, but they appear in the SDO, with the difference

I. Kovacic, *Nonlinear Oscillations*,
https://doi.org/10.1007/978-3-030-53172-0_5

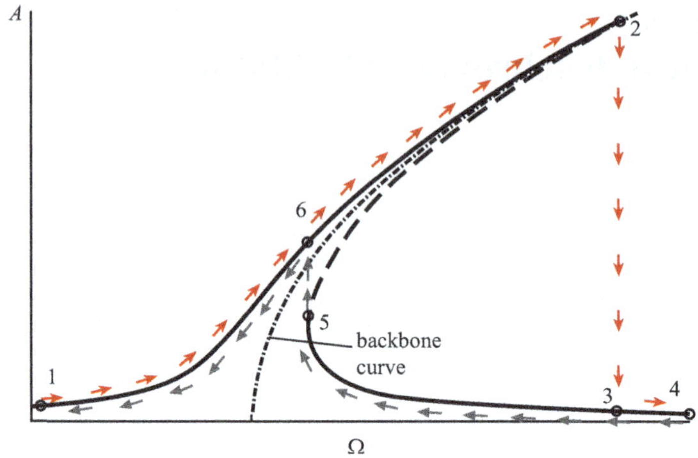

Fig. 5.1 A sketch of the characteristic amplitude-frequency response curve of the damped, forced Duffing oscillator with hysteresis and jump phenomenon

related to the bending of the backbone curve and the frequency-amplitude curve to the left-hand side [1, 2].

It may happen that in an engineering application, the system in question is exposed to repeated changes in forcing frequency so that the system is exposed to repeated jumps. Each of these jumps is not periodic and reproduced exposure to such a loading situation may be objectionable (see, for example, [3]).

It is in such a scenario that an engineering designer could wish for a nonlinear system which does not exhibit jump phenomenon. One scheme for attaining such a goal would be if the associated backbone curve was a vertical straight line, i.e. its period/frequency of vibration is fixed, amplitude-independent. The Simple Harmonic Oscillator (SHO) with the Lagrangian

$$L_{\mathrm{SHO}} = \frac{\dot{X}^2}{2} - \frac{X^2}{2}, \tag{5.3}$$

and the equation of motion

$$\ddot{X} + X = 0, \tag{5.4}$$

is the archetypal example of the system with this property. This property can actually be related to *isochronicity/isochrony*—the characteristic of oscillatory systems which have a fixed, amplitude-independent frequency/period [4]. The SHO is said to have an isochronous centre as the period is constant in the neighborhood of the centre.

First results in the field of isochronous oscillators are believed to date back to Galileo Galilei and Christian Huygens [5]. Although Galileo did not live to complete his design, he had thought that a pendulum is isochronous in the sense that the time it takes to complete one full swing is the same regardless of the size of the swing. Huygens, however, pushed this matter further, noting that this is true for pendulums that swing only a few degrees. He pursued the question of achieving perfect isochronicity and showed that it can be realized in a simple pendulum that wraps around the cycloid [6, 7] (see Example 5.5 below).

More recent investigations of isochronicity have been directed towards nonlinear oscillators, which are in general known to have a frequency that depends on their amplitude. Thus, some Liénard-type equations

$$\ddot{x} + u(x)\dot{x} + v(x) = 0, \tag{5.5}$$

are found to exhibit the isochronicity characteristic. Sabatini [8] gave necessary and sufficient mathematical conditions for isochronicity of the differential equation (5.5) in terms of the coefficient functions $u(x)$ and $v(x)$: Let $u(x)$, $v(x)$ be analytic, $v(x)$ odd, $u(0) = v(0) = 0$, $v'(0)>0$. Then the origin O is a centre if and only if $u(x)$ is odd, and O is an isochronous centre if and only if

$$\left(\int_0^x su(s)\,\mathrm{d}s\right)^2 - x^3\left(v(x) - v'(0)x\right) \equiv 0. \tag{5.6}$$

He illustrated the existence of this behaviour in the system (5.5) with

$$u(x) = (2n+3)x^{2n+1}, \quad v(x) = x + x^{4n+3}, \tag{5.7}$$

where n is a non-negative integer. Iacono and Russo showed that this system can be explicitly solved [9]. Necessary and sufficient mathematical conditions for the isochronicity of the differential equation (5.5) have also been provided by Christopher and Devlin [10]: the system (5.5) with $u(x)$ and $v(x)$ being analytic, has an isochronous centre at the origin if and only if

$$v(x) = ww'\left(1 + \frac{1}{w^4}\left(\int_0^x w(s)u(s)\,\mathrm{d}s\right)^2\right), \tag{5.8}$$

where $w(x)$ solves the functional equation in $F(x) = \int_0^x u(s)\mathrm{d}s$:

$$F(x - 2w(x)) = F(x), \quad w(0) = 0, \quad w'(0) = 1. \tag{5.9}$$

Chandrasekar et al. [11] investigated in detail the so-called modified Emden equation, which is a Liénard-type nonlinear oscillator (5.5) with

$$u(x) = kx, \quad v(x) = \lambda_1 x + \frac{k^2}{9}x^3, \tag{5.10}$$

and determined the conditions under which it can yield isochronous oscillations. Chandrasekar et al. [12] found a class of coupled Liénard-type equations that exhibits the isochronicity property.

It should be noted that Liénard-type equations (5.5) have a term linear in the generalized velocity. Sabatini [13] studied the equation analogous to (5.5), but with $\dot{x}$ replaced by $\dot{x}^2$

$$\ddot{x} + p(x)\dot{x}^2 + q(x) = 0, \tag{5.11}$$

deriving a sufficient condition for its solution to be oscillatory, i.e. for the origin to be a centre:

$$xq(x) > 0. \tag{5.12}$$

Based on his previous results [8], Sabatini also proved in [13] that when $p(x)$ and $q(x)$ are odd and analytic, and Eq. (5.12) is satisfied for small values of $x \neq 0$, the origin is an isochronous centre if and only if the following expression is equal to zero in the whole domain

$$x\left[q(x)\Phi'(x) - \Phi(x)q'(x) - \Phi(x)q(x)p(x)\right] \equiv 0, \tag{5.13}$$

where

$$\Phi(x) = \int_0^x \exp(P(s))\,\mathrm{d}s, \quad P(x) = \int_0^x p(s)\,\mathrm{d}s. \tag{5.14}$$

Sabatini further gave a characterization of isochronous centres: when $p(x)$ and $q(x)$ are polynomials and the condition (5.12) is satisfied, the origin represents a global isochronous centre if and only if both $p(x)$ and $q(x)$ have an odd degree and $p(x)$ has a positive leading coefficient [13].

This theory related to the oscillators modelled by Eq. (5.11) neither links the equation of motion with mechanical models, nor provide general solution for their isochronous motion. The results proposed in this chapter resolves this shortage presenting a family of oscillators, answering a fundamental question of designing a differential equation which is similar to the Duffing equation (5.1) in that it is conservative, but for which the backbone curve is a straight vertical line in the $\omega - A$ plane, like for the linear oscillator. The answers to this question are obtained in several different ways: via perturbations methods in Sect. 5.2 and via energy considerations in Sects. 5.3 and 5.4.

5.2 Derivations Based on Perturbation Methods

Let us start from equations of the form [14]:

$$\ddot{x} + x + x\dot{x}^2 + f(x) = 0, \tag{5.15}$$

where $f(x)$ is odd and nonlinear in x:

$$f(x) = a_3x^3 + a_5x^5 + a_7x^7 + \cdots . \tag{5.16}$$

We ask how one can define the coefficients a_i in $f(x)$ such that Eq. (5.15) will exhibit a straight-line backbone curve, and thus when forced and possibly damped, will not exhibit jumps.

To begin with, one can note that Eq. (5.15) is conservative and may be derived using Lagrange's equation. We take the Lagrangian to be of the form

$$L = \exp(x^2)\left(\frac{1}{2}\dot{x}^2 - g(x)\right), \tag{5.17}$$

where $g(x)$ is to be determined. Lagrange's equation becomes

$$\ddot{x} + x\dot{x}^2 + \frac{\mathrm{d}g(x)}{\mathrm{d}x} + 2xg(x) = 0. \tag{5.18}$$

Comparing Eqs. (5.15), (5.16), (5.18) one can see that $g(x)$ must satisfy the equation:

$$\frac{\mathrm{d}g(x)}{\mathrm{d}x} + 2xg(x) = x + a_3x^3 + a_5x^5 + a_7x^7 + \cdots . \tag{5.19}$$

This may be solved for $g(x)$ by taking $g(x)$ in the form of a power series with even-powered terms:

$$g(x) = b_2x^2 + b_4x^4 + b_6x^6 + \cdots . \tag{5.20}$$

Substitution of (5.20) into (5.19) leads to expressions for the b_i coefficients, the first few of which are

$$b_2 = \frac{1}{2}, \quad b_4 = \frac{a_3 - 1}{4}, \quad b_6 = \frac{-2a_5 + a_3 - 1}{12}, \quad \cdots . \tag{5.21}$$

Thus the Lagrangian (5.17) produces the differential equation of motion (5.15) which exhibits the first integral

$$\exp(x^2)\left(\frac{1}{2}\dot{x}^2 + g(x)\right) = \text{constant}. \tag{5.22}$$

5.2.1 Straight-Line Backbone Curve

We set $x = \sqrt{\epsilon}\,\tilde{x}$ in Eqs. (5.15), (5.16) and drop the tilde for convenience, giving

$$\ddot{x} + x + \epsilon x\dot{x}^2 + \epsilon a_3 x^3 + \epsilon^2 a_5 x^5 + \epsilon^3 a_7 x^7 + \cdots = 0. \tag{5.23}$$

In order to obtain an approximate solution to Eq. (5.23), one can expand x in a power series in ϵ:

$$x = x_0 + \epsilon x_1 + \epsilon^2 x_2 + \cdots . \tag{5.24}$$

Note that since we are after a straight-line backbone curve, there is no need to expand frequency in a power series in ϵ as is usual in Lindstedt's method [1]. Substituting (5.24) into (5.23) and collecting terms one can get a sequence of equations of which the first few are given by

$$\ddot{x}_0 + x_0 = 0, \tag{5.25}$$

$$\ddot{x}_1 + x_1 = -x_0\dot{x}_0^2 - a_3 x_0^3, \tag{5.26}$$

$$\ddot{x}_2 + x_2 = -2\ x_0\ \dot{x}_0\dot{x}_1 - \dot{x}_0^2\ x_1 - 3\ a_3\ x_0{}^2\ x_1 -\ a_5\ x_0{}^5. \tag{5.27}$$

We take the solution to Eq. (5.25) to be

$$x_0 = A\cos t, \tag{5.28}$$

whereupon (5.26) becomes, after some trigonometric reduction:

$$\ddot{x}_1 + x_1 = \frac{a_3 - 1}{4}A^3\cos 3t + \frac{3a_3 + 1}{4}A^3\cos t. \tag{5.29}$$

We take $a_3 = -1/3$ to remove resonance (secular) terms, and obtain

$$\ddot{x}_1 + x_1 = -\frac{1}{3}A^3\cos 3t, \tag{5.30}$$

which gives the particular solution:

$$x_1 = -\frac{1}{24}A^3\cos 3t. \tag{5.31}$$

Substituting (5.28), (5.31) into (5.27) gives

$$\ddot{x}_2 + x_2 = \frac{15a_5 - 1}{24}R^5\cos t + \text{NRT}, \tag{5.32}$$

where NRT stands for non-resonant terms. For no resonance, one can choose $a_5 = 1/15$. Proceeding in this way, one can obtain the following values for the coefficients a_i in Eq. (5.16):

$$\begin{aligned}
a_3 &= -1/3, \\
a_5 &= 1/15 = 1/(3*5), \\
a_7 &= -1/105 = -1/(3*5*7), \\
a_9 &= 1/945 = 1/(3^3*5*7), \\
a_{11} &= -1/10395 = -1/(3^3*5*7*11), \\
a_{13} &= 1/135135 = 1/(3^3*5*7*11*13), \\
a_{15} &= -1/2027025 = -1/(3^4*5^2*7*11*13), \\
a_{17} &= 1/34459425 = 1/(3^4*5^2*7*11*13*17).
\end{aligned}$$

Now it is a remarkable fact that the typical term in the foregoing list of coefficients may be written in the following compact form:

$$a_{2n+1} = \frac{(-1)^n}{(2n+1)!!}. \tag{5.33}$$

Thus, the straight-line backbone differential equations (5.15), (5.16) may be written

$$\ddot{x} + x + x\dot{x}^2 + \sum_{n=1}^{\infty} \frac{(-1)^n x^{2n+1}}{(2n+1)!!} = 0, \tag{5.34}$$

or absorbing the x term into the sum, cf. Eqs. (5.18), (5.19),

$$\ddot{x} + x\dot{x}^2 + \sum_{n=0}^{\infty} \frac{(-1)^n x^{2n+1}}{(2n+1)!!} = 0. \tag{5.35}$$

In order to check this analytical result, Eq. (5.35) was solved numerically and the frequency was extracted from the time response for different values of the initial amplitude A. This was done for a different number of odd-powered polynomial terms in the sum from Eq. (5.35). The backbone curves of the corresponding oscillators (O_j) are shown in Fig. 5.2, where the subscript j denotes the highest power included into the sum. Thus, the oscillator O_5 has the backbone of the softening type. Additional terms change the way how the backbone curve is bent, alternating it between hardening and softening, making it be straighter for higher amplitudes. The oscillator O_{15} has a backbone curve that is straight on the region of A considered. The backbone curve of the oscillators with higher powers of nonlinearity remains straight on that region, too.

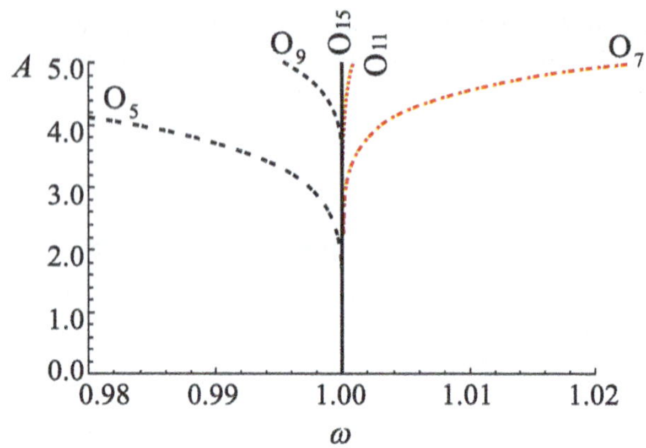

Fig. 5.2 Numerically obtained backbone curves of the oscillator (5.35) for a different number of odd-powered terms (the highest power is indicated in the subscript)

5.2.2 Closed-Form Solution

It is another remarkable fact that the sum in Eq. (5.35) can be written in the following closed form:

$$\sum_{n=0}^{\infty} \frac{(-1)^n x^{2n+1}}{(2n+1)!!} = \sqrt{\frac{\pi}{2}}\, e^{-\frac{x^2}{2}}\, \mathrm{erfi}\left(\frac{x}{\sqrt{2}}\right), \tag{5.36}$$

where erfi denotes the 'imaginary error function' defined as [15]

$$\mathrm{erfi}(z) = -i\, \mathrm{erf}\,(iz)\,, \tag{5.37}$$

where erf represents the error function [15, 16],

$$\mathrm{erf(z)} = \frac{2}{\sqrt{\pi}} \int_0^z e^{-t^2} dt. \tag{5.38}$$

We note that $\mathrm{erfi}(z)$ satisfies the equation [15]:

$$\frac{d}{dz}\mathrm{erfi}(z) = \frac{2}{\sqrt{\pi}} \exp z^2. \tag{5.39}$$

Thus, starting from a local perturbation analysis we have been able to obtain an expression for the straight-line backbone differential equation which is valid for all x, namely [14]

$$\ddot{x} + x\dot{x}^2 + \sqrt{\frac{\pi}{2}}\, e^{-\frac{x^2}{2}}\, \mathrm{erfi}\left(\frac{x}{\sqrt{2}}\right) = 0. \tag{5.40}$$

The terms a_i obtained by use of the perturbation series (5.24) were derived with the requirement that the frequency of oscillations be unity, this being equivalent to requiring that the period of oscillations be 2π. Because of the local nature of such a procedure, results may be expected to be valid only for small amplitudes of vibration. Since the sum has been obtained as the series in closed form, and since it is easy to show using the ratio test that the resulting series converges for all x, it should be possible to show directly the straight-line backbone property, i.e., that all motions of Eq. (5.40) have period 2π, regardless of amplitude.

From Eq. (5.22) with the initial condition $x(0) = A$, $\dot{x}(0) = 0$, one can obtain

$$\exp(x^2)\left(\frac{1}{2}\dot{x}^2 + g(x)\right) = \exp(A^2)g(A). \tag{5.41}$$

We may use this equation to compute the period of oscillation once $g(x)$ is known. From Eqs. (5.18), (5.19), (5.40), one has

$$\frac{dg(x)}{dx} + 2xg(x) = \sqrt{\frac{\pi}{2}}\, e^{-\frac{x^2}{2}} \operatorname{erfi}\left(\frac{x}{\sqrt{2}}\right). \tag{5.42}$$

Multiplying by $\exp(x^2)$, an integrating factor, one obtains

$$\frac{d}{dx}\left(e^{x^2} g(x)\right) = \sqrt{\frac{\pi}{2}}\, e^{\frac{x^2}{2}} \operatorname{erfi}\left(\frac{x}{\sqrt{2}}\right). \tag{5.43}$$

Now from Eq. (5.39) one has

$$\frac{d}{dx}\operatorname{erfi}^2\left(\frac{x}{\sqrt{2}}\right) = 2\sqrt{\frac{2}{\pi}}\, e^{\frac{x^2}{2}} \operatorname{erfi}\left(\frac{x}{\sqrt{2}}\right). \tag{5.44}$$

Comparison of Eqs. (5.43), (5.44) gives that one can may take $g(x)$ as

$$g(x) = \frac{\pi}{4} e^{-x^2} \operatorname{erfi}^2\left(\frac{x}{\sqrt{2}}\right). \tag{5.45}$$

Having found $g(x)$, one can may now return to Eq. (5.41), which may be solved for $\dot{x}$, leading to the following expression for period of oscillation T:

$$T = 4\int_0^R \frac{\mathrm{d}x}{|\dot{x}|} = 4\int_0^R \frac{\mathrm{d}x}{\sqrt{-2g(x) + 2\exp(R^2 - x^2)g(R)}}. \tag{5.46}$$

Substitution of Eq. (5.45) in (5.46) gives

$$T = 4\sqrt{\frac{2}{\pi}}\int_0^R \frac{\exp(x^2/2)\mathrm{d}x}{\sqrt{\operatorname{erfi}^2(R/\sqrt{2}) - \operatorname{erfi}^2(x/\sqrt{2})}}. \tag{5.47}$$

Careful differentiation gives the result that

$$\frac{\mathrm{d}}{\mathrm{d}x}\arctan\frac{\mathrm{erfi}(x/\sqrt{2})}{\sqrt{\mathrm{erfi}^2(R/\sqrt{2})-\mathrm{erfi}^2(x/\sqrt{2})}}=\sqrt{\frac{2}{\pi}}\frac{\exp(x^2/2)dx}{\sqrt{\mathrm{erfi}^2(R/\sqrt{2})-\mathrm{erfi}^2(x/\sqrt{2})}}, \tag{5.48}$$

whereupon one has

$$T=4\arctan\left.\frac{\mathrm{erfi}(x/\sqrt{2})}{\sqrt{\mathrm{erfi}^2(R/\sqrt{2})-\mathrm{erfi}^2(x/\sqrt{2})}}\right|_0^R=4\,\frac{\pi}{2}=2\pi. \tag{5.49}$$

So, the period of oscillations is 2π.

5.2.3 Restoring Force

There are two ways of approaching the concept of restoring force for the foregoing oscillator. On the one hand, one may consider the Lagrangian (5.17) in which the potential energy is given by $E_p=\exp(x^2)g(x)$, where $g(x)$ is defined by Eq. (5.45), which gives

$$E_p=\frac{\pi}{4}\mathrm{erfi}^2\left(\frac{x}{\sqrt{2}}\right). \tag{5.50}$$

The potential energy E_p is plotted as a solid line in Fig. 5.3a. In this case the force that corresponds to it is $F_{r1}=-\partial E_p/\partial x$. Omitting the minus sign, one can recognize that the restoring force-displacement law is given by the right-hand side of Eq. (5.43), which is shown as a solid line in Fig. 5.3b. For comparison, these same figures

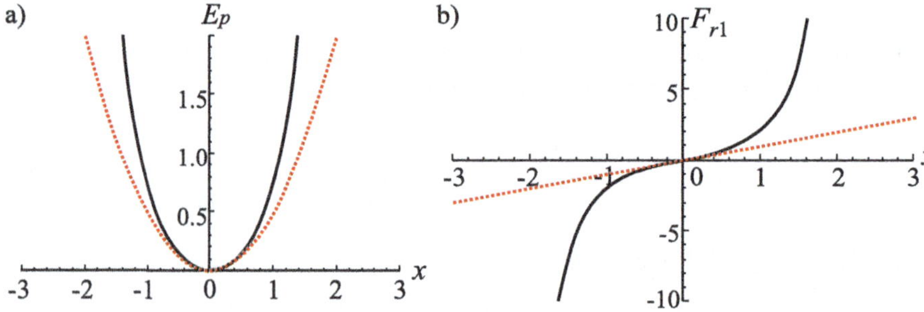

Fig. 5.3 **a** Potential energy E_p defined by Eq. (5.50) (solid line); **b** the corresponding force-displacement law $F_{r1}-x$, given by the right-hand side of Eq. (5.43) (solid line). The potential energy and force-displacement law of an harmonic oscillator with period 2π are plotted as dotted lines

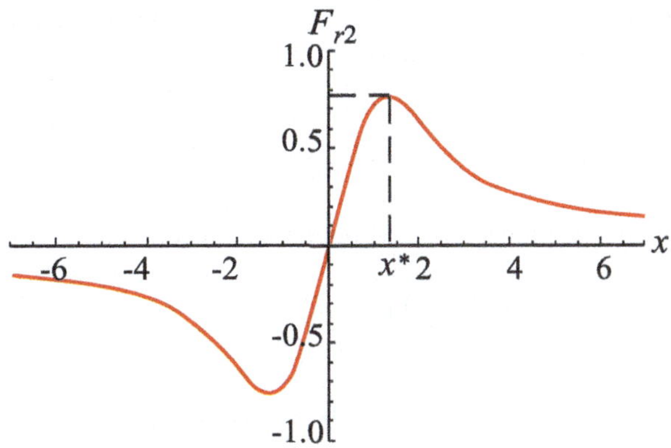

Fig. 5.4 F_{r2} given by the right-hand side of Eq. (5.36) plotted as a function of x

contain plots of the potential energy $E_p = x^2/2$ and restoring force x of an harmonic oscillator which has the same period 2π. Note that both the nonlinear oscillator under consideration and the corresponding harmonic oscillator have a single-well potential. An example of a mechanical system that mimics this behaviour is a particle whose mass changes exponentially with its position, and which moves along a fixed smooth horizontal surface, being connected to the spring whose potential energy and the force-displacement law correspond to those given in Fig. 5.3.

Another approach to treating the restoring force is to take the nonlinear oscillator as being given in the form (5.40). In this case one may identify the restoring force as F_{r2}, its value being given by the right-hand side of (5.36). This force-displacement law is plotted in Fig. 5.4. It is seen that F_{r2} has a limited codomain, with a maximum at $x^* \approx 1.307$, for which $F_{r2}(x^*) = 1/x^* \approx 0.765$. As x further increases, F_{r2} approaches zero monotonically.

5.2.4 Related Oscillators

The oscillator given by Eqs. (5.15), (5.16) may be generalized by including even-powered terms in the expansion for $f(x)$:

$$\ddot{x} + x + a_0 x \dot{x}^2 + a_2 x^2 + a_3 x^3 + a_4 x^4 + a_5 x^5 + \cdots = 0. \tag{5.51}$$

In this case one may simplify the a_0 coefficient by stretching x. We set $x = \mu y$ and obtain

$$\ddot{y} + y + a_0 \mu^2 y \dot{y}^2 + a_2 \mu y^2 + a_3 \mu^2 y^3 + \cdots = 0. \tag{5.52}$$

Thus by choosing $\mu^2 = 1/|a_0|$ one can make the coefficient of the $y\dot{y}^2$ term equal to 1 or to −1, i.e. to sgn (a_0).

Returning to (5.51), one can prepare it for perturbation treatment by setting $x = \epsilon\,\tilde{x}$:

$$\ddot{x} + x + a_0\epsilon^2 x\dot{x}^2 + a_2\epsilon x^2 + a_3\epsilon^2 x^3 + a_4\epsilon^3 x^4 + \cdots = 0, \tag{5.53}$$

where the tilde has been dropped for convenience. Proceeding as before, requiring the solution to have frequency $\omega = 1$, collecting terms, solving equations and removing resonances, one can obtain

$$9\,a_3 = 10\,a_2^2 - 3\,a_0, \tag{5.54}$$

$$135\,a_5 = 9\,a_0^2 + 63\,a_2^2\,a_0 + 378\,a_2\,a_4 - 280\,a_2^4, \tag{5.55}$$

$$\begin{aligned}14175\,a_7 = {} & -135\,a_0^3 + 192\,a_2^2\,a_0^2 + 11934 a_2\,a_4\,a_0 - 40880\,a_2^4\,a_0 \\ & + 48600\,a_2\,a_6 + 20412\,a_4^2 - 186480\,a_2^3\,a_4 + 148400\,a_2^6,\end{aligned} \tag{5.56}$$

$$\begin{aligned}1148175\,a_9 = {} & 1215\,a_0^4 + 6993\,a_2^2\,a_0^3 + 73386\,a_2\,a_4\,a_0^2 - 1216755\,a_2^4\,a_0^2 \\ & + 886950\,a_2\,a_6\,a_0 + 500823\,a_4^2\,a_0 - 15790140\,a_2^3\,a_4\,a_0 + 29912400\,a_2^6\,a_0 \\ & + 4677750\,a_2\,a_8 + 3608550\,a_4\,a_6 - 24057000\,a_2^3\,a_6 - 29189160\,a_2^2\,a_4^2 \\ & + 131392800\,a_2^5\,a_4 - 93940000\,a_2^8,\end{aligned} \tag{5.57}$$

$$\begin{aligned}442047375\,a_{11} = {} & -42525\,a_0^5 - 435024\,a_2^2\,a_0^4 + 2671542\,a_2\,a_4\,a_0^3 - 112405293\,a_2^4\,a_0^3 \\ & + 38321100\,a_2\,a_6\,a_0^2 + 17272926\,a_4^2\,a_0^2 - 2156954184\,a_2^3\,a_4\,a_0^2 \\ & + 8833359150\,a_2^6\,a_0^2 + 314940150\,a_2\,a_8\,a_0 + 388520550\,a_4\,a_6\,a_0 \\ & - 7348433400\,a_2^3\,a_6\,a_0 - 10286259984\,a_2^2\,a_4^2\,a_0 + 97229424600\,a_2^5\,a_4\,a_0 \\ & - 133948183600\,a_2^8\,a_0 + 2089678500\,a_2\,a_{10} + 1532430900\,a_4\,a_8 \\ & - 14189175000\,a_2^3\,a_8 + 703667250\,a_6^2 - 31086455400\,a_2^2\,a_4\,a_6 \\ & + 97459362000\,a_2^5\,a_6 8459018568\,a_2\,a_4^3 + 188659270800\,a_2^4\,a_4^2 \\ & - 586687701600\,a_2^7\,a_4 + 379443064000\,a_2^{10}.\end{aligned} \tag{5.58}$$

We now discuss a number of special cases of Eq. (5.51):

Equation (5.51) *with* $a_0 = 0$

In this case the formulas (5.54)–(5.58) show that the coefficients may be chosen so that the resulting equation has a straight-line backbone curve. The first few of them are

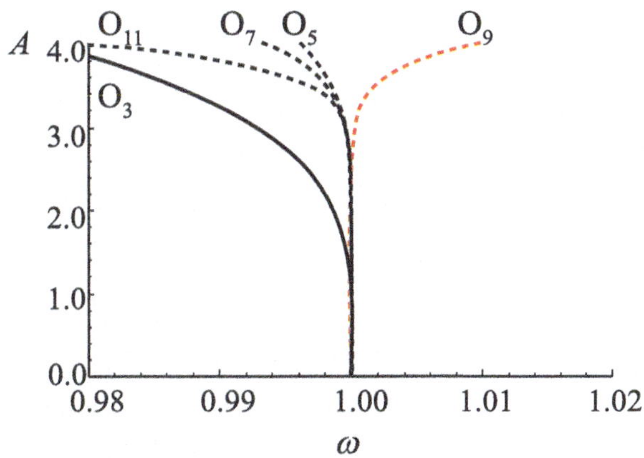

Fig. 5.5 Numerically obtained backbone curves of the oscillator (5.53) with $a_0 = 0$ and $\epsilon = 0.1$ for a different number of even and odd-powered terms

$$a_3 = \frac{10}{9}\, a_2^2, \tag{5.59}$$

$$a_5 = \frac{378\, a_2\, a_4 - 280\, a_2^4}{135}, \tag{5.60}$$

$$a_7 = \frac{48600\, a_2\, a_6 + 20412\, a_4^2 - 186480\, a_2^3\, a_4 + 148400\, a_2^6}{14175}. \tag{5.61}$$

The corresponding backbone curves obtained numerically from the differential equations of motion are shown in Fig. 5.5 for $\epsilon = 0.1$ and for the coefficients of the even powers equal to unity. The oscillator O_3 with a quadratic and cubic nonlinear term corresponds to the so-called Helmholtz–Duffing oscillator [2]. The relationship between a_3 and a_2 in Eq. (5.59) agrees with the well-know result yielding the frequency independent of amplitude, which is in the literature obtained for the calculation procedure when terms of $O\left(\varepsilon^3\right)$ are neglected (see, for example, [17], pp. 55 or pp. 198). As seen from Fig. 5.5, an increase in the number of terms in the truncation ceases to produce an increase in the straightness of the backbone curve for values of amplitude A which are larger than about 2.5. We conjecture that this effect is due to divergence of the associated perturbation series.

Equation (5.51) *with* $a_1 = a_3 = a_5 = \cdots = 0$

When all the odd-powered terms are absent, the formulas (5.54)–(5.58) determine the following coefficients of the even-powered terms such that the backbone curve is a straight line:

$$a_2 = \pm 3\sqrt{\frac{a_0}{30}}, \tag{5.62}$$

$$a_4 = \mp \frac{1}{14}\sqrt{\frac{a_0^3}{30}}, \tag{5.63}$$

$$a_6 = \mp \frac{2209}{18900}\sqrt{\frac{a_0^5}{30}}, \tag{5.64}$$

$$a_8 = \pm \frac{91673}{1157625}\sqrt{\frac{a_0^7}{30}}, \tag{5.65}$$

$$a_{10} = \pm \frac{2014958909}{35006580000}\sqrt{\frac{a_0^9}{30}}. \tag{5.66}$$

Note that there are two sets of these coefficients (with the upper and lower sign), i.e. two types of oscillators with even-powered terms that have a straight-line backbone curve. Note however that solutions exist only for $a_0 > 0$. By solving Eq. (5.53) with even-powered terms for the coefficients (5.62)–(5.66) with the upper signs and for $a_0 = 1$, $\epsilon = 0.1$, the backbone curves presented in Fig. 5.6 are obtained.

Equation (5.51) *with* $a_2 = a_4 = a_6 = \cdots = 0$
This case has been the subject of the earlier sections based on Eq. (5.23) where $a_0 = 1$. However now one can note that the form of the closed-form solution is dependent on the sign of a_0. When $a_0 > 0$, the coefficients a_{2n+1} are defined by Eq. (5.33) and have an alternating sign. When $a_0 < 0$, they are all positive and defined by

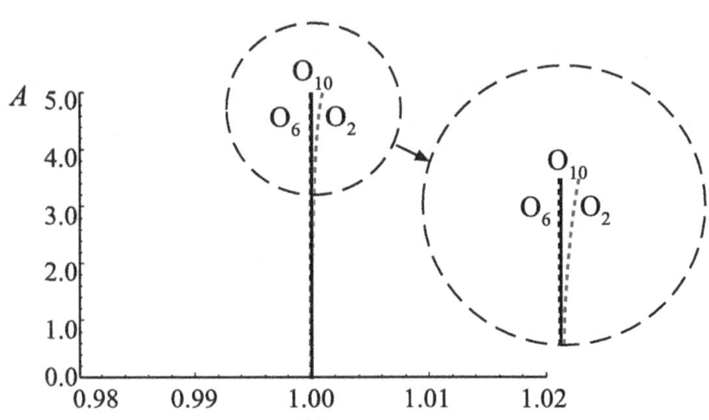

Fig. 5.6 Numerically obtained backbone curves of the oscillator (5.53), (5.62)–(5.66) with $a_0 = 1$ and $\epsilon = 0.1$ for a different number of even-powered terms

$$a_{2n+1} = \frac{|a_0|^n}{(2n+1)!!}. \tag{5.67}$$

In order to show how additional terms change the shape of the corresponding backbone curve, Eq. (5.51) with no even-powered terms can be first rescaled to correspond to Eq. (5.23) and then solved numerically for $a_0 = -1$, $\epsilon = 0.1$ and for a_{2n+1} defined by Eq. (5.67). The backbone curves are of the hardening type, but every additional even-powered term unbends it more and more, producing the desirable effect of a straight-line backbone curve on the region of A considered.

In the case of Eq. (5.51) with no even-powered terms but with $a_0 < 0$, the sum can be expressed in the following closed form:

$$\sum_{n=0}^{\infty} \frac{|a_0|^n x^{2n+1}}{(2n+1)!!} = \sqrt{\frac{\pi}{2|a_0|}} e^{\frac{a_0 x^2}{2}} \operatorname{erf}\left(\sqrt{\frac{|a_0|}{2}} x\right), \tag{5.68}$$

so that Eq. (5.51) becomes

$$\ddot{x} - |a_0| x \dot{x}^2 + \sqrt{\frac{\pi}{2|a_0|}} e^{\frac{|a_0| x^2}{2}} \operatorname{erf}\left(\sqrt{\frac{|a_0|}{2}} x\right) = 0. \tag{5.69}$$

5.3 Energy Considerations I

The aim of this section is to present a general procedure for defining isochronous mechanical systems that have the equation of motion in the form

$$\ddot{x} + G(x)\dot{x}^2 + F_r(x) = 0, \tag{5.70}$$

based on the system's kinetic and potential energy.

5.3.1 Oscillators with Position-Dependent Mass

Let us consider conservative oscillators whose Lagrangian has the form

$$L = \frac{1}{2} m(x)\dot{x}^2 - E_p(x), \tag{5.71}$$

where $m(x)$ is a mass that changes with the displacement x and $E_p(x)$ is the potential energy that is required to be positive definite and to yield the amplitude-independent frequency.

Lagrange's equation corresponding to the Lagrangian (5.71) is

$$\ddot{x} + \frac{m'}{2m}\dot{x}^2 + \frac{E_p'}{m} = 0, \tag{5.72}$$

where $m' = \mathrm{d}m/\mathrm{d}x$.

Now, putting the requirement of the equivalence between the Lagrangian of the oscillator under consideration (5.17) and the Lagrangian of the SHO with the isochronous centre at the origin (5.3), one can conclude that the following should be satisfied

$$\dot{X} = \sqrt{m(x)}\dot{x}, \tag{5.73}$$

$$E_p\,(x) = \frac{X^2}{2}. \tag{5.74}$$

Equation (5.73) gives

$$X = \int_0^x \sqrt{m(s)}\mathrm{d}s, \tag{5.75}$$

and Eq. (5.74) defines the potential energy, so that the equation of motion (5.72) becomes

$$\ddot{x} + \frac{m'}{2m}\dot{x}^2 + \frac{1}{\sqrt{m}}\left[\int_0^x \sqrt{m(s)}\mathrm{d}s\right] = 0. \tag{5.76}$$

Given the fact that classical mechanical systems are such that $m > 0$, the denominators in Eq. (5.76) are not zero and the radicand is always positive. Equation (5.76) can be related to Eq. (5.11) by identifying

$$p\,(x) = \frac{m'}{2m}, \quad q\,(x) = \frac{1}{\sqrt{m}}\left[\int_0^x \sqrt{m(s)}\mathrm{d}s\right] = \frac{X}{\sqrt{m}}, \tag{5.77}$$

where the latter term plays the role of the restoring force $F_r\,(x) \equiv q\,(x)$.

Comparing (5.77) and (5.14), one can obtain that $P = \ln\sqrt{m}$ and that $\Phi\,(x) = \int_0^x \sqrt{m\,(s)}\mathrm{d}s = X$ from Eq. (5.75). Then the condition derived by Sabatini in (5.13) is satisfied, since the expression in square brackets in Eq. (5.13) becomes

$$\frac{X}{\sqrt{m}}\sqrt{m} - X\frac{\sqrt{m}\sqrt{m} - X\frac{m'}{2\sqrt{m}}}{m} - X\frac{X}{\sqrt{m}}\frac{m'}{2m} \equiv 0. \tag{5.78}$$

In addition, the form of $m(x)$ should be such that $p\,(x)$ and $q\,(x)$ are odd and analytic. To determine the form of $m(x)$, one can analyse $p\,(x)$ in Eq. (5.77): the ratio of $m'(x)$ and $m(x)$ needs to be odd. Given the properties of mass in classical mechanical systems, one concludes that $m\,(x)$ should be even.

Noting that $m'/\,(2m) = \left(\sqrt{m}\right)'/\sqrt{m}$, Eq. (5.76) can be expressed as [18]

$$\ddot{x} + \frac{M'}{M}\dot{x}^2 + \frac{1}{M}\left[\int_0^x M(s)\mathrm{d}s\right] = 0, \tag{5.79}$$

where $M(x) = \sqrt{m(x)}$. So, by choosing $M(x)$ as an even analytic function, and performing its differentiation and integration with respect to x, one can find the differential equation (5.79) with an isochronous centre. Thus, Eq. (5.79) and its other version (5.76) yield a family of conservative isochronous oscillators.

Since the solution for motion of the SHO (5.4) has a general form $A\cos(t+\alpha)$, Eq. (5.75) also defines how x changes with time

$$X = \int_0^x \sqrt{m(s)}\mathrm{d}s = A\cos(t+\alpha), \tag{5.80}$$

where A and α can be found from the initial conditions $x(0)$ and $\dot{x}(0)$. So, not only does this approach yield mechanical and mathematical models of isochronous oscillators, but it also enables one to find their isochronous motion (note that this solution can be implicit). In addition, in case when the isochronous motion exists and when the corresponding initial energy level is h, the energy-conservation law can be used

$$\frac{1}{2}m(x)\dot{x}^2 + E_p(x) = \frac{1}{2}m(x)\dot{x}^2 + \frac{\left[\int_0^x \sqrt{m(s)}\mathrm{d}s\right]^2}{2} = h, \tag{5.81}$$

to define this motion in the phase plane, i.e. to obtain the phase trajectory

$$\dot{x}^2 = \frac{2h - \left[\int_0^x \sqrt{m(s)}\mathrm{d}s\right]^2}{m(x)}. \tag{5.82}$$

A few following examples illustrate potential use and benefits of the theoretical findings presented above.

5.3.1.1 Example 5.1: Oscillators with a Known Mass-Displacement Law

The first example is related to the problem in which the form of the mass-displacement law is known, and the corresponding potential energy and the equation of motion that result in isochronous oscillations are obtained. Let the mass changes in accordance with

$$m(x) = \exp\left(x^{2k}\right), \tag{5.83}$$

where k is a positive integer. This position-dependent mass will result in an odd-powered monomial term in front of $\dot{x}^2$ in Eq. (5.76).

To find the restoring force, the expression (5.75) is solved in terms of the lower incomplete gamma function labelled by γ subsequently [19]

$$X = \int_0^x \exp\left(\frac{s^{2k}}{2}\right) \mathrm{d}s = \frac{(-1)^{-\frac{1}{2k}} 2^{\frac{1}{2k}}}{2k} \gamma\left(\frac{1}{2k}, -\frac{x^{2k}}{2}\right). \tag{5.84}$$

The associate potential energy (5.74) is

$$E_p(x) = \frac{\left[\frac{(-1)^{-\frac{1}{2k}} 2^{\frac{1}{2k}}}{2k} \gamma\left(\frac{1}{2k}, -\frac{x^{2k}}{2}\right)\right]^2}{2}. \tag{5.85}$$

Equation (5.76) now transforms to a closed-form differential equation:

$$\ddot{x} + k x^{2k-1} \dot{x}^2 + \frac{(-1)^{-\frac{1}{2k}} 2^{\frac{1}{2k}}}{2k} \gamma\left(\frac{1}{2k}, -\frac{x^{2k}}{2}\right) \exp\left(-\frac{x^{2k}}{2}\right) = 0. \tag{5.86}$$

Based on Eqs. (5.74) and (5.80), the isochronous oscillations are found to be defined by

$$\frac{(-1)^{-\frac{1}{2k}} 2^{\frac{1}{2k}}}{2k} \gamma\left(\frac{1}{2k}, -\frac{x^{2k}}{2}\right) = A \cos(t + \alpha). \tag{5.87}$$

Further, using the series representation for the lower incomplete gamma function [19, 20]

$$\gamma\left(\frac{1}{2k}, -\frac{x^{2k}}{2}\right) = \exp\left(\frac{x^{2k}}{2}\right) \left(-\frac{x^{2k}}{2}\right)^{\frac{1}{2k}} \Gamma\left(\frac{1}{2k}\right) \sum_{n=0}^{\infty} \frac{\left(-\frac{x^{2k}}{2}\right)^n}{\Gamma\left(\frac{1}{2k} + n + 1\right)}, \tag{5.88}$$

the restoring force can be expressed in a polynomial form, so that the equation yielding isochronous oscillations (5.86) becomes

$$\ddot{x} + k x^{2k-1} \dot{x}^2 + \frac{\Gamma\left(\frac{1}{2k}\right)}{2k} \sum_{n=0}^{\infty} \frac{(-1)^n x^{2k \cdot n+1}}{2^n \ \Gamma\left(\frac{1}{2k} + n + 1\right)} = 0. \tag{5.89}$$

Now, it is easy to compare Eq. (5.89) with (5.11) and to recognize that $p(x)$ and $q(x)$ are polynomials with odd degree and that $p(x)$ has a positive leading coefficient because $k>0$. Therefore, the isochronous centre is global.

To illustrate these results, the mass variation (5.83) is plotted in Fig. 5.7a for $k = 2$. The associated potential energy (5.85) is shown in Fig. 5.7b, from which it is seen that the potential energy is single welled. The solution for time response obtained numerically by integrating directly Eq. (5.86) for $x(0) = 1$ and $\dot{x}(0) = 0$ is plotted in Fig. 5.7c in red dots. The analytical solution corresponding to these initial conditions is given by Eq. (5.87) with $A = 1.1156579646$, $\alpha = 0$ and is plotted as a solid black line. These two types of solutions coincide. In addition, phase trajectories obtained numerically from Eq. (5.86) and based on Eq. (5.82) for $h = 0.622346347$ are plotted in Fig. 5.7d and demonstrate perfect matching. Figure 5.7c also contains the solutions shown for two additional pairs of the initial conditions $x(0) = 0.1; 0.5$

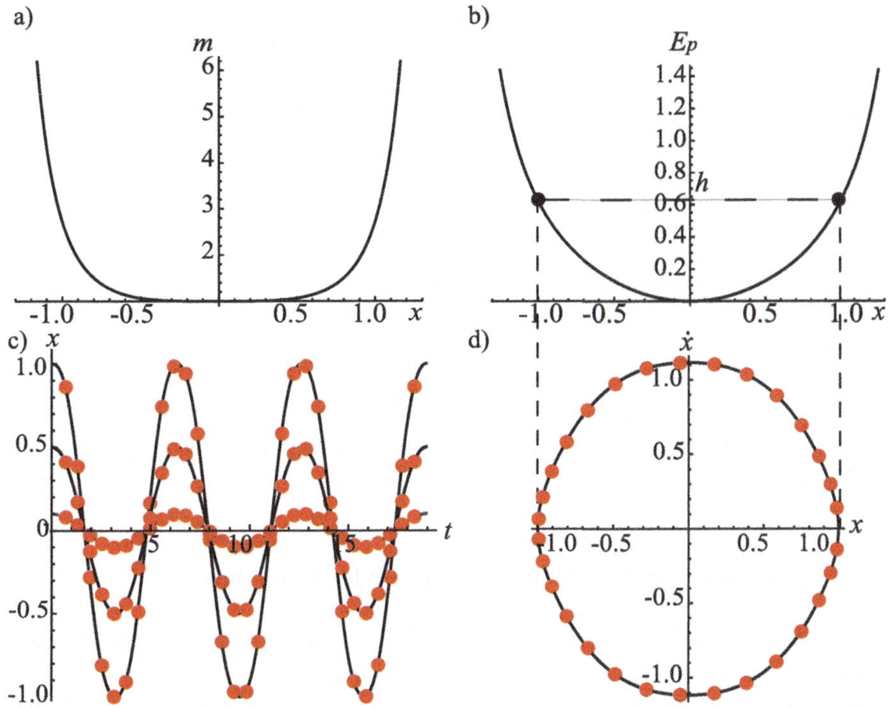

Fig. 5.7 Example 5.1 for $k = 2$: **a** mass variation (5.83); **b** potential energy (5.85); **c** time response obtained numerically from Eq. (5.86) for $x(0) = 0.1; 0.5; 1$, $\dot{x}(0) = 0$ (red dots), and the analytical solution Eq. (5.87) with $A = 0.100001; 0.503152; 1.1156579646$, $\alpha = 0$ (solid black line); phase trajectories obtained numerically from Eq. (5.86) (red dots) and based on Eq. (5.82) for $h = 0.622346347$ (solid black line)

and $\dot{x}(0) = 0$. As it can be seen, all the resulting time histories have the same, constant period.

It is interesting to note that when $k = 1$, the lower incomplete gamma function turns into the imaginary error function [21], i.e.

$$X = \int_0^x \exp\left(\frac{s^2}{2}\right) \mathrm{d}s = \frac{-i}{\sqrt{2}} \gamma\left(\frac{1}{2}, -\frac{x^2}{2}\right) = \sqrt{\frac{\pi}{2}} \operatorname{erfi}\left(\frac{x}{\sqrt{2}}\right), \tag{5.90}$$

with the potential energy

$$E_p(x) = \frac{\pi}{4} \operatorname{erfi}^2\left(\frac{x}{\sqrt{2}}\right), \tag{5.91}$$

and the equation of motion

$$\ddot{x} + x\dot{x}^2 + \sqrt{\frac{\pi}{2}} \exp\left(-\frac{x^2}{2}\right) \operatorname{erfi}\left(\frac{x}{\sqrt{2}}\right) = 0. \tag{5.92}$$

Given Eq. (5.88), the expression for the restoring force can be written down as

$$F_r = \frac{\Gamma\left(\frac{1}{2}\right)}{2} \sum_{n=0}^{\infty} \frac{(-1)^n x^{2n+1}}{2^n \quad \Gamma\left(\frac{1}{2} + n + 1\right)} = \sum_{n=0}^{\infty} \frac{(-1)^n x^{2n+1}}{(2n+1)!!}, \tag{5.93}$$

and the corresponding equation of motion becomes

$$\ddot{x} + x\dot{x}^2 + \sum_{n=0}^{\infty} \frac{(-1)^n x^{2n+1}}{(2n+1)!!} = 0. \tag{5.94}$$

The results (5.91)–(5.94) agree with the results obtained in Sect. 5.2.2 and [14] by using a perturbation method.

5.3.1.2 Example 5.2: Oscillators with a Fixed Restoring Force

Equations (5.76) and (5.79) can be used to find the mass-displacement characteristic of an isochronous oscillator with a fixed nonlinearity, i.e. with the known form of the restoring force. To illustrate this, it is assumed that the restoring force has the following form:

$$F_r\,(x) \equiv q\,(x) = x + x^{2n+1}, \tag{5.95}$$

where n is a positive integer.

Identifying from Eq. (5.79) that

$$\frac{1}{M}\left[\int_0^x M(s)\mathrm{d}s\right] = x + x^{2n+1}, \tag{5.96}$$

and solving this for $M\,(x)$, one can obtain

$$M\,(x) = \left(1 + x^{2n}\right)^{-\frac{2n+1}{2n}}. \tag{5.97}$$

Equation (5.79) gives

$$\ddot{x} - (2n+1)\frac{x^{2n-1}}{1 + x^{2n}}\dot{x}^2 + x + x^{2n+1} = 0. \tag{5.98}$$

When $n = 1$, the restoring force (5.95) is of the Duffing hardening type (see Chap. 2). The mass should change in accordance with

$$m(x) = \frac{1}{(1+x^2)^3}, \tag{5.99}$$

and the potential energy should be

$$E_p(x) = \frac{x^2}{2(1+x^2)}. \tag{5.100}$$

Then, the equation of motion is

$$\ddot{x} - \frac{3x}{1+x^2}\dot{x}^2 + x + x^3 = 0, \tag{5.101}$$

with the isochronous oscillations being defined by

$$\frac{x}{\sqrt{1+x^2}} = A\cos(t+\alpha). \tag{5.102}$$

Figure 5.8a, b shows, respectively, the mass-displacement expression (5.99) and the potential energy (5.100). Note that for larger $|x|$, mass decreases considerably and

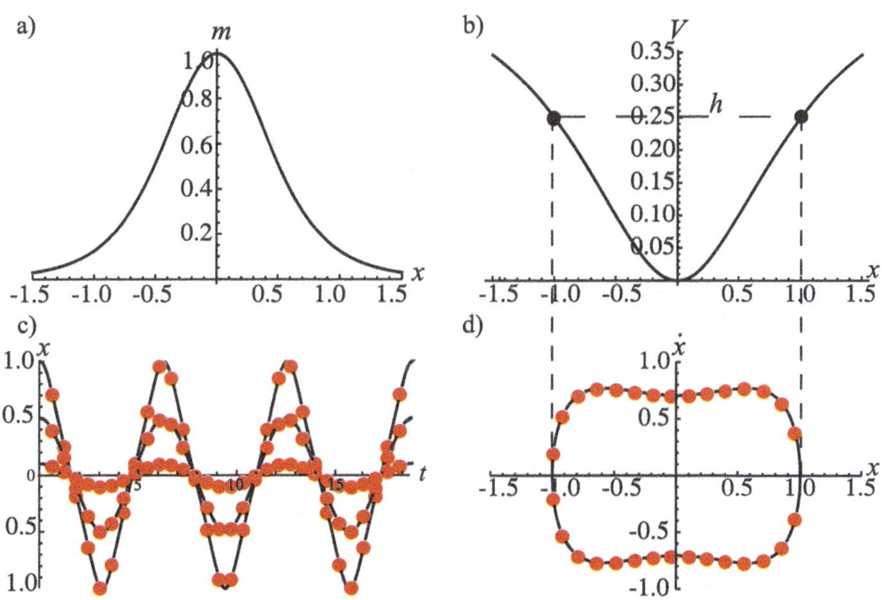

Fig. 5.8 Example 5.2: **a** the mass-displacement expression (5.99); **b** potential energy (5.100); **c** numerically obtained time response from Eq. (5.101) for $x(0) = 1$ and $\dot{x}(0) = 0$ (red dots), and the analytical solution given by Eq. (5.102) with $A = 0.0995037; 0.447214; 1/\sqrt{2}$, $\alpha = 0$ (black solid line); **d** phase trajectories obtained numerically from Eq. (5.101) and based on Eq. (5.82) for $h = 1/4$

becomes very small. A numerically obtained time response from Eq. (5.101) is plotted in Fig. 5.8c in red dots for $x(0) = 1$ and $\dot{x}(0) = 0$, while the analytical solution given by Eq. (5.102) with $A = 1/\sqrt{2}$, $\alpha = 0$ is depicted by a solid black line. Phase trajectories obtained numerically from Eq. (5.101) and based on Eq. (5.82) for $h = 1/4$ and plotted in Fig. 5.8d. Both Fig. 5.8c, d validates the analytical results derived. Figure 5.8c also includes the analytical and numerical solutions shown for two additional pairs of the initial conditions $x(0) = 0.1; 0.5$ and $\dot{x}(0) = 0$. All these time histories have the same, constant period. This confirms that the period/frequency are amplitude-independent.

5.3.1.3 Example 5.3: Oscillators with Prescribed Motion

The results presented in Sect. 5.3.1 can also be used to determine if there is an isochronous oscillator having the motion of the given form and, if there is, to find its mechanical and mathematical model. Such situation arises, for example, if the prescribed motion is given by

$$x = \sinh(\cos(t)). \tag{5.103}$$

Based on Eq. (5.80), one follows

$$X = \operatorname{arc\,sinh}(x). \tag{5.104}$$

The mass-displacement law is obtained first by differentiating the right-hand side of Eq. (5.104) and then squaring what has been obtained

$$m(x) = \frac{1}{1+x^2}, \tag{5.105}$$

while the potential energy should be

$$E_p(x) = \frac{(\operatorname{arc\,sinh}(x))^2}{2}. \tag{5.106}$$

The corresponding equation of motion is

$$\ddot{x} - \frac{x}{1+x^2}\dot{x}^2 + \sqrt{1+x^2}\operatorname{arc\,sinh}(x) = 0. \tag{5.107}$$

The time response obtained numerically by integrating directly Eq. (5.107) for $x(0) = \sinh 1 = 1.1752011936$ and $\dot{x}(0) = 0$ is shown in Fig. 5.9a in red dots, while the analytical solution (5.103) is plotted as a solid black line. A perfect match is seen. Figure 5.9b shows how the mass (5.105) changes with time, illustrating that it corresponds to periodically varying, i.e. pulsating mass with a period smaller than the one of the response.

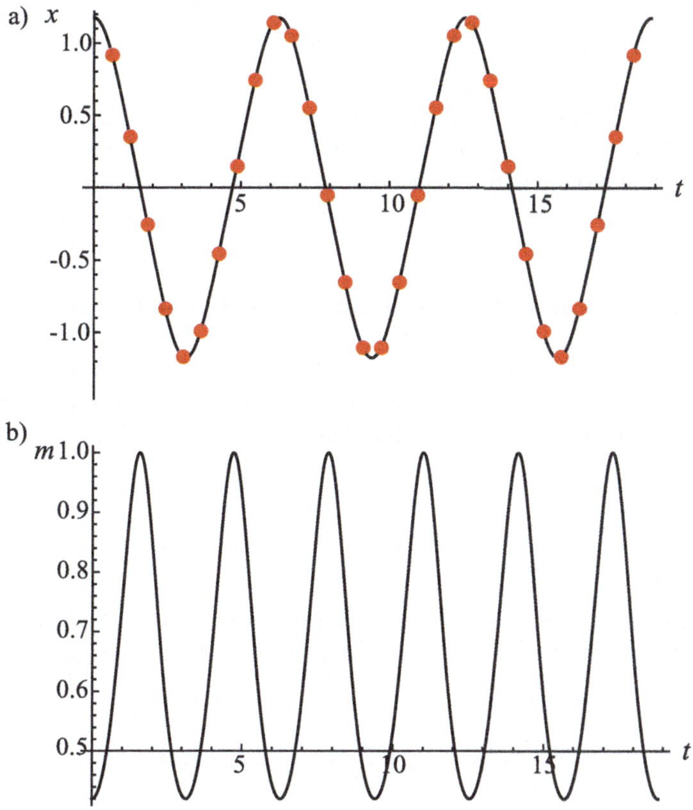

Fig. 5.9 Example 5.3: **a** time response obtained numerically from Eq. (5.107) for $x\,(0) = 1.1752011936$ and $\dot{x}\,(0) = 0$ (red dots), and the analytical solution (5.103) (solid black line); **b** time-varying mass plotted based on Eq. (5.105)

5.3.2 *On Some Other Oscillators with Position-Dependent Coefficient of the Kinetic Energy*

Besides conservative oscillators with the position-dependent mass analysed previously, there is another family of conservative oscillators that can be considered in the same context as above. Their Lagrangian has the form

$$L = \frac{1}{2}\tilde{E}_k\,(x)\,\dot{x}^2 - E_p(x), \tag{5.108}$$

where $\tilde{E}_k\,(x)$ is the coefficient of the kinetic energy $E_k(x, \dot{x}) = \frac{1}{2}\tilde{E}_k\,(x)\,\dot{x}^2$. Note that here one does not associate $\tilde{E}_k\,(x)$ with the mass of the system under consideration, but assume that it exists due to the geometry of motion.

The Lagrange's equation corresponding to the Lagrangian (5.108) is

$$\ddot{x} + \frac{\tilde{E}_k'}{2\tilde{E}_k}\dot{x}^2 + \frac{E_p'}{\tilde{E}_k} = 0. \tag{5.109}$$

The form of this equation can directly be related to Eq. (5.11) and Sabatini's results [13]. Thus, the cases with $\tilde{E}_k$ and E_p being even functions in x are seen as leading to the equation of motion with an isochronous solution. The following examples are to illustrate two cases with such properties and to demonstrate some further benefits.

5.3.2.1 Example 5.4: A Mechanism with Two Sliders and a Spring

Let us consider the mechanism shown in Fig. 5.10: the sliders A and B of equal mass m are connected by a light rigid bar of length l and move with negligible friction in the slots shown, both of which are in a horizontal plane; the slider A is also connected with a spring. The kinetic energy of this system is given by

$$E_k(x, \dot{x}) = \frac{1}{2}\tilde{E}_k(x)\,\dot{x}^2, \quad \tilde{E}_k(x) = \frac{ml^2}{l^2 - x^2}. \tag{5.110}$$

Assuming that the spring is linear and has a stiffness k, and introducing the non-dimensional variables $\bar{x} = x/l$, $\bar{t} = \sqrt{k/m}t$, the corresponding equation of motion is

$$\frac{\mathrm{d}^2\bar{x}}{\mathrm{d}\bar{t}^2} + \frac{\bar{x}}{1 - \bar{x}^2}\left(\frac{\mathrm{d}\bar{x}}{\mathrm{d}\bar{t}}\right)^2 + \bar{x} - \bar{x}^3 = 0. \tag{5.111}$$

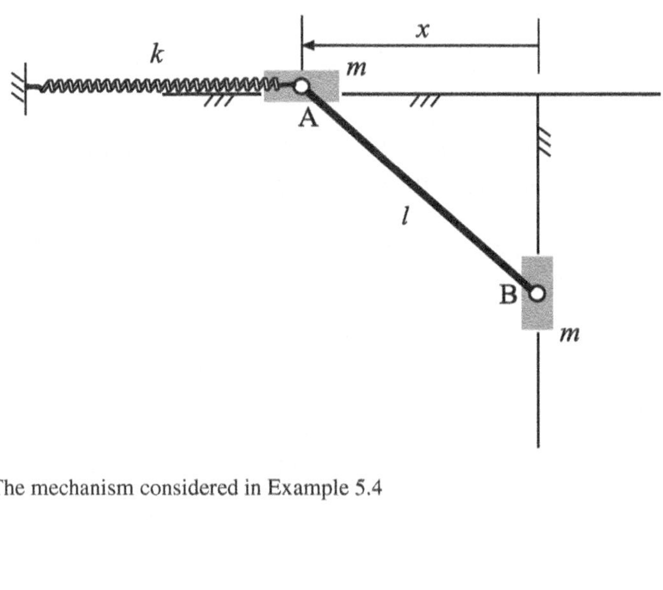

Fig. 5.10 The mechanism considered in Example 5.4

Note that unlike Example 5.2, but here one deals with the softening Duffing nonlinearity. For $\bar{x} \neq 1$, Eq. (5.111) satisfies the conditions for the existence of isochronous oscillations. However, the restoring force and the coefficient in front of the square of the generalized velocity do not match the form (5.79) with the solution (5.80). Here one can pose a question of the form of the potential energy and the restoring force for which the equation of motion will correspond to Eq. (5.79). Thus, by calculating $M(\bar{x})$ from $M'/M = \bar{x}/(1-\bar{x}^2)$, Eq. (5.79) leads to

$$\frac{\mathrm{d}^2\bar{x}}{\mathrm{d}\bar{t}^2} + \frac{\bar{x}}{1-\bar{x}^2}\left(\frac{\mathrm{d}\bar{x}}{\mathrm{d}\bar{t}}\right)^2 + \sqrt{1-\bar{x}^2}\arcsin\bar{x} = 0. \tag{5.112}$$

Equation (5.80) yields its isochronous solution

$$\arcsin\bar{x} = A\cos\left(\bar{t}+\alpha\right). \tag{5.113}$$

It should be noted that the restoring force in Eq. (5.112) can be approximated as follows:

$$\sqrt{1-\bar{x}^2}\arcsin\bar{x} \approx \bar{x} - \frac{\bar{x}^3}{3}, \tag{5.114}$$

so that Eq. (5.112) becomes

$$\frac{\mathrm{d}^2\bar{x}}{\mathrm{d}\bar{t}^2} + \frac{\bar{x}}{1-\bar{x}^2}\left(\frac{\mathrm{d}\bar{x}}{\mathrm{d}\bar{t}}\right)^2 + \bar{x} - \frac{\bar{x}^3}{3} = 0. \tag{5.115}$$

Comparing Eq. (5.115) with (5.111) one concludes that in case of small oscillations, the isochronous solution of the former can be taken as a good approximate solution of the latter. To confirm this, Eqs. (5.111) and (5.115) are solved numerically for $\bar{x}\,(0) = 0.1$ and $\frac{d\bar{x}}{d\bar{t}}\,(0) = 0$ and plotted in Fig. 5.11. In addition, the solution of Eq. (5.113) with $A = \arcsin(0.1)$, $\alpha = 0$ is also shown. It is seen that these solutions agree well.

5.3.2.2 Example 5.5: Huygens' Isochronous Pendulum

It was Huygens [6, 7] who showed that if a pendulum of length l and mass m wraps around a cycloid [7]

$$x = \frac{l}{4}\left(\theta - \sin\theta\right),$$
$$y = -\frac{l}{4}\left(\cos\theta - 1\right),$$

it performs isochronous oscillations. The corresponding parametric equations of motion of the bob are [7]

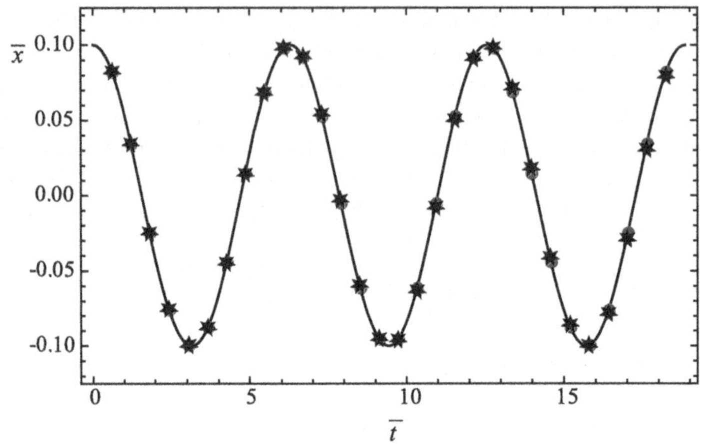

Fig. 5.11 Example 5.4: **a** time response obtained numerically from Eq. (5.115) for $\bar{x}(0) = 0.1$ and $\frac{d\bar{x}}{d\bar{t}}(0) = 0$ (red dots) and from Eq. (5.111) (black stars); analytical solution (5.113) with $A = \arcsin(0.1)$, $\alpha = 0$ (solid black line)

$$X = \frac{l}{4}(\theta + \sin\theta), \tag{5.116}$$

$$Y = -\frac{l}{4}(3 + \cos\theta), \tag{5.117}$$

Both of these cycloids are shown in Fig. 5.12 for l=4 and for $-\pi \leq \theta \leq \pi$.

The corresponding kinetic energy

$$E_k = \frac{1}{2}m(\dot{X}^2 + \dot{Y}^2), \tag{5.118}$$

is

$$E_k = \frac{1}{2}m\left(\frac{l}{2}\cos\frac{\theta}{2}\right)^2\dot{\theta}^2. \tag{5.119}$$

It is seen that the kinetic energy (5.119) of Huygens' pendulum has a position-dependent coefficient.

Its potential energy

$$E_p = mgY, \tag{5.120}$$

has the form

$$E_p = -mg\frac{l}{4}(3 + \cos\theta). \tag{5.121}$$

On the other hand, a Simple Pendulum (SP) of the same length and the same mass has a constant period if it performs small oscillations. Its kinetic energy is of the form

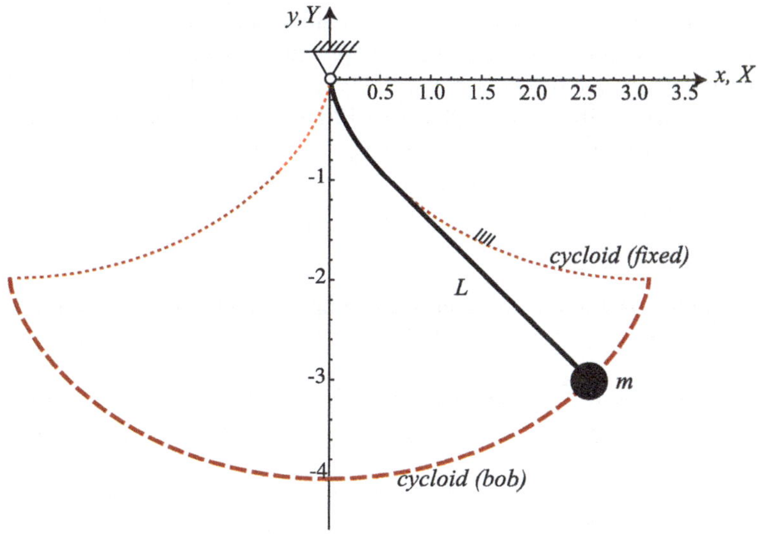

Fig. 5.12 Huygens' isochronous pendulum in motion

$$E_{k\mathrm{SP}} = \frac{1}{2}m(l\dot{\varphi})^2, \tag{5.122}$$

and the potential energy is

$$E_{p\mathrm{SP}} = -mgl\left(1 - \frac{\varphi^2}{2}\right). \tag{5.123}$$

The equality between two expressions for the kinetic energy (5.119) and (5.122) yields

$$\dot{\varphi} = \frac{1}{2}\dot{\theta}\cos\frac{\theta}{2}. \tag{5.124}$$

This is satisfied for

$$\varphi = \sin\frac{\theta}{2}. \tag{5.125}$$

By using (5.125), the expression for the potential energy E_p (5.121) becomes

$$E_p = -mgl\left(1 - \frac{1}{2}\sin^2\frac{\theta}{2}\right) = -mgl\left(1 - \frac{\varphi^2}{2}\right), \tag{5.126}$$

i.e. it transforms to the potential energy of the simple pendulum (5.123), which confirms that Huygens' pendulum belongs to a wide class of isochronous oscillators that can be transformed into simple harmonic oscillators by establishing the equivalence between their Lagrangians or the kinetic and potential energies.

Let us now define the problem in a deductive way: find the form of X and Y for which the system described by (5.118) and (5.120) performs isochronous oscillations. To that end, one can make (5.120) equal to (5.123), which leads to

$$Y = l\left(\frac{\varphi^2}{2} - 1\right). \tag{5.127}$$

Knowing that the solution for motion of the simple pendulum has the form

$$\varphi = A\sin\left(\sqrt{\frac{g}{l}}t + \varphi\right) = A\sin\psi, \tag{5.128}$$

one has

$$Y = l\left(\frac{(A\sin\psi)^2}{2} - 1\right). \tag{5.129}$$

We also make (5.118) equal to (5.122), deriving

$$\dot{X}^2 = l^2\dot{\varphi}^2 - \dot{Y}^2 = l^2\dot{\varphi}^2\left(1 - \varphi^2\right), \tag{5.130}$$

or

$$X = l\int_0^{\varphi}\sqrt{1 - \varphi^2}\mathrm{d}\varphi. \tag{5.131}$$

Integrating (5.131), one obtains

$$X = \frac{l}{2}\left(\varphi\sqrt{1 - \varphi^2} + \arcsin\varphi\right). \tag{5.132}$$

This is a conditional expression for which $-1 \leq A \leq 1$. Taking, $A = 1$, one follows

$$X = \frac{l}{2}\left(\sin\psi\sqrt{1 - (\sin\psi)^2} + \arcsin\left(\sin\psi\right)\right), \tag{5.133}$$

so that the solutions for X and Y become equal to Eqs. (5.116) and (5.117) with $2\psi = \theta$. Thus, this transformation approach presented in this paper obviously gives the same results as derived in [7].

5.4 Energy Considerations II

All the equations of motion of isochronous systems considered above contain the term proportional to the square of the generalized velocity. The goal of this section is to use the energy approach to derive those equations of motion that contain the term linear in the generalized velocity:

$$\ddot{x} + G(x)\dot{x} + F_r(x) = 0, \tag{5.134}$$

and to find the corresponding Lagrangians, conservation laws, as well as solutions for motion.

In this case one can consider the system whose potential and kinetic energy are

$$E_p = \frac{1}{2}X^2 = \frac{1}{2}(x\,f)^2, \quad E_k = \frac{1}{2}\dot{X}^2 = \frac{1}{2}\left(\dot{x}\,f + x^2\,f\,'\right)^2, \tag{5.135}$$

where $f \equiv f(I)$, $I = \int_0^t x(t)\mathrm{d}t$ and $f\,' = \mathrm{d}f/\mathrm{d}I$.

The corresponding Lagrange's equation is

$$\ddot{x} + 3x\dot{x}\frac{f\,'}{f} + x + \frac{f\,''}{f}x^3 = 0. \tag{5.136}$$

This system has two independent first integrals. The first one is the energy conservation law stemming from $\dot{X}^2/2 + X^2/2 = const.$:

$$\left(\dot{x}\,f + x^2\,f\,'\right)^2 + (x\,f)^2 = h_1, \quad h_1 = const. \tag{5.137}$$

The other first integral is related to the principle of conservation of momentum for the SHO: $\dot{X} + \int_0^t X\mathrm{d}t = \dot{X}(0) = const.$ By using X and $\dot{X}$ from (5.135), and knowing that $dI/dt = x$, one can obtain

$$\dot{x}\,f + x^2\,f\,' + \int f(I)\,\mathrm{d}I = h_2, \quad h_2 = const. \tag{5.138}$$

In addition, as the solution for motion for the SHO can be written down as $X = a\sin(t+\alpha)$, the following should be satisfied:

$$a\sin(t+\alpha) = x\,f, \quad a\cos(t+\alpha) = \dot{x}\,f + x^2\,f\,'. \tag{5.139}$$

For the hardening-type nonlinearity in Eq. (5.136), one requires $f\,''/f = 1$, which leads to

$$f_{\mathrm{HDO}} = \exp\left(\int_0^t x(t)\,\mathrm{d}t\right), \tag{5.140}$$

and the equation of motion takes the form [22]

$$\ddot{x} + 3x\dot{x} + x + x^3 = 0. \tag{5.141}$$

Two first integrals (5.137) and (5.138) are

$$\exp(2x_1)\left[\left(x_3 + x_2^2\right)^2 + x_2^2\right] = h_1, \tag{5.142}$$

and

$$\exp(x_1)\left(x_3 + x_2^2 + 1\right) = h_2, \tag{5.143}$$

where

$$x_1 = \int_0^t x(t)\,\mathrm{d}t, \quad x_2 = \dot{x}_1 = x, \quad x_3 = \dot{x}_2 = \dot{x}, \tag{5.144}$$

with initial conditions being:

$$x_1(0) = 0, \quad x_2(0) = \dot{x}_1(0) = A, \quad x_3(0) = \ddot{x}_1(0) = 0. \tag{5.145}$$

Equation (5.142) is plotted in Fig. 5.13a for $h_1 = 1$. To analyse phase trajectories in more detail, Eq. (5.143) is squared and divided by Eq. (5.142) to obtain

$$\frac{\left(x_3 + x_2^2 + 1\right)^2}{\left(x_3 + x_2^2\right)^2 + x_2^2} = B, \quad B = const. \tag{5.146}$$

This expression agrees with the first integral obtained and studied in [9], and is plotted in Fig. 5.13b, where periodic solutions correspond to the case $B > 1$. Note that for the initial conditions (5.145), one has $B = 1 + 1/A^2$, which implies that B is always higher than unity.

By using (5.139) and (5.140) one can derive

$$\dot{x}\,\sin(t+\alpha) + x^2\sin(t+\alpha) - x\cos(t+\alpha) = 0\,. \tag{5.147}$$

Its solution satisfying the initial conditions is

$$x = \frac{\sin\left(t + \arctan\frac{1}{A}\right)}{\sqrt{1 + \frac{1}{A^2}} - \cos\left(t + \arctan\frac{1}{A}\right)}\,. \tag{5.148}$$

To compare the analytical solution for motion (5.148) with a numerically obtained solution of the equation of motion (5.141), the latter is written down in the form [22]

$$\dddot{x}_1 + 3\dot{x}_1\ddot{x}_1 + \dot{x}_1 + \dot{x}_1^3 = 0, \tag{5.149}$$

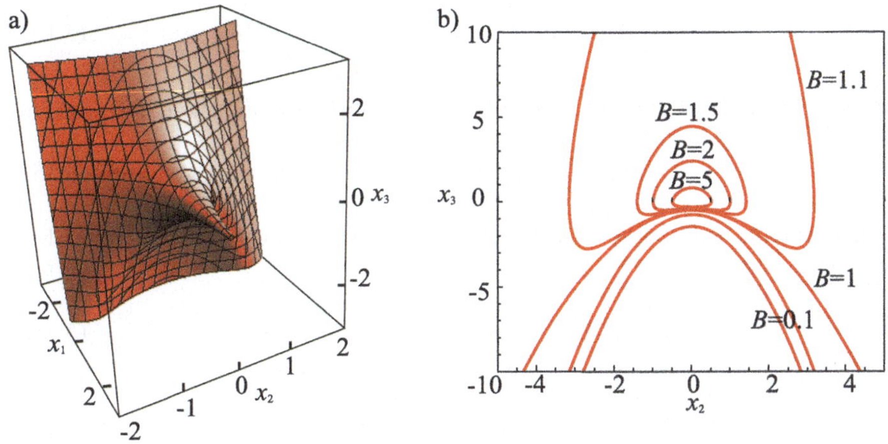

Fig. 5.13 **a** 3D plot of Eq. (5.142); **b** phase trajectories obtained from Eq. (5.146) for different values of B

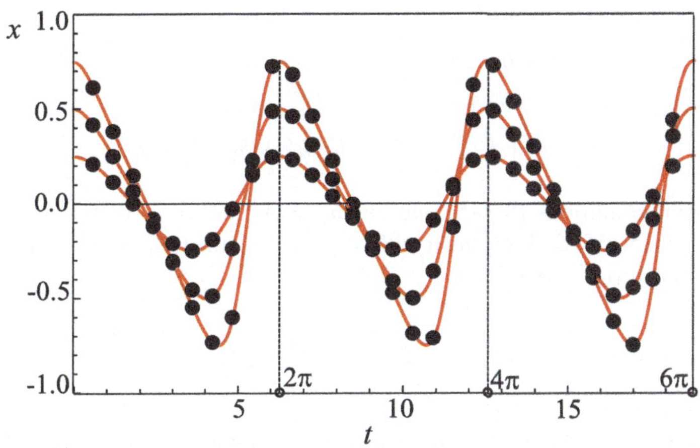

Fig. 5.14 Time response of the HDO, Eq. (5.141) for $A = 0.25; 0.5; 0.75$: numerically obtained solution from Eq. (5.149) (black dots) and from Eq. (5.148) (blue solid line)

and numerically integrated by using the initial conditions (5.145). This comparison, plotted in Fig. 5.14, shows that these two types of solution are in full agreement as well as that the period is amplitude-independent.

The equation of motion (5.136) corresponds to the SDO if $f''/f = -1$, which is satisfied for

$$f_{\mathrm{SDO}} = \cos\left(\int_0^t x(t)\,\mathrm{d}t\right). \tag{5.150}$$

This equation of motion is now given by

$$\ddot{x} - 3x\dot{x}\tan\left(\int_0^t x(t)\,\mathrm{d}t\right) + x - x^3 = 0. \tag{5.151}$$

By using the notation given in Eq. (5.144), the equation of motion (5.151) transforms to

$$\dddot{x}_1 - 3\dot{x}_1\ddot{x}_1\tan x_1 + \dot{x}_1 - \dot{x}_1^3 = 0, \tag{5.152}$$

with the initial conditions given in Eq. (5.145). Two first integrals (5.137) and (5.138) become

$$\left(x_3\,\cos x_1 - x_2^2\,\sin x_1\right)^2 + x_2^2\,\cos^2 x_1 = h_1, \tag{5.153}$$

and

$$x_3\,\cos x_1 - x_2^2\,\sin x_1 + \sin x_1 = h_2. \tag{5.154}$$

These two integrals can be manipulated to exclude x_1 and to derive

$$\begin{aligned}
&(x_2^2 - 1)^2(h_1 + (-1 - h_1 + h_2^2)x_2^2 + x_2^4)^2 + \\
&\quad 2(1 - h_1 - h_2^2 + (-1 + h_1 - h_2^2)x_2^2) \\
&\quad (h_1(x_2^2 - 1) - x_2^2(h_2^2 + x_2^2 - 1))x_3^2 + \\
&\quad ((-1 + h_1)^2 - 2(1 + h_1)h_2^2 + h_2^4)x_3^4 = 0.
\end{aligned} \tag{5.155}$$

For the initial conditions (5.145) one has $h_1 = A^2$ and $h_2 = 0$. Introducing these values into Eq. (5.155) and solving it with respect to x_3, the following explicit solution for phase trajectories is obtained:

$$x_3 = \pm(x_2^2 - 1)\sqrt{\frac{A^2 - x_2^2}{1 - A^2}}. \tag{5.156}$$

Combining equations in (5.139) and using $a = A$ and $\alpha = \pi/2$, one can derive

$$\dot{x}\,A\cos t - x^2\sqrt{x^2 - A^2\cos^2 t} + x\,A\,\sin t = 0\,. \tag{5.157}$$

Its solution satisfying the initial conditions is

$$x = \frac{A\cos t}{\sqrt{1 - A^2\sin^2 t}}. \tag{5.158}$$

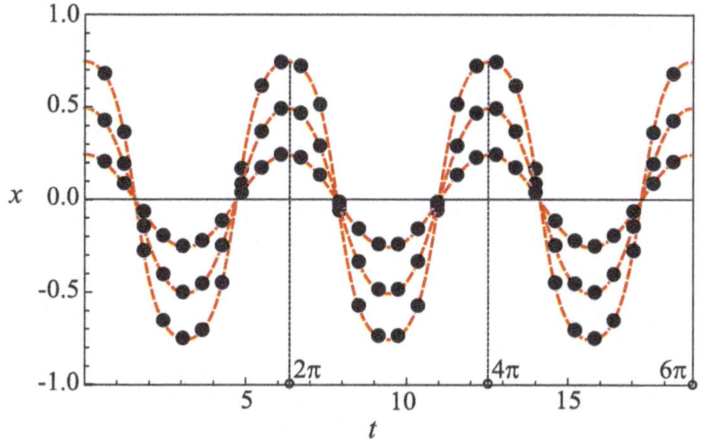

Fig. 5.15 Time response of the SDO, Eq. (5.151) with $A = 0.25; 0.5; 0.75$: numerically obtained solution from Eq. (5.152) (black dots) and from Eq. (5.158) (red dashed line)

This solution is plotted in Fig. 5.15 together with the numerical solution of Eq. (5.151) with (5.145) for different values of A. These solutions coincide and confirm isochronicity.

References

1. Rand RH (2012) Lecture notes on nonlinear vibrations (version 53). http://dspace.library.cornell.edu/handle/1813/28989. Accessed 12 Aug 2012
2. Kovacic I, Brennan MJ (eds) (2011) The Duffing equation: nonlinear oscillators and their behaviour. Wiley, Chichester
3. Kluger JM, Moon FC, Rand RH (2013) Shape optimization of a blunt body vibro-wind galloping oscillator. J Fluids Struct 40:185–200
4. Calogero F (2008) Isochronous systems. Oxford University Press, Oxford
5. http://math.jbpub.com/advancedengineering/docs/Project2.8_TrickyTiming.pdf. Accessed 15 Apr 2020
6. Knoebel A, Laubenbacher R, Lodder J, Pengelley D (2007) Mathematical masterpieces: further chronicles by the explorers. Springer, Berlin
7. http://thep.housing.rug.nl/sites/default/files/users/user12/Huygens_pendulum.pdf. Accessed 15 Feb 2013
8. Sabatini M (1999) On the period function of Liénard systems. J Differ Equ 152:467–487
9. Iacono R, Russo F (2011) Class of solvable nonlinear oscillators with isochronous orbits. Phys Rev E 83, Art. No. 027601
10. Christopher C, Devlin J (2004) On the classification of Liénard systems with amplitude-independent periods. J Differ Equ 200:1–17
11. Chandrasekar VK, Senthilvelan M, Lakshmanan M (2005) Unusual Liénard-type nonlinear oscillator. Phys Rev E 72, Art. No. 066203

12. Chandrasekar VK, Sheeba JH, Pradeep RG, Divyasree RS, Lakshmanan M (2012) A class of solvable coupled nonlinear oscillators with amplitude independent frequencies. Phys Lett A 376:2188–2194
13. Sabatini M (2004) On the period function of $x'' + f(x)x'^2 + g(x) = 0$. J Differ Equ 196:151–168
14. Kovacic I, Rand R (2013) Straight line backbone curve. Commun Nonlinear Sci Numer Simul 18:2281–2288
15. http://mathworld.wolfram.com/Erf.html. Accessed 15 Apr 2020
16. Korn GA, Korn TM (2000) Mathematical handbook for scientists and engineers. Dover Publications Inc., Mineola
17. Nayfeh AH, Mook DT (1995) Nonlinear oscillations. Wiley, New York
18. Kovacic I, Rand R (2013) About a class of nonlinear oscillators with amplitude-independent frequency. Nonlinear Dyn 74:455–465
19. Temme NM (1975) Uniform asymptotic expansions of the incomplete gamma functions and the incomplete beta functions. Math Comput 29:1109–1114
20. Gradshteyn IS, Ryzhik IM (2000) Tables of integrals, series and products. Academic, New York
21. Abramowitz M, Stegun IA (1964) Handbook of mathematical functions: with formulas, graphs, and mathematical tables. Dover Publication, New York
22. Kovacic IN, Rand RH (2014) Duffing-type oscillators with amplitude-independent period. In: Awrejcewicz J (ed) Applied non-linear dynamical systems. Springer proceedings in mathematics and statistics, vol 93. Springer, Cham

Chapter 6
From Chains of Nonlinear Oscillators to Continuous Nonlinear Systems

6.1 Introduction

The previous chapters have been concerned with single degree-of-freedom systems. This chapter extends the considerations related to obtaining the exact or approximate solutions for their oscillatory response to obtaining exact solutions for the systems with more than one degree of freedom, ending up with those having an infinite number of them, i.e. with continuous systems. Thus, the chains of nonlinear oscillators are analysed first from the viewpoint of similar normal modes and the related exact solutions. The number of masses is increased gradually from two to three and then infinitely, so that the longitudinal vibration of continuous systems are dealt with at the end. The exact solutions are based on the considerations presented in Chap. 2 for purely nonlinear and pure cubic oscillators.

6.2 Chains of Purely Nonlinear Oscillators

Chains consisting of coupled oscillators, i.e. spring-mass systems, are considered. Coupled oscillators model a variety of systems, ranging from mathematical biology to physics and engineering [1]. As a results of this fact, they have attracted considerable interest of researchers during the past several decades. However, emerging micro- and nano-electro-mechanical systems have recently renewed this interest [2–5]. In order to model these systems in accordance with experimental results, strong nonlinearity with real-valued powers should be taken into consideration [6–9], which makes the existing approaches less adequate or inadequate. This is related especially to the possibility to determine exact solutions for their free and forced response.

It was Rosenberg who defined the concept of the 'exact steady state' of a strongly nonlinear, undamped, discrete system [10, 11]: the ratio of the response and the amplitude in the steady-state forced response of a single degree of freedom is 'cosine-like' [11] and of the same period as the periodic forcing function. Hsu [12] considered

I. Kovacic, *Nonlinear Oscillations*,
https://doi.org/10.1007/978-3-030-53172-0_6

Duffing-type oscillators and used the fact that these undamped oscillators have exact closed-form solutions in terms of Jacobi elliptic functions. Hsu designed the external excitation to be proportional to the forced displacement and, thus, expressed it also in terms of Jacobi elliptic functions. This idea has been extended to some other undamped nonlinear oscillators with cubic or quadratic nonlinearities that have exact closed-form solutions [13]. Caughey and Vakakis [14] examined two-degree-of-freedom strongly nonlinear systems, where the power of nonlinearity is an odd integer. To find the exact forced response, they expressed the forcing as a certain power function of the forced displacement. This study was concerned with homogenous systems (nonlinearizable, essentially nonlinear), with stiffness nonlinearities proportional to the same power of displacement. Rosenberg [10] found that a system with a homogeneous potential function possesses similar normal mode oscillations, i.e. normal modes with straight-line trajectories in the configuration space. The existence of at least n similar normal modes in n-degree-of-freedom homogeneous systems was proven by van Groesen [15]. The number of similar normal modes, associated bifurcations and stability were addressed in [16–19] (other references can also be found in [19]).

Similar normal modes are investigated in the following section. In Sect. 6.2, the case of pure nonlinearity is treated, where the power of nonlinearity is assumed to be any real number higher than unity, including non-integer and integer numbers [20, 21]. First, two-mass chains of purely nonlinear oscillators are investigated. Exact solutions for the corresponding responses are obtained for free and then also for forced vibrations with a specially designed external excitation. Then, a chain with three and more masses purely nonlinear springs is considered and the corresponding similar normal modes are found and illustrated.

6.2.1 Two-Mass Chain: Free Vibrations

The system of coupled oscillators with two masses ($N = 2$) shown in Fig. 6.1a, where all the springs are purely nonlinear, is governed by

$$\ddot{x} + K_1\, x\, |x|^{\alpha-1} + K_2\, (x-y)\, |x-y|^{\alpha-1} = 0, \tag{6.1}$$

$$\ddot{y} + K_3\, y\, |y|^{\alpha-1} + K_2\, (y-x)\, |y-x|^{\alpha-1} = 0, \tag{6.2}$$

where K_1, K_2 and K_3 are constants, while α is assumed to be a real positive number higher than unity. The absolute value function is used to make the restoring forces be an odd function for all the values of α defined. The two masses are assumed to be equal to the unit mass, but in case this does not hold, an appropriate normalization can be carried out.

Similar normal modes [10] for the system of equations (6.1), (6.2) can exist if $x(t)$ and $y(t)$ are related to each other as follows:

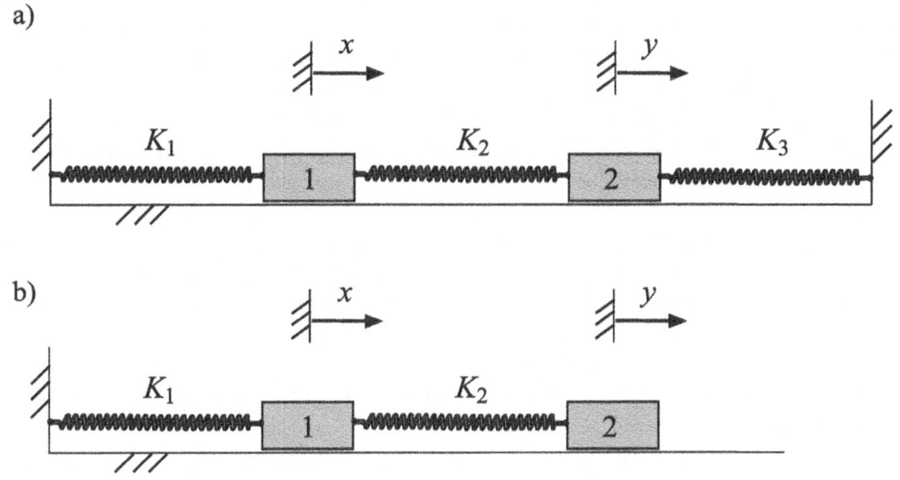

Fig. 6.1 **a** Coupled two-mass system in Case I; **b** Coupled two-mass system in Case II

$$y = Cx, \tag{6.3}$$

where C is a constant parameter to be determined. This yields

$$\ddot{x} + \left[K_1 + K_2 (1 - C) |1 - C|^{\alpha-1}\right] x |x|^{\alpha-1} = 0, \tag{6.4}$$

$$\ddot{x} + \left[K_3 |C|^{\alpha-1} + K_2 \frac{C-1}{C} |C - 1|^{\alpha-1}\right] x |x|^{\alpha-1} = 0. \tag{6.5}$$

The problem can be simplified to a one-degree-of-freedom Purely Nonlinear Oscillator (PNO), discussed in detail in Chap. 2, Sect. 2.5:

$$\ddot{x} + c_\alpha x |x|^{\alpha-1} = 0, \tag{6.6}$$

with c_α being here:

$$c_\alpha = K_1 + K_2 (1 - C) |1 - C|^{\alpha-1}, \tag{6.7}$$

provided that the coefficients in front of the nonlinear terms in Eqs. (6.4), (6.5) are equal, which leads to

$$K_1 + K_2 |1 - C|^{\alpha-1} (1 - C) \frac{1 + C}{C} - K_3 |C|^{\alpha-1} = 0. \tag{6.8}$$

As presented in Chap. 2, Sect. 2.5.2.3, the solution of Eq. (6.6) can be expressed in the form

$$x = A \,\mathrm{ca}\left(\alpha, 1, |A|^{(\alpha-1)/2} \sqrt{\frac{c_\alpha (\alpha + 1)}{2}} t\right), \tag{6.9}$$

where A is the amplitude of vibrations. The first argument in the ca solution (6.9) is the power of nonlinearity α, the second argument is always unity, while the coefficient in front of t in the third argument represents the frequency of the ca function: $\omega_{ca} = |A|^{(\alpha-1)/2}\sqrt{c_\alpha(\alpha+1)/2}$.

In addition, by considering the first integral of Eq. (6.6), one can derive the following expression for the period of vibrations, which is discussed in details in Chap. 2, Sect. 2.5.1:

$$T_{ex} = \frac{4B\left(\frac{1}{\alpha+1}, \frac{1}{2}\right)}{\sqrt{2c_\alpha(\alpha+1)}}|A|^{(1-\alpha)/2} = \sqrt{\frac{8\pi}{c_\alpha(\alpha+1)}}\frac{\Gamma\left(\frac{1}{\alpha+1}\right)}{\Gamma\left(\frac{\alpha+3}{2(\alpha+1)}\right)}|A|^{(1-\alpha)/2}, \tag{6.10}$$

where B is the complete beta function and Γ is the Euler gamma function (see Appendix A).

Note that the expression for the frequency ω_{ca} is the consequence of the fact that the Ateb function $\mathrm{ca}(\alpha, 1, z)$ has the exact period $T_{\mathrm{ca}(\alpha,1,z)} = 2B\left(\frac{1}{\alpha+1}, \frac{1}{2}\right)$ [22], so that one can use $\omega_{ca} = T_{\mathrm{ca}(\alpha,1,z)}/T_{ex}$ and obtain ω_{ca} as given above.

Further, for the Pure Cubic Oscillator (PCO) $\alpha = 3$, and its exact solution for motion can be expressed in the form of Jacobi cn function (see Chap. 2, Sect. 2.3.3 or [23]):

$$x = A\,\mathrm{cn}\left(\sqrt{c_3}At\left|\frac{1}{2}\right.\right), \tag{6.11}$$

$$c_3 = K_1 + K_2(1-C)^3. \tag{6.12}$$

The frequency of the cn function is the coefficient in front of t in the first argument $\omega_{cn} = \sqrt{c_3}A$, while the second argument is the elliptic parameter $m = 1/2$. The frequency ω_{cn} was obtained from the fact that the cn function $\mathrm{cn}(z\,|1/2)$ has the exact period $T_{\mathrm{cn}(z|1/2)} = 4K(1/2)$, where K stands for the complete elliptic integral of the first kind. Consequently, one can derive (after some transformations): $\omega_{cn} = T_{\mathrm{cn}(z|1/2)}/T_{ex} = \sqrt{c_3}A$.

When $\alpha = 1$, one deals with the Simple Harmonic Oscillator (SHO), for which $x = A\cos\left(\sqrt{c_1}t\right), c_1 = K_1 + K_2(1-C)$ and $T_{ex} = 2\pi/\sqrt{c_1}$. All the solutions mentioned hold for the initial conditions $x(0) = A, \dot{x}(0) = 0$ as they are of importance for the subsequent investigations.

In order to determine the solutions for all α and detect how they differ from the one corresponding to $\alpha = 1$, one should calculate the values of C, and this is done herein for two cases: Case I refers to the equal stiffness coefficients of the anchoring springs 1 and 3 (Fig. 6.1a), while Case II is related to the system without the spring connecting the second mass with the base/wall on the right-hand side (Fig. 6.1b).

6.2.1.1 Case I: Equal Stiffness Coefficients of Anchoring Springs

If $K_3 = K_1$, the condition given by Eq. (6.8) leads to

$$1 + \frac{K_2}{K_1} |1 - C|^{\alpha-1} (1 - C) \frac{1 + C}{C} - |C|^{\alpha-1} = 0. \tag{6.13}$$

This expression is used to plot the diagrams shown in Fig. 6.2, which depict how the value of C changes with the ratio $K = K_2/K_1$ for different values of the power α. As each of these diagrams represents the values of C that balance purely nonlinear term in Eqs. (6.4), (6.5), they will be denoted as 'balancing diagrams' (this terminology is taken from [14, 24]).

In case of the linear spring $\alpha = 1$, two values of C exist: $C = 1$ and $C = -1$. The former corresponds to the case when the masses oscillate with the same amplitude and in the same direction (in phase) as they were connected with a rigid link. The latter corresponds to the case when they oscillate with the same amplitude but in the opposite directions (out–of-phase). It is easy to see from Eq. (6.13) that the values $C = \pm 1$ exist also for all α. However, for some other values of α, a subcritical pitchfork bifurcation occurs at the value K^*(Fig. 6.2b, c, d), as a result of which two additional modes bifurcate from $C = -1$, and both of them correspond to a negative C (note that the unstable solutions are labelled by the dashed line, while the stable ones are labelled by the solid line; the stability analyses is carried out by using the methodology and results from [24], but all the details are omitted here). For $K > K^*$, the modes are the same as in the linear system and the mode corresponding to $C = -1$ changes from unstable to stable. Other modes existing for higher values of the power of nonlinearity α are presented in Fig. 6.2e, f, where the stable and unstable solutions are also indicated.

In order to investigate the characteristics of these bifurcations, the value of K^* is examined first. To that end, one can use Eq. (6.13) to express K and calculate

$$K^* = \lim_{C->-1} \frac{|C|^{\alpha-1} - 1}{|1 - C|^{\alpha-1} (1 - C) (1 + C)} C = \frac{\alpha - 1}{2^\alpha}. \tag{6.14}$$

This expression is plotted in Fig. 6.3a. It is interesting to note that the value of $K^* = 1/4$ corresponding to $\alpha = 2$ and $\alpha = 3$ is the same. In addition, the parameter values corresponding to the maximum are

$$\bar{\alpha} = \frac{1 + \log 2}{\log 2} = 2.4427, \quad \bar{K}^* = \frac{1}{e \log 4} = 0.265369. \tag{6.15}$$

When $\alpha < \bar{\alpha}$, the value of K^* increases with α; if $\alpha > \bar{\alpha}$, the higher the value of α, the lower the value of K^*.

The loci of the bifurcation points K^* and K^{**} from Fig. 6.2 are shown in Fig. 6.3b.

The exact solutions for motion can now be obtained by using the expression given in Sect. 6.2.1. For example, taking $\alpha = 5/3$, $K = K_2 = 0.2$, $K_1 = 1$, two bifurcating values of C (see Fig. 6.3b) are calculated from Eq. (6.13): $C = -0.6308$

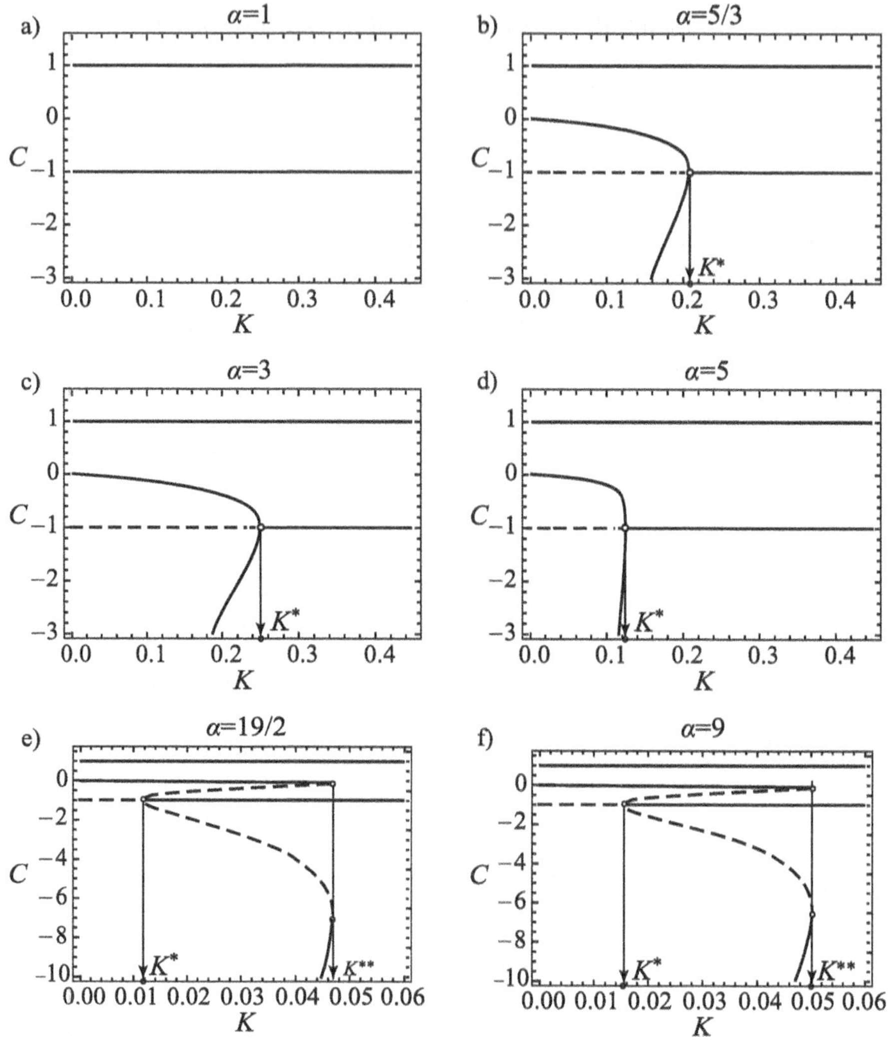

Fig. 6.2 Balancing diagrams $C - K$ for a fixed value of the power α and different values of the ratio $K = K_2/K_1$ in Case I for: **a** $\alpha = 1$; **b** $\alpha = 5/3$; **c** $\alpha = 3$; **d** $\alpha = 5$; **e** $\alpha = 19/2$; **f** $\alpha = 9$

and $C = -1.5851$. Equation (6.9) now gives

$$x = A \operatorname{ca}\left(\frac{5}{3}, 1, |A|^{1/3} \sqrt{\frac{4}{3}\left(1 + 0.2\,(1 - C)\,|1 - C|^{2/3}\right)t}\right). \tag{6.16}$$

The solution for y is defined by Eq. (6.3). The resulting responses are presented in Fig. 6.4a, b for both modes and for $A = 1$. To provide a further insight into the

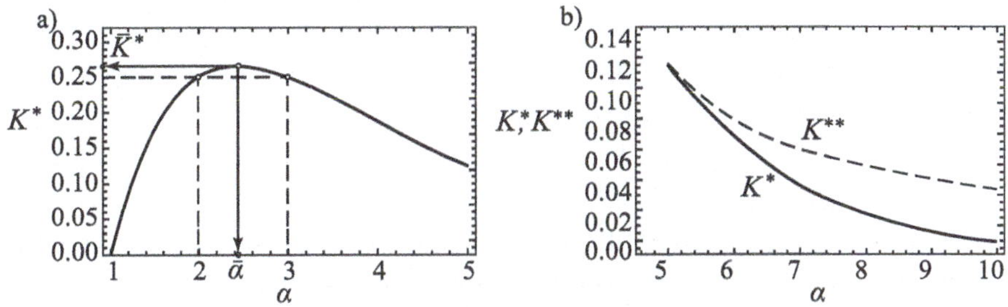

Fig. 6.3 **a** Bifurcation point K^* as a function of the power α in Case I; **b** Curves corresponding to K^* and K^{**} meeting at a cusp

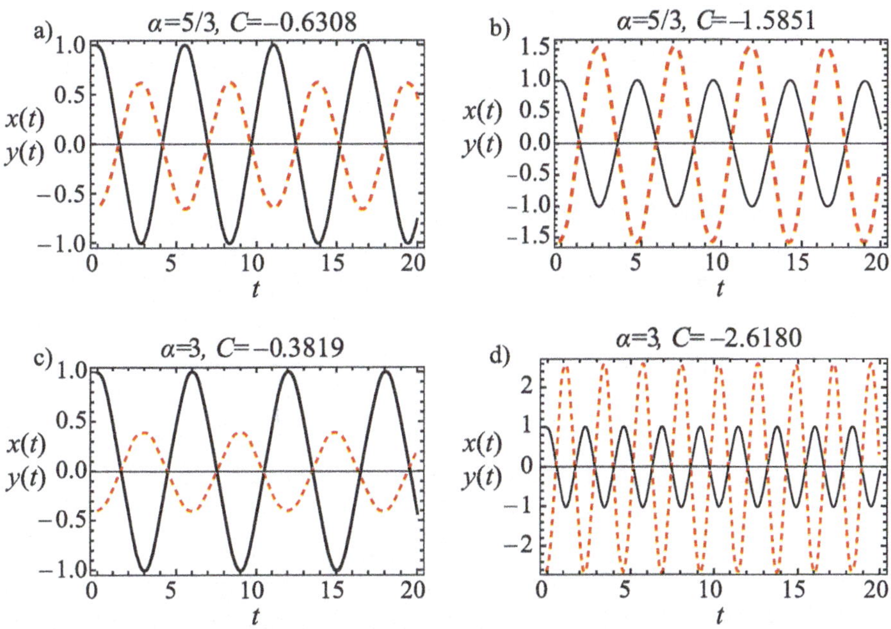

Fig. 6.4 Time-history diagrams for x (black solid line) and y (red dashed line) corresponding to $K = K_2 = 0.2$, $K_1 = 1$, $A = 1$ in Case I for: **a** $\alpha = 5/3$, $C = -0.6308$; **b** $\alpha = 5/3$, $C = -1.5851$; **c** $\alpha = 3$, $C = -0.3819$; **d** $\alpha = 7/2$, $C = -2.6180$

physical meaning of this solution, one can use the Fourier series expansion from Appendix C and [25] to find out, for example, that the solution (6.16) corresponding to Fig. 6.4a consists of the following harmonics:

$$\begin{aligned} x \approx\ & 0.982522\cos(1.13364t) + 0.0182686\cos(3.40092t) - \\ & 0.000939578\cos(5.66819t) + 0.000190823\cos(7.93547t) - \\ & 0.0000583094\cos(10.2027t)\,. \end{aligned} \tag{6.17}$$

In the second example, it is assumed that $\alpha = 3$, $K = K_2 = 0.2$, $K_1 = 1$. Equation (6.13) gives $C = -0.3819$ and $C = -2.6180$. Equation (6.11) now becomes

$$x = A \operatorname{cn}\left(\sqrt{1 + 0.2\,(1 - C)^3}\,At \left| \frac{1}{2} \right.\right). \tag{6.18}$$

The corresponding time-history diagrams for x and y are shown in Fig. 6.4c, d for both modes and for $A = 1$. The Fourier series expansion from Appendix C can be used to present the solution (6.18) corresponding to Fig. 6.4c as follows

$$\begin{aligned} x \approx\; & 0.955006\cos(1.04721t) + 0.0430495\cos(3.14164t) + \\ & 0.0018605\cos(5.23606t) + 0.0000804069\cos(7.33049t) + \\ & 3.4794 \times 10^{-6}\cos(9.42492t). \end{aligned} \tag{6.19}$$

6.2.1.2 Case II: No Right Anchoring Spring

If the right-hand side spring is removed (Fig. 6.1b), the equations of motion have the form

$$\ddot{x} + K_1\, x\, |x|^{\alpha-1} + K_2\, (x - y)\, |x - y|^{\alpha-1} = 0, \tag{6.20}$$

$$\ddot{y} + K_2\, (y - x)\, |y - x|^{\alpha-1} = 0, \tag{6.21}$$

which together with Eq. (6.3), yields the following condition:

$$1 + \frac{K_2}{K_1}\, |1 - C|^{\alpha-1}\, (1 - C)\, \frac{1 + C}{C} = 0. \tag{6.22}$$

This expression is used to plot the $K - C$ balancing diagrams in Fig. 6.5a, b, c, d, where again $K = K_2/K_1$. There are several distinctive and different features with respect to Case I. Namely, in Case II, there is no fixed value of C, but it changes both with K and α. The type of bifurcation is also different. As seen in Fig. 6.5c, d, a saddle-node bifurcation appears for higher values of α in a certain narrow range of K.

The bifurcation points are defined by

$$\hat{K} = -\frac{2^{\alpha}\,(\alpha - 1 + \Lambda)}{(-1 - \alpha + \Lambda)} \left(\frac{\alpha}{-1 + 3\alpha + \Lambda}\right)^{\alpha}, \tag{6.23}$$

$$\check{K} = \frac{2^{\alpha}\,(\alpha - 1 - \Lambda)}{(1 + \alpha + \Lambda)} \left(-\frac{\alpha}{1 - 3\alpha + \Lambda}\right)^{\alpha}, \tag{6.24}$$

where

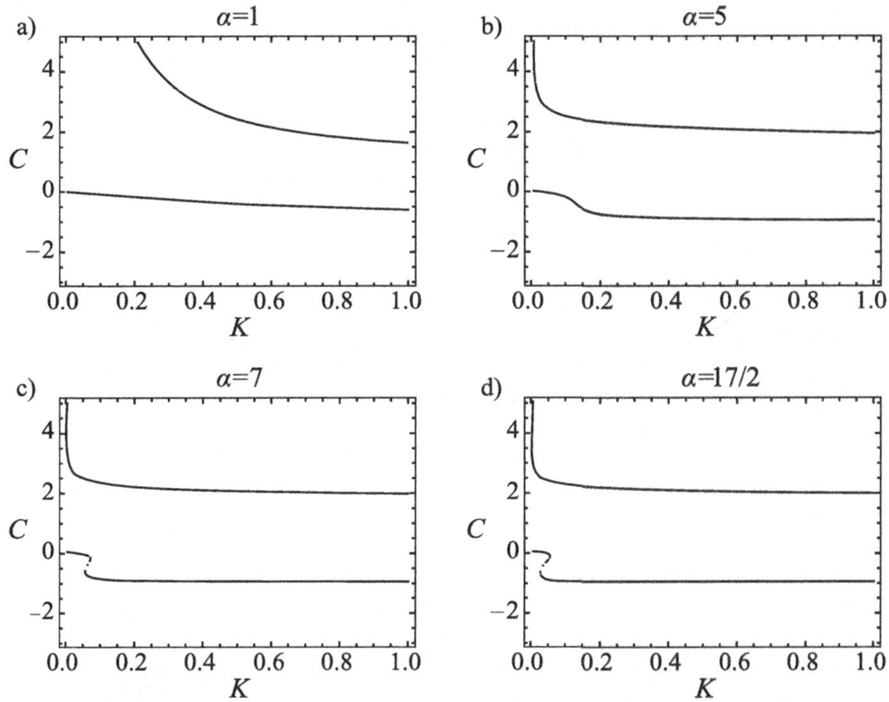

Fig. 6.5 Balancing diagrams $C - K$ for a fixed value of the power α and different values of the ratio $K = K_2/K_1$ in Case II: **a** $\alpha = 1$; **b** $\alpha = 5$; **c** $\alpha = 7$; **d** $\alpha = 17/2$

$$\Lambda = \sqrt{\alpha^2 - 6\alpha + 1}. \tag{6.25}$$

They are depicted in Fig. 6.6a, together with the following associated values

$$\hat{C} = \frac{1 - \alpha - \Lambda}{2\alpha}, \tag{6.26}$$

$$\check{C} = \frac{1 - \alpha + \Lambda}{2\alpha}. \tag{6.27}$$

They exist for $\Lambda > 0$, i.e. $\alpha > 3 + 2\sqrt{2}$ ($\alpha > 5.82843$), which corresponds to $\bar{K} = 2^{-2-\sqrt{2}} = 0.0938036$. These two values are labelled in Fig. 6.6b, which shows two curves meeting at the cusp in terms of the power α. The way how $\hat{C}$ and $\check{C}$ change with α is presented in Fig. 6.6c, where it is also noted that $\bar{C} = 1 - \sqrt{2} = -0.414214$.

To illustrate how the exact solution for motion can be obtained, it is assumed that $\alpha = 9$, $K = K_2 = 0.04$, $K_1 = 1$. Two bifurcating values of C (see Fig. 6.6a) are

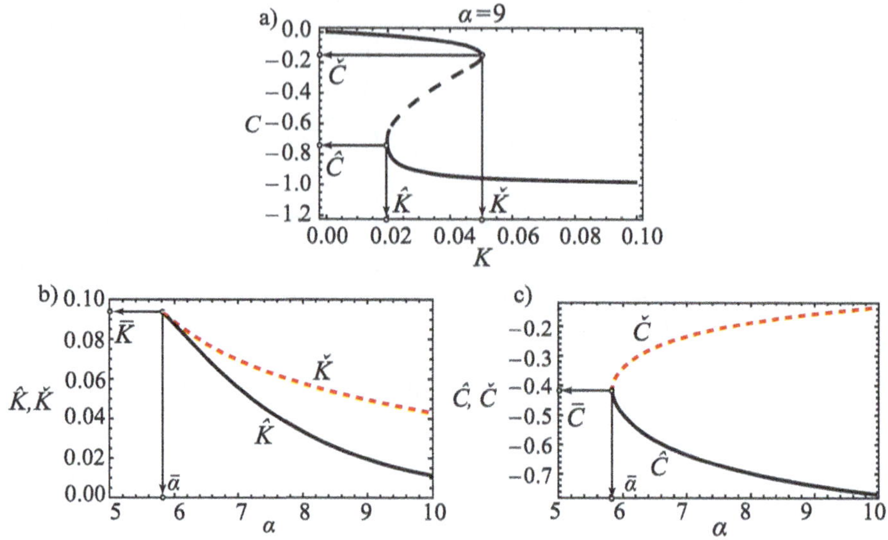

Fig. 6.6 **a** Enlarged balancing diagrams $C - K$ diagram for $\alpha = 9$; **b** Bifurcation values of K as a function of α; **c** Bifurcation values of C as a function of α

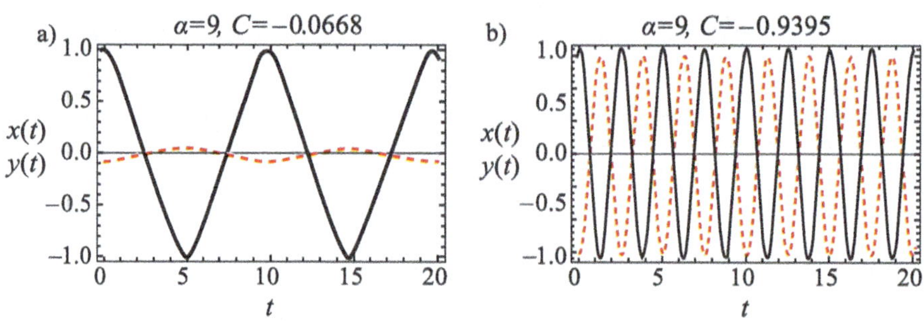

Fig. 6.7 Time-history diagrams for x (black solid line) and y (red dashed line) corresponding to $\alpha = 9$, $K = K_2 = 0.04$, $K_1 = 1$, $A = 1$ in Case I for: **a** $C = -0.0668$; **b** $C = -0.9395$

calculated from Eq. (6.22): $C = -0.0668$ and $C = -0.9395$. Equation (6.9) now gives

$$x = A\,\mathrm{ca}\left(9, 1, A^4\sqrt{5\left(1 + 0.04\,(1 - C)^5\right)}t\right). \tag{6.28}$$

Figure 6.7 contains this response for $A = 1$ as well as the one corresponding to $y(t)$. One can use the Fourier series expansion from Appendix C and [25] to find out that the solution (6.28) corresponding to Fig. 6.7a can be represented as the sum of the following harmonics

$$x \approx 0.893115 \cos(0.642219t) + 0.0800142 \cos(1.92666t) + 0.0190361 \cos(3.21109t) + 0.00544285 \cos(4.49553t) + 0.00164111 \cos(5.77997t). \quad (6.29)$$

6.2.2 Two-Mass Chain: Forced Vibrations

This section is concerned with the system of coupled oscillators from Fig. 6.1a, where the first mass is excited by the forcing f, so that the equations of motion have the form:

$$\ddot{x} + K_1\, x\, |x|^{\alpha-1} + K_2\, (x - y)\, |x - y|^{\alpha-1} = f, \quad (6.30)$$

$$\ddot{y} + K_3\, y\, |y|^{\alpha-1} + K_2\, (y - x)\, |y - x|^{\alpha-1} = 0. \quad (6.31)$$

In order to find the forced response in an exact form, the expression for the external excitation is designed in a special way: it is assumed in the form related to the form of the purely nonlinear restoring force as follows

$$f = \frac{F_0}{A} x \left| \frac{F_0}{A} x \right|^{\alpha-1}, \quad (6.32)$$

where F_0 is constant. Now, assuming also that Eq. (6.3) holds, the excitation (6.32) becomes the part of the stiffness coefficient in front of the nonlinear term and the equations of motion become uncoupled:

$$\ddot{x} + \left[K_1 + K_2 (1 - C) |1 - C|^{\alpha-1} - \frac{F_0}{A} \left| \frac{F_0}{A} \right|^{\alpha-1} \right] x\, |x|^{\alpha-1} = 0, \quad (6.33)$$

$$\ddot{x} + \left[K_3\, |C|^{\alpha-1} + K_2 \frac{C-1}{C} |C - 1|^{\alpha-1} \right] x\, |x|^{\alpha-1} = 0. \quad (6.34)$$

Proceeding as previously, i.e. equating the coefficients in front of the nonlinear terms, one derives

$$K_1 + K_2\, |1 - C|^{\alpha-1} (1 - C) \frac{1 + C}{C} - K_3\, |C|^{\alpha-1} = \frac{F_0}{A} \left| \frac{F_0}{A} \right|^{\alpha-1}. \quad (6.35)$$

One can use this expression to investigate how the ratios F_0/A and the stiffness coefficients influence the value of C for a fixed value of α. This is done separately for both Cases considered in the previous section.

Note also that now the solution for x is defined by

$$\ddot{x} + c_{\alpha,F}\, x\, |x|^{\alpha-1} = 0, \quad (6.36)$$

$$c_{\alpha,F} = K_1 + K_2 (1 - C) |1 - C|^{\alpha-1} - \frac{F_0}{A} \left| \frac{F_0}{A} \right|^{\alpha-1}, \quad (6.37)$$

but it can easily be related to the form given by Eqs. (6.6) and (6.9). The expression for the force follows then from Eq. (6.32).

6.2.2.1 Case I: Equal Stiffness Coefficients of Anchoring Springs

Equation (6.35) is used to plot Fig. 6.8 for the forced response as the extension of the one shown in Fig. 6.2a, b, c, d for the same parameter values, two different values of the excitation force and $A = 1$. As it is seen, the branches in some regions tend to those corresponding to the free response, which are plotted as solid line and represent their backbone curves.

To illustrate how x, y and the force f change in time, Fig. 6.9 is created for $\alpha = 5/3$, $K = 0.2$, $A = 1$ and two negative values of C. The force is in phase with x in Fig. 6.9a, and in phase with y in Fig. 6.9b.

Note that frequency-response curves are not shown here, but can be obtained by using Eq. (6.35) in association with the relationship for the frequency $\omega = 2\pi/T_{\text{ex}}$, where T_{ex} is defined by Eq. (6.10).

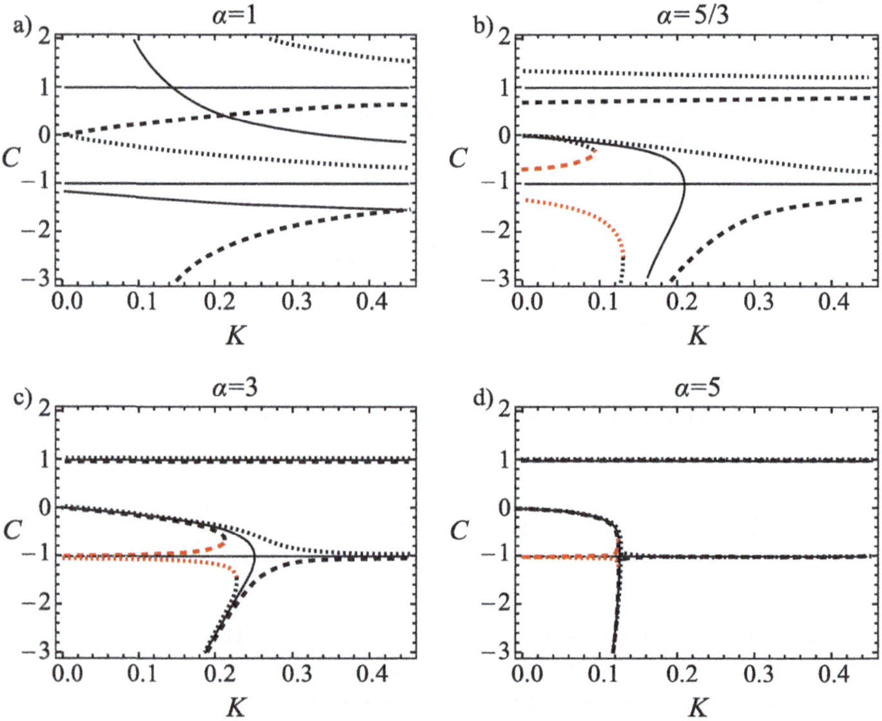

Fig. 6.8 Balancing diagrams $C - K$ diagrams for forced vibrations in Case I for $A = 1$ and: **a** $\alpha = 1$; **b** $\alpha = 5/3$; **c** $\alpha = 3$; **d** $\alpha = 5$ (solid line: $F_0 = 0$, dashed line: $F_0 = 0.4$, dotted line $F_0 = -0.4$)

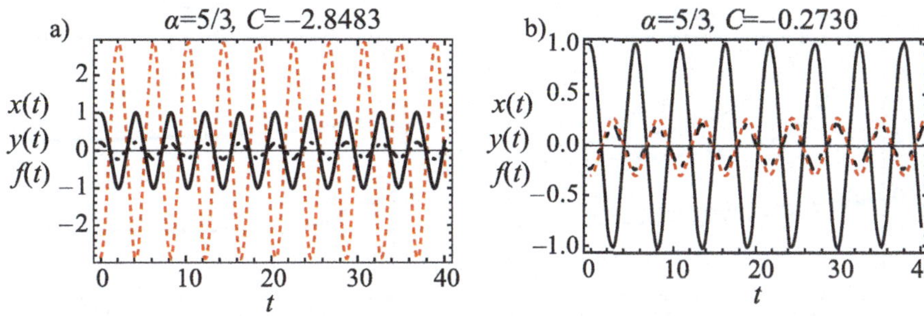

Fig. 6.9 Diagrams for x (black solid line), y (red dashed line) and f (black dashed-dotted line) corresponding to Fig. 6.8b, $\alpha = 5/3$, $K = K_2 = 0.2$, $K_1 = 1$, $A = 1$ in Case I for: **a** $C = -2.84836$; **b** C =0.273026

6.2.2.2 Case II: No Right Anchoring Spring

Equation (6.35) is now used to create Fig. 6.10 for the forced response in Case II as an extension of the one shown in Fig. 6.5 for the same parameter values, two values of the excitation force and $A = 1$. Unlike the linear case, for nonlinear cases the curves of a forced system are very close to those corresponding to the free system.

6.2.3 Three-Mass Chains

If the system consists of three masses ($N = 3$) attached mutually as well as with the base with purely nonlinear springs of equal stiffness coefficients, the equations of motion are

$$\ddot{x} + K\ x\,|x|^{\alpha-1} + K\ (x-y)\,|x-y|^{\alpha-1} = 0, \tag{6.38}$$

$$\ddot{y} + K\ (y-x)\,|y-x|^{\alpha-1} + K\ (y-z)\,|y-z|^{\alpha-1} = 0, \tag{6.39}$$

$$\ddot{z} + K\ z\,|z|^{\alpha-1} + K\ (z-y)\,|z-y|^{\alpha-1} = 0, \tag{6.40}$$

where K is constant. Of interest is again to find similar normal modes, i.e. the solutions satisfying Eq. (6.3) and

$$z = D\,x, \tag{6.41}$$

where C and D are now constant parameters to be calculated. This yields

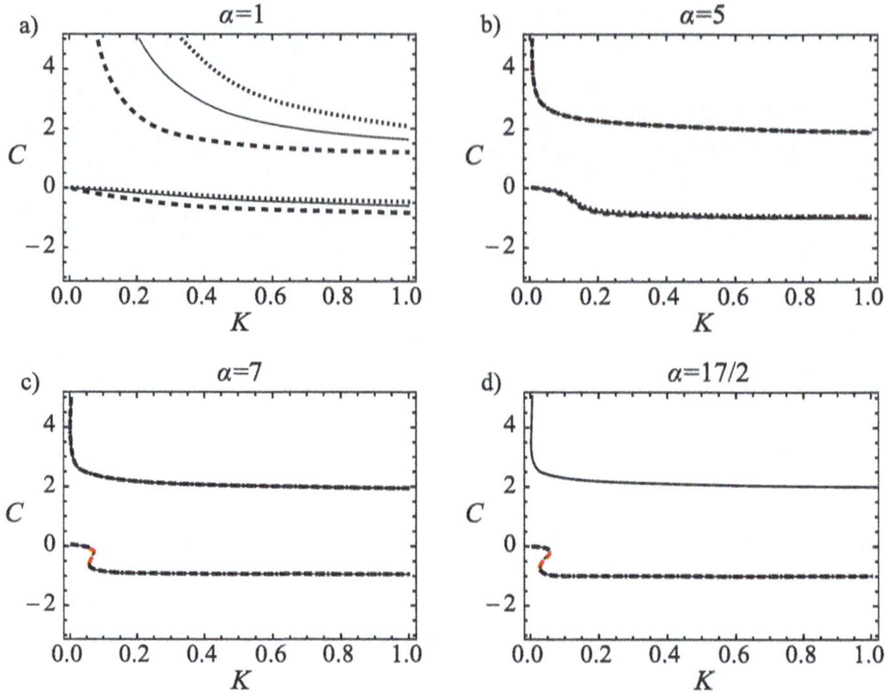

Fig. 6.10 Balancing diagrams $C-K$ diagrams for forced vibrations in Case II for $A=1$ and: **a** $\alpha=1$; **b** $\alpha=5$; **c** $\alpha=7$; **d** $\alpha=17/2$ (solid line: $F_0=0$, dashed line: $F_0=0.4$, dotted line $F_0=-0.4$)

$$\ddot{x}+K\left[1+(1-C)\left|1-C\right|^{\alpha-1}\right]x\left|x\right|^{\alpha-1}=0, \tag{6.42}$$

$$\ddot{x}+K\left[\frac{C-1}{C}\left|C-1\right|^{\alpha-1}+\frac{C-D}{C}\left|C-D\right|^{\alpha-1}\right]x\left|x\right|^{\alpha-1}=0, \tag{6.43}$$

$$\ddot{x}+K\left[\left|D\right|^{\alpha-1}+\frac{D-C}{D}\left|D-C\right|^{\alpha-1}\right]x\left|x\right|^{\alpha-1}=0. \tag{6.44}$$

To simplify this to a one-degree-of-freedom system, the following system of equation needs to be satisfied:

$$1+(1-C)\left|1-C\right|^{\alpha-1}=\frac{C-1}{C}\left|C-1\right|^{\alpha-1}+\frac{C-D}{C}\left|C-D\right|^{\alpha-1}, \tag{6.45}$$

$$1+(1-C)\left|1-C\right|^{\alpha-1}=\left|D\right|^{\alpha-1}+\frac{D-C}{D}\left|D-C\right|^{\alpha-1}. \tag{6.46}$$

Solving it for a fixed α, one can obtain the values of C and D, and then calculate the corresponding frequency $\omega=2\pi/T_{\text{ex}}$, where T_{ex} is defined by Eq. (6.10), and c_α is obtained from Eq. (6.7) with $K_1=K_2=K$. These results are presented in Table 6.1

Table 6.1 Values of C and D obtained from Eqs. (6.46), (6.45), the corresponding modal frequency ω obtained from Eqs. (6.10), (6.7) with $K_1 = K_2 = K = 1$; n is the mode number/modal index associated with each of them

	$\alpha = 1$		$\alpha = 5/3$		$\alpha = 3$	
	(C,D)	ω	(C,D)	ω	(C,D)	ω
$n = 1$	$(\sqrt{2},1)$	0.76537	(1.61689,1)	0.699603	(1.77783,1)	0.61643
$n = 2$	(0,−1)	1.41421	(0,1)	1.33051	(0,−1)	1.19814
$n = 3$	$(-\sqrt{2},1)$	1.84776	(−1.67506,1)	2.33408	(−1.18117,0.27081)	2.85762
$n = 4$	–	–	–	–	(−1.923,1)	4.31779
$n = 5$	–	–	–	–	(−4.36161,3.69263)	10.55212

for $K = 1$. They are listed with respect to the increasing ω. Each pair (C, D) and the corresponding ω have the mode number/modal index n associated with them.

These results are also plotted in Fig. 6.11 for $A = 1$ and for all the values of α from Table 6.1 as well as for $\alpha = 5$ (due to a large value, one solution for $\alpha = 5$, which is $\omega = 3168.06$, is not shown). The mode number/modal index n is shown on the horizontal axis and the associated frequency ω on the vertical axis.

One can further use the results presented in Table 6.1 and illustrate them graphically by rotating the displacements for 90°clockwise. The results obtained are shown for $\alpha = 5/3$ in Fig. 6.12a and for $\alpha = 3$ in Fig. 6.12b (these displacements are plotted by using the same scale to make them be comparable). While the results shown in Fig. 6.12a resemble the well-known modes corresponding to the linear system, those in Fig. 6.12b illustrate the appearance of two additional asymmetric modes, the existence of which is also seen in Fig. 6.11.

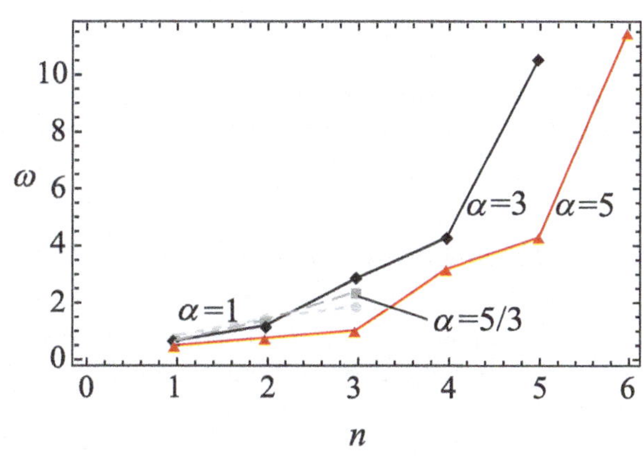

Fig. 6.11 Modal frequencies ω versus mode number n for different values of the power of nonlinearity

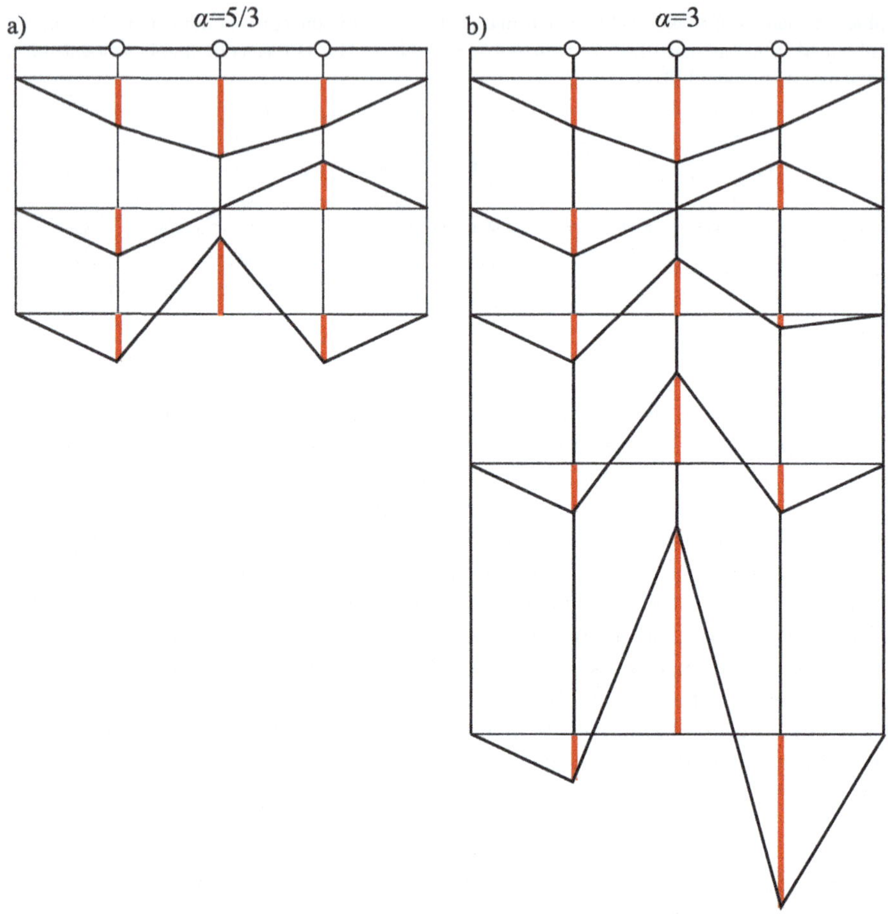

Fig. 6.12 Modal responses of a three-mass chain of oscillators rotated for 90°, plotted proportionally with respect to each other, based on the calculations carried out for $K = 1$, $A = 1$ and: **a** $\alpha = 5/3$, **b** $\alpha = 3$

6.2.4 From Purely Nonlinear Chains to Continuous Systems

In order to examine the case of a continuous system—a bar exhibiting longitudinal vibrations, one can use the previously derived results letting the number of masses tend to infinity ($N \to \infty$). For that purpose, we focus on Eq. (6.39), which can be used for an arbitrary mass in the chain, but change the notation. First, we will introduce the mass M and use $y \equiv u = u(x_i)$ (u is the displacement of the ith mass and x_i is the equilibrium position of the ith mass), $z \equiv u(x_i + \Delta x_i)$ with Δx_i being the equal spacing between the equilibrium positions of all the masses, $x \equiv u(x_i - \Delta x_i)$. This gives

$$M\frac{\partial^2 u(x_i)}{\partial t^2} = K\ (u(x_i + \Delta x_i) - u(x_i))\,|u(x_i + \Delta x_i) - u(x_i)|^{\alpha-1} -$$
$$K\ (u(x_i) - u(x_i - \Delta x_i))\,|u(x_i) - u(x_i - \Delta x_i)|^{\alpha-1}. \quad (6.47)$$

Now, by dividing the equation with Δx_i and dropping the subscript i, one has

$$\frac{M}{\Delta x}\frac{\partial^2 u(x)}{\partial t^2} = K\left\{\frac{(u(x+\Delta x) - u(x))\,|u(x+\Delta x) - u(x)|^{\alpha-1}}{\Delta x}\right.$$
$$\left. - \frac{(u(x) - u(x-\Delta x))\,|u(x) - u(x-\Delta x)|^{\alpha-1}}{\Delta x}\right\}. \quad (6.48)$$

Considering the case $\alpha = 1$, it is easy to recognize that letting $\Delta x \to 0$ (which is the case when $N \to \infty$), the ratios on the right-hand side represents the first derivative with respect to x:

$$\frac{M}{\Delta x}\frac{\partial^2 u}{\partial t^2} = K\ \Delta x \frac{u'(x) -\ u'(x-\Delta x)}{\Delta x}, \quad (6.49)$$

where the primes denote spatial derivatives. For $M/\Delta x = \rho$ and $K\ \Delta x = E$, this further simplifies to

$$\rho\frac{\partial^2 u}{\partial t^2} = E\frac{\partial^2 u}{\partial x^2}, \quad (6.50)$$

which corresponds to the governing equation for longitudinal vibrations with $E/\rho = c^2$ as well as to the form of a wave equation in one dimension.

Following the analogous procedure, the Right-Hand Side (RHS) of Eq. (6.48) in a general case can be written down as

$$\mathrm{RHS(Eq.(48))} = K\Delta x\,|\Delta x|^{\alpha-1}\left\{\frac{\frac{(u(x+\Delta x)-u(x))}{\Delta x}\left|\frac{u(x+\Delta x)-u(x)}{\Delta x}\right|^{\alpha-1}}{\Delta x}\right.$$
$$\left. - \frac{\frac{(u(x)-u(x-\Delta x))}{\Delta x}\left|\frac{u(x)-u(x-\Delta x)}{\Delta x}\right|^{\alpha-1}}{\Delta x}\right\}$$
$$= K\Delta x\,|\Delta x|^{\alpha-1}\frac{\partial}{\partial x}\left[\left(\frac{\partial u}{\partial x}\right)\left|\frac{\partial u}{\partial x}\right|^{\alpha-1}\right]$$
$$= \alpha K\Delta x\,|\Delta x|^{\alpha-1}\frac{\partial^2 u}{\partial x^2}\left|\frac{\partial u}{\partial x}\right|^{\alpha-1}. \quad (6.51)$$

Equation (6.48) now becomes

$$\rho\frac{\partial^2 u}{\partial t^2} = \alpha E\frac{\partial^2 u}{\partial x^2}\left|\frac{\partial u}{\partial x}\right|^{\alpha-1}, \quad (6.52)$$

with $E = K \Delta x \, |\Delta x|^{\alpha-1}$. This generalizes Eq. (6.50) to a nonlinear case when the bar is made of a material characterized by a nonlinear stress–strain relationship (some examples of the systems/materials for which this relationship is nonlinear can be found, for instance, in [26]).

6.3 Chains of Pure Cubic Oscillators

6.3.1 Two-Mass Chain

A one degree-of-freedom free oscillator consisting of a mass attached to a pure cubic spring has the exact closed-form solution for its response (see [27, 28] or Sect. 2.3.3 of Chap. 2 in this book). It is shown therein that a one degree-of-freedom oscillator with pure cubic nonlinearity is governed by

$$\ddot{x} + c_3 \, x^3 = 0, \tag{6.53}$$

where c_3 is a constant and the initial conditions are: $x\,(0) = A$, $\dot{x}\,(0) = 0$, an exact solution for its response exists in closed form that involves the Jacobi cn function:

$$x = A \, \mathrm{cn}\left(\omega \, t \left| \frac{1}{2} \right.\right), \quad \omega = \sqrt{c_3} A, \tag{6.54}$$

where the angular frequency of the cn function ω is the coefficient in front of t in the first argument (note that it depends on the amplitude A), while the second argument is the elliptic parameter $m = 1/2$. This Jacobi cn function can be interpreted as multi-harmonic. To illustrate this fact, its Fourier series expansion is given in Appendix C. This series implies that the first harmonic dominates, but the whole response also includes higher, odd harmonics, which are the consequence of the odd-powered restoring force.

Oscillatory systems considered subsequently in this section represent a direct extension of the oscillator presented in Sect. 6.2: the same unit masses and the same pure cubic springs are added in line to the pure cubic oscillator, thus forming a chain [29]. They are categorized in two ways: with respect to the number of masses that they contain (two masses or more than two masses), and depending on whether they are anchored at both ends (Fig. 6.13a) or just at one end (Fig. 6.13b). The number of masses is labelled by N, and the generalized coordinates are chosen to be the absolute coordinates x_i, $i = 1, \ldots, N$.

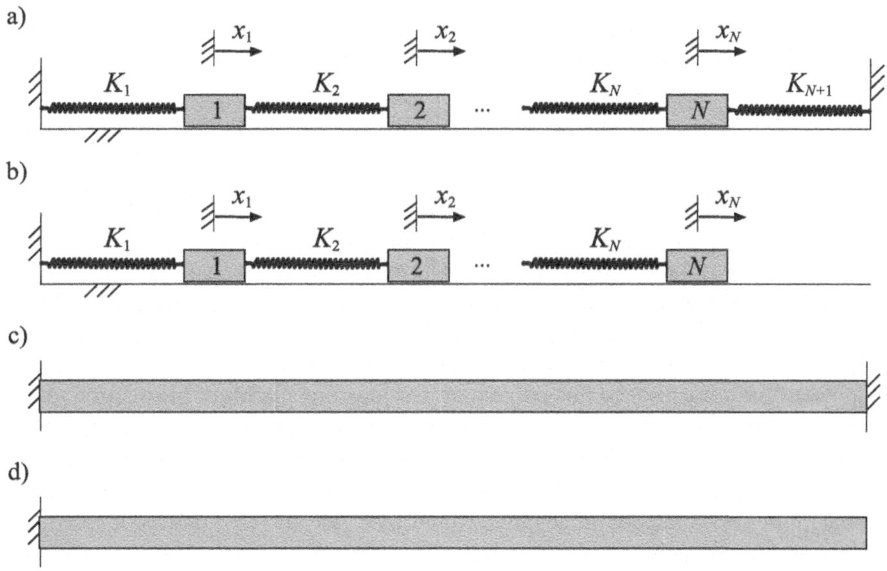

Fig. 6.13 **a** Chain of oscillators attached at both ends; **b** Chain of oscillators attached at one end; **c** Clamped–clamped bar; **d** Clamped-free bar

6.3.1.1 Chains Anchored at Both Ends

The equations of motion of the system with two discrete masses ($N = 2$) from Fig. 6.13a are given by

$$\ddot{x}_1 + K_1\, x_1^3 + K_2\, (x_1 - x_2)^3 = 0, \tag{6.55}$$

$$\ddot{x}_2 + K_3\, x_2^3 + K_2\, (x_2 - x_1)^3 = 0, \tag{6.56}$$

where $K_1,\ K_2$ and K_3 stand for the stiffness coefficient of the springs. The two masses are assumed to be equal to a unit mass.

Of interest here is to find similar normal modes [10], which relate $x_1$and x_2 in accordance with the following linear relationship:

$$x_2 = C\, x_1, \tag{6.57}$$

where C is a constant to be calculated. With this substitution, Eqs. (6.55), and (6.56) become

$$\ddot{x}_1 + \left[K_1 + K_2\,(1 - C)^3\right]\, x_1^3 = 0, \tag{6.58}$$

$$\ddot{x}_1 + \left[K_3 C^2 + K_2 \frac{(C-1)^3}{C}\right]\, x_1^3 = 0. \tag{6.59}$$

These two equations are equivalent mutually and simplify to Eq. (6.55) if the coefficients of the cubic terms are equal, which can be written down as

$$1+\frac{K_2}{K_1}(1-C)^3\,\frac{1+C}{C}-\frac{K_3}{K_1}C^2=0. \tag{6.60}$$

This expression (6.60) defines the value of C as a function of the ratios between the stiffness of springs 2 and 3 with respect to the first one: $K = K_2/K_1$ and $\tilde{K} = K_3/K_1$. Different cases are possible and they are shown in Fig. 6.14a, b, c, d, e for certain increasing values of $\tilde{K}$ ranging from $\tilde{K} = 0.75$ to $\tilde{K} = 1.25$. The central figure corresponding to $\tilde{K} = 1$ (Fig. 6.14c) shows a subcritical pitchfork bifurcation exists at $K^* = 1/4$, which is already shown in Fig. 6.2c in the context of the influence of the power of nonlinearity to this behaviour. Besides this pitchfork bifurcation, there is always a constant value of C that appears for all K. Additional $K - C$ diagrams in Fig. 6.14 show how this bifurcation is generated when $\tilde{K} < 1$ (Fig. 6.14a, b) and destroyed when $\tilde{K} > 1$ (Fig. 6.14d and e). Besides affecting fundamentally similar normal modes associated with negative values of C, the stiffness ratio $\tilde{K}$ also affects the one associated with a positive value of C: when $\tilde{K} < 1$, one has $C > 1$, while for $\tilde{K} > 1$, one has $C < 1$.

Let us now explain the behaviour shown in Fig. 6.14c in more detail and illustrate it. Below K^*, four distinct similar normal modes exist, three of which are stable, labelled by the solid line (one solution is $C = 1$, and there are two negative solutions $C < 0$). Above K^*, there are two of them and both are stable ($C = 1$ and $C = -1$). Taking, for example, $K = 0.2$, one can calculate the values of C from Eq. (6.60): $C = 1$; $C = -0.38$; $C = -1$; $C = -2.62$. Analyzing their stability, one can determine that the first, second and fourth are stable. They are labelled in Fig. 6.15 as Cases I_1 (in phase oscillations), II_1 (out–of-phase oscillations) and III_1 (out–of-phase oscillations) and the corresponding mechanical model and its characteristic mode is plotted for each of them as well. The stable mode that appears for $K > K^*$ together with case I_1 is labelled as II_2 ($C = -1$) and the corresponding mechanical model is shown with the characteristic mode, too.

It should be noted that after calculating the constant C, one can compare Eq. (6.58) with (6.53) and use its solution to write down the exact solution for motion of x_1 in the form of Eq. (6.54), but with $\omega = \sqrt{K_1 + K_2(1-C)^3 A}$, where $A = x_1(0)$. Given Eq. (6.57), the exact closed-form solution for x_2 is then known as well.

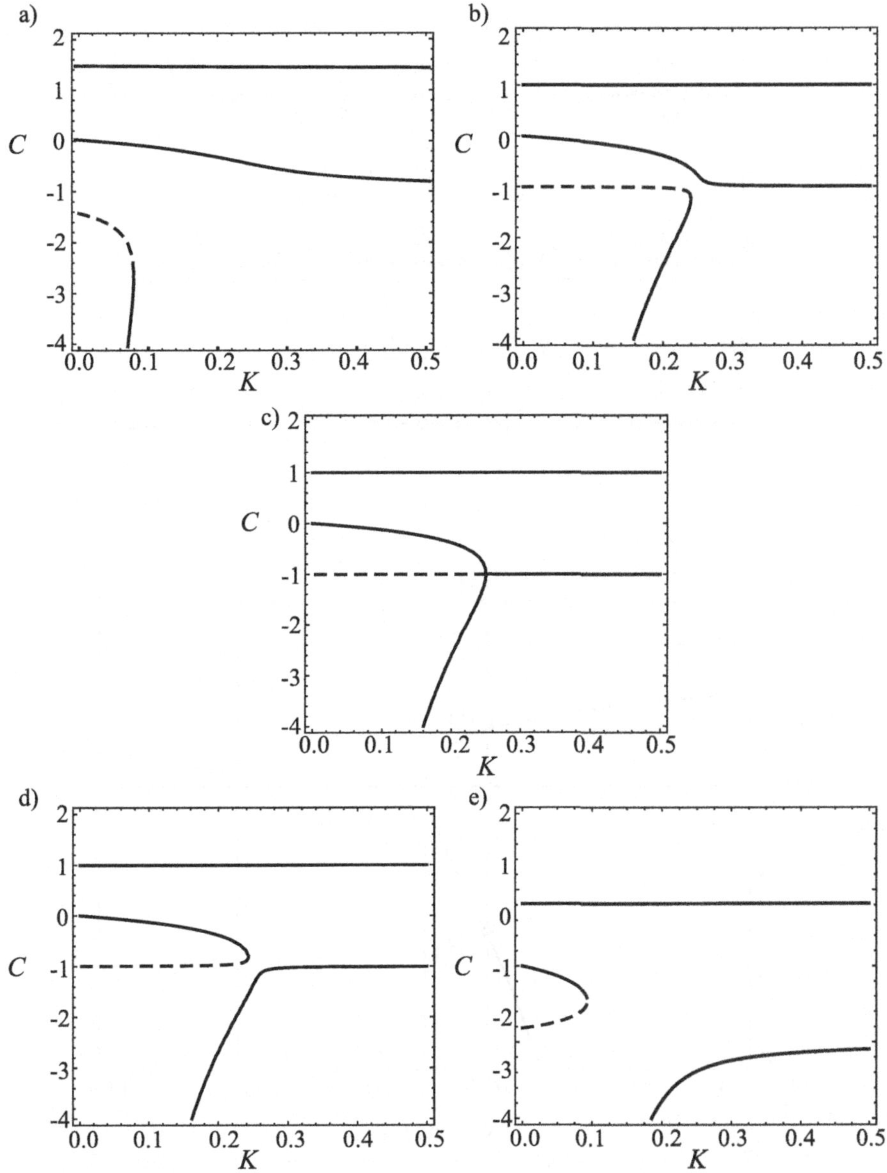

Fig. 6.14 Values of C, Eq. (6.60) as a function of the ratio $K = K_2/K_1$ for different values of $\tilde{K} = K_3/K_1$: **a** $\tilde{K} = 0.75$; **b** $\tilde{K} = 0.99$; **c** $\tilde{K} = 1$; **d** $\tilde{K} = 1.01$; **e** $\tilde{K} = 1.25$

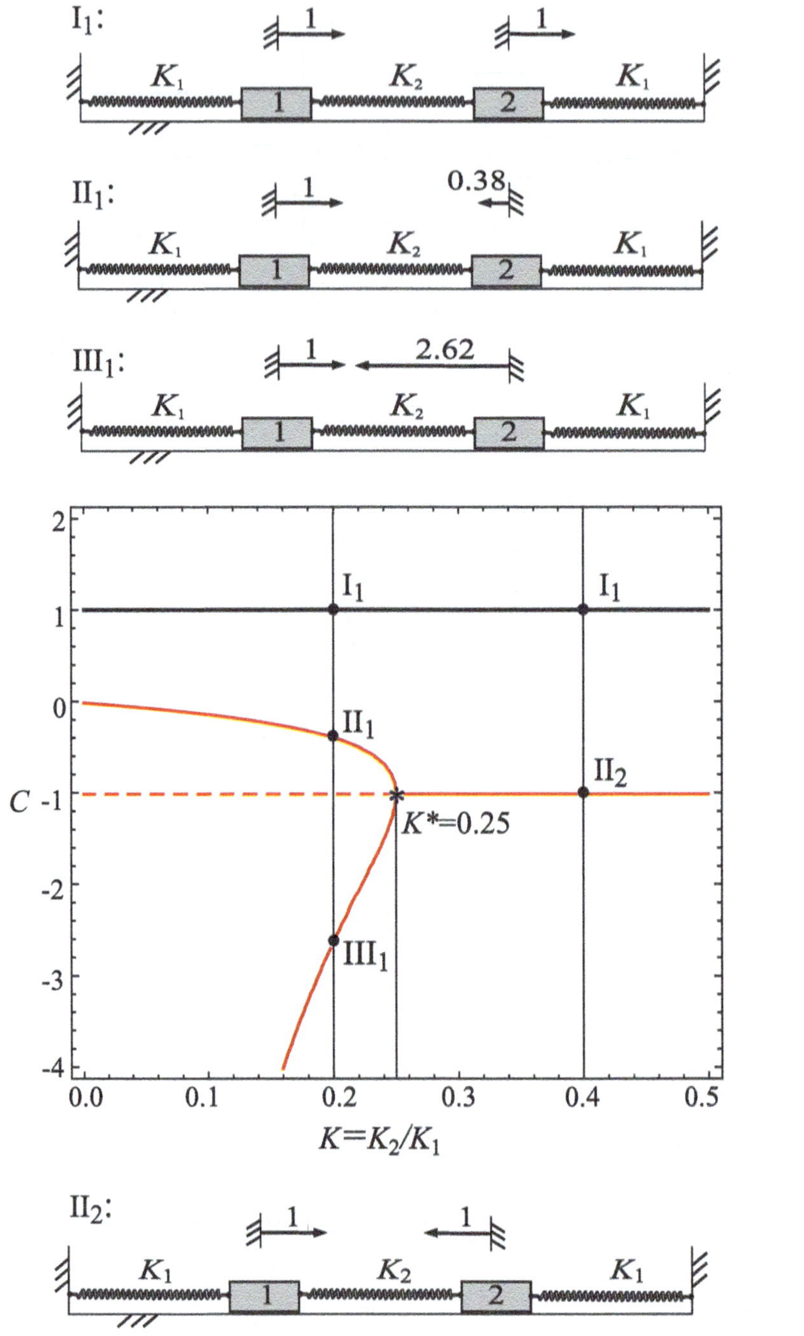

Fig. 6.15 Values of C, Eq. (6.60) as a function of the ratio $K = K_2/K_1$ with the mechanical models of the stable similar normal modes

6.3.1.2 Chains Anchored at One End

The equations of motion of the system with two lumped masses ($N = 2$) from Fig. 6.13b are given by (6.55), (6.56) with $K_3 = 0$. The procedure presented previously yields the following condition for similar normal modes:

$$1 + K\,(1 - C)^3\,\frac{1 + C}{C} = 0. \tag{6.61}$$

This expression (6.61) is used to plot the $K - C$ diagram in Fig. 6.16. Unlike the chain anchored at both ends, this chain anchored at one end always exhibits two similar normal modes regardless of the values of K, and no bifurcation is found to exist. One of these modes corresponds to in phase oscillations ($C > 0$) and the other one to out-of-phase oscillations ($C < 0$).

6.3.2 Chains of Pure Cubic Oscillators with More Than Two Masses

6.3.2.1 Chains Anchored at Both Ends

The system consisting of N masses is considered now, where N is an odd integer higher than unity. Note that the analysis holds for even N as well, but to establish some links with continuous systems later on, the symmetry requires N to be odd. The equations of motion are

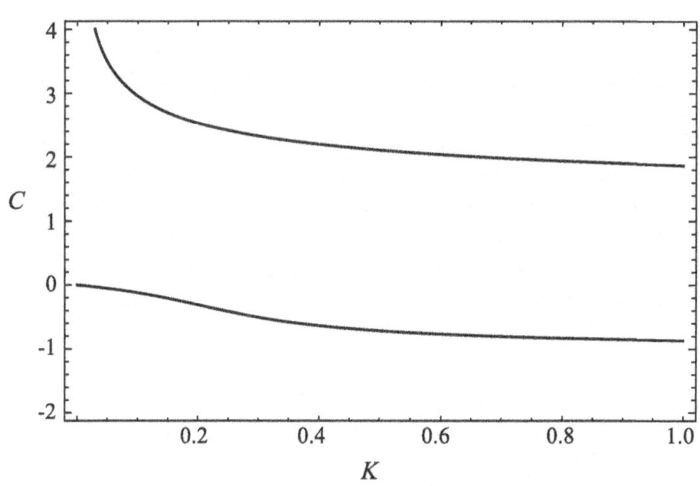

Fig. 6.16 Values of C, Eq. (6.61) as a function of the ratio $K = K_2/K_1$

$$\ddot{x}_1 + K\, x_1^3 + K\, (x_1 - x_2)^3 = 0, \tag{6.62}$$

$$\ddot{x}_j + K\, \left(x_j - x_{j-1}\right)^3 + K\, \left(x_j - x_{j+1}\right)^3 = 0, \quad j = 2, \ldots, N-1, \tag{6.63}$$

$$\ddot{x}_N + K\, x_N^3 + K\, (x_N - x_{N-1})^3 = 0. \tag{6.64}$$

To find similar normal modes, the solutions of Eqs. (6.62)–(6.64) are required to be related mutually as follows:

$$x_i = C_i\, x_1, i = 1, \ldots, N \tag{6.65}$$

where C_i are constant parameters to be calculated (note that this notation implies $C_1 = 1$). Substituting (6.65) into Eqs. (6.62)–(6.64), yields

$$\ddot{x}_1 + K\left[1 + (1 - C_2)^3\right] x_1^3 = 0, \tag{6.66}$$

$$\ddot{x}_1 + K\left[\frac{\left(C_j - C_{j-1}\right)^3}{C_j} + \frac{\left(C_j - C_{j+1}\right)^3}{C_j}\right] x_1^3 = 0, \tag{6.67}$$

$$\ddot{x}_1 + K\left[C_N^2 + \frac{(C_N - C_{N-1})^3}{C_N}\right] x_1^3 = 0. \tag{6.68}$$

Following the principle of the equivalence of these equations, i.e. equating the coefficients in front of the cubic terms, gives

$$1 + (1 - C_2)^3 = \frac{\left(C_j - C_{j-1}\right)^3}{C_j} + \frac{\left(C_j - C_{j+1}\right)^3}{C_j}, \tag{6.69}$$

$$1 + (1 - C_2)^3 = C_N^2 + \frac{(C_N - C_{N-1})^3}{C_N}. \tag{6.70}$$

Solving this system of algebraic equations, one can obtain the combinations of $C_2 - C_N$, and then calculate the corresponding mode frequency. The modal shapes obtained in this way are presented in Fig. 6.17 for $N = 3$ and in Fig. 6.18 for $N = 5$. It is seen that their number is higher than the number of degrees of freedom. Figure 6.19 shows the corresponding angular frequencies ω associated with each mode for $N = 3$ (Fig. 6.19a) and for $N = 5$ (Fig. 6.19b).

Note that in this system, the motion of each mass has again the exact closed-form solution. Namely, comparing Eq. (6.66) with Eq. (6.53), one can use Eq. (6.54) to obtain

$$x_1 = A\, \mathrm{cn}\left(\sqrt{K\left[1 + (1 - C_2)^3\right]}At \,\middle|\, \frac{1}{2}\right), \tag{6.71}$$

$$x_{2,\ldots N} = C_{2,\ldots N}\, A\, \mathrm{cn}\left(\sqrt{K\left[1 + (1 - C_2)^3\right]}At \,\middle|\, \frac{1}{2}\right). \tag{6.72}$$

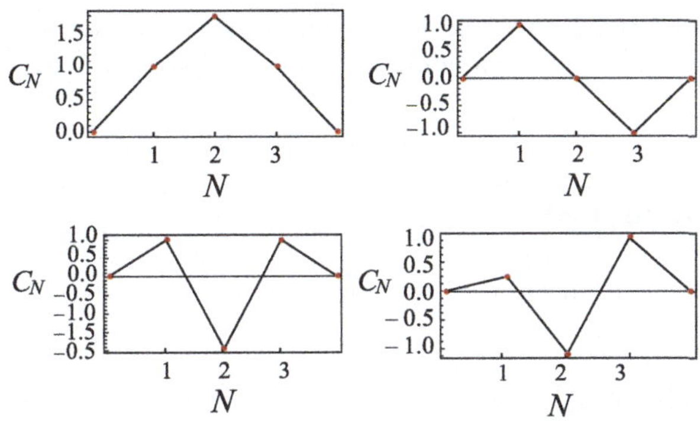

Fig. 6.17 Mode shapes of a chain anchored at both ends with $N = 3$

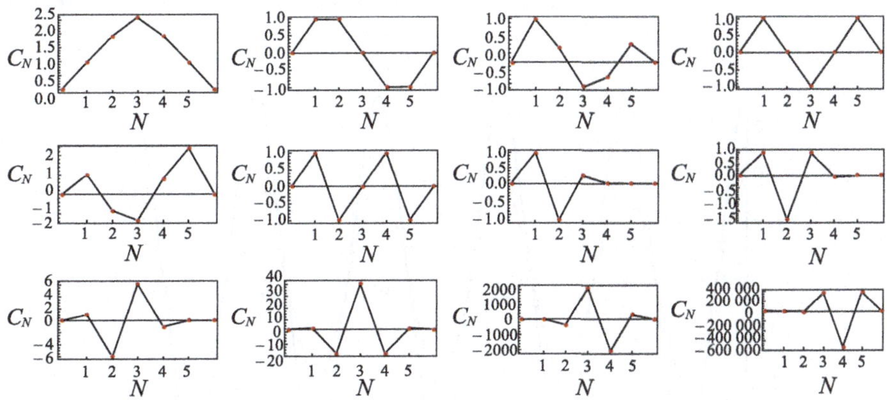

Fig. 6.18 Mode shapes of a chain anchored at both ends with $N = 5$

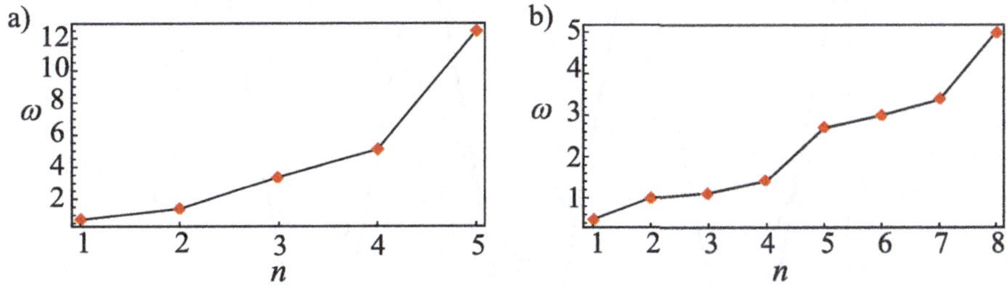

Fig. 6.19 Angular mode frequencies ω versus the mode number n of a chain anchored at both ends with $K = 1$: **a** $N = 3$; **b** $N = 5$ (not all calculated frequencies are shown; there are four more calculated, but they are significantly higher than the ones included and could not be presented to achieve a clear and visible overview)

These expressions (6.71) and (6.72) are used to plot the responses of the first and middle mass in the first mode in the chain with $N = 3$ (Fig. 6.20a) and $N = 5$ (Fig. 6.20b).

6.3.2.2 Chains Anchored at One End

The equations of motion of chains from Fig. 6.13b are given by

$$\ddot{x}_1 + K\ x_1^3 + K\ (x_1 - x_2)^3 = 0, \tag{6.73}$$

$$\ddot{x}_j + K\ \left(x_j - x_{j-1}\right)^3 + K\ \left(x_j - x_{j+1}\right)^3 = 0, \quad j = 2, \ldots, N-1, \tag{6.74}$$

$$\ddot{x}_N + K\ (x_N - x_{N-1})^3 = 0. \tag{6.75}$$

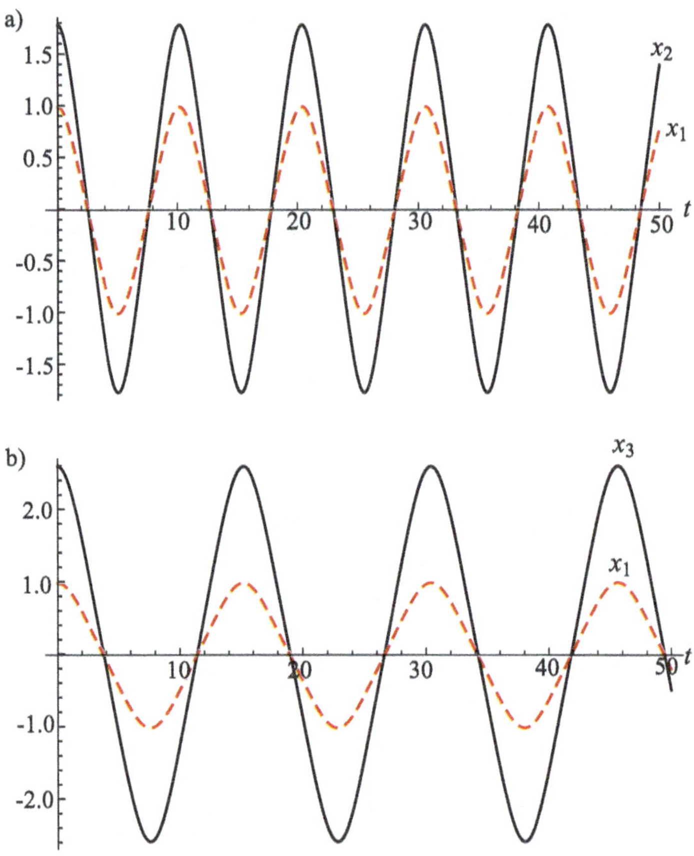

Fig. 6.20 Exact solutions for the time response of the first and middle mass in the chain, Eqs. (6.71) and (6.72) with $K = 1$: **a** $N = 3$; **b** $N = 5$

To find similar normal modes, the expressions (6.65) are substituted into Eqs. (6.73)–(6.75) and the procedure from Sect. 6.3.1 is repeated, yielding the following system of algebraic equations for the constants C_2-C_N:

$$1 + (1 - C_2)^3 = \frac{\left(C_j - C_{j-1}\right)^3}{C_j} + \frac{\left(C_j - C_{j+1}\right)^3}{C_j}, \tag{6.76}$$

$$1 + (1 - C_2)^3 = \frac{(C_N - C_{N-1})^3}{C_N}. \tag{6.77}$$

To make a parallel with the results from the previous section, similar results as those presented in Figs. 6.17, 6.18, 6.19 and 6.20 are obtained for this type of chains as well. The mode shapes are presented in Fig. 6.21 for $N = 3$ and in Fig. 6.22 for $N = 5$. Figure 6.23 shows the mode frequencies ω that correspond to the modes found.

The expressions given by Eqs. (6.71) and (6.72), but with C_2 calculated from Eqs. (6.76) and (6.77) are used to plot the responses of the first and ending mass in the chain with $N = 3$ (Fig. 6.24a) and $N = 5$ (Fig. 6.24b).

6.4 Continuous Systems with Pure Cubic Nonlinearity

It is derived in Sect. 6.2.4 that the equation of motion of the internal mass from a chain of oscillators with purely nonlinear power form of nonlinearity in springs of equal stiffness can be used to derive the governing equation for longitudinal vibrations of a bar with pure power-form nonlinearity between the axial stress and axial strain. The results presented therein holds for any real power of nonlinearity higher than unity. However, the nonlinear governing equation is only derived, but not solved. Thus, the aim here is to obtain an exact analytical solution for the first mode for the case of pure cubic nonlinearity. Based on Eq. (6.52), the corresponding governing equation takes the form:

$$\frac{\partial^2 u}{\partial t^2} = \varkappa^2 \frac{\partial^2 u}{\partial x^2} \left(\frac{\partial u}{\partial x}\right)^2, \tag{6.78}$$

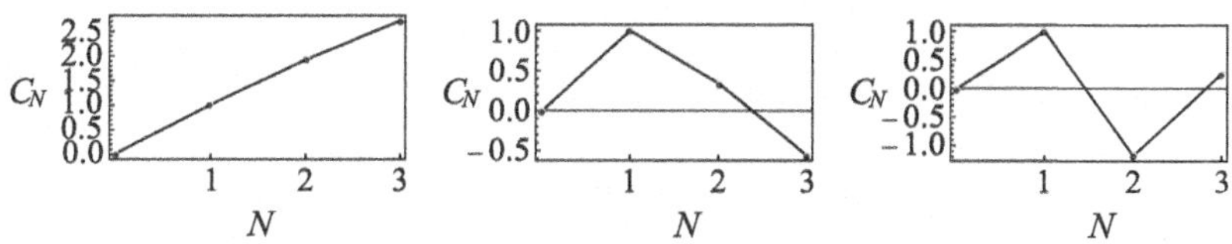

Fig. 6.21 Mode shapes of a chain anchored at one end with $N = 3$

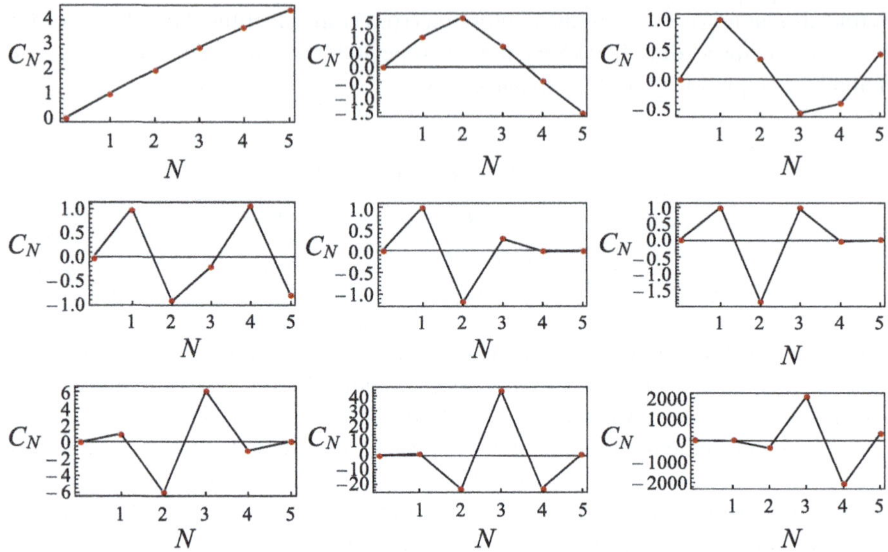

Fig. 6.22 Mode shapes of a chain anchored at one end with $N = 5$

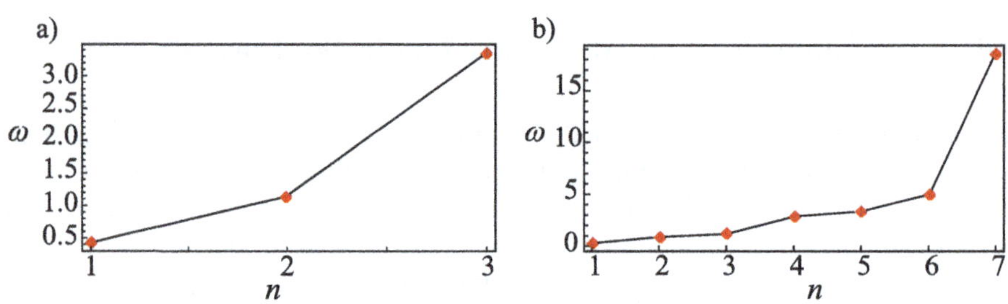

Fig. 6.23 Angular mode frequencies ω versus the mode number n of a chain anchored at one end with $K = 1$: **a** $N = 3$; **b** $N = 5$ (not all calculated frequencies are shown; there are four more calculated, but they are significantly higher than the ones included and could not be presented to achieve a clear and visible overview)

where $u = u(x, t)$ and $\varkappa = const$. Rescaling $U = u/l$, $X = x/l$, $T = \varkappa t/l$, one obtains

$$\frac{\partial^2 U}{\partial T^2} = \frac{\partial^2 U}{\partial X^2}\left(\frac{\partial U}{\partial X}\right)^2, \tag{6.79}$$

which is a simplified version of Eq. (6.78) in which $\varkappa$ is omitted.

To determine the first mode, the variables X and T are separated as follows $U(X, T) = g(X)\, f(T)$, yielding the following relationship between the unknown spatial function g and the time function f:

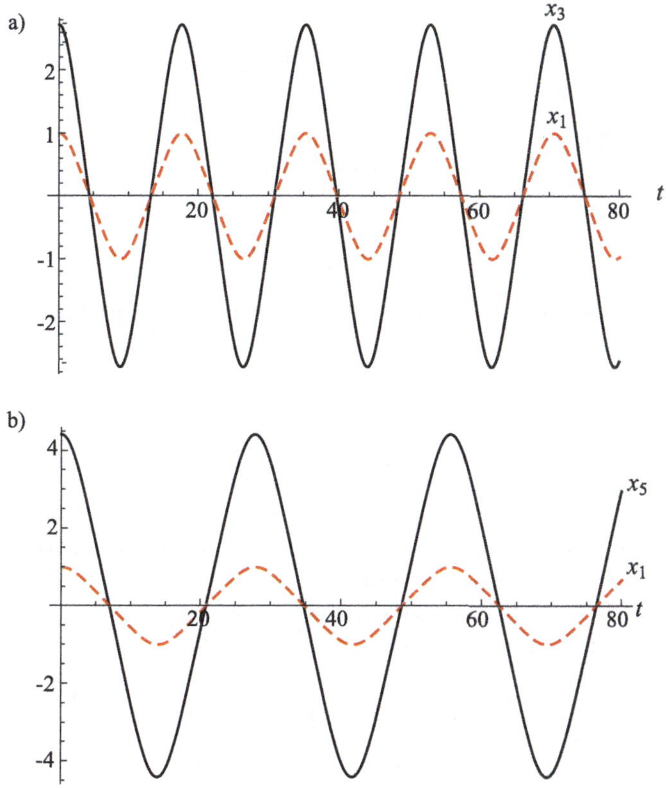

Fig. 6.24 Exact solutions for the time response of the first and ending mass in the chain, Eqs. (6.71) and (6.72) with $K = 1$: **a** $N = 3$; **b** $N = 5$

$$\frac{\ddot{f}}{f^3} = \frac{g''g'^2}{g} = -\omega^2, \tag{6.80}$$

where ω^2 must be constant; the dots stand for the differentiation with respect to T and the primes with respect to X. This expression can be separated further into two ordinary differential equations, one for the time function f:

$$\ddot{f} + \omega^2 f^3 = 0, \tag{6.81}$$

and the other one for the spatial displacement g:

$$g''g'^2 + \omega^2 g = 0. \tag{6.82}$$

Comparing Eq. (6.81) to Eq. (6.53) and assuming that initial velocities are zero, one can write down the solution for $f(T)$ as

$$f(T) = A_t \operatorname{cn}\left(\omega A_t T \,\middle|\, \frac{1}{2}\right), \tag{6.83}$$

where A_t is a constant.

Equation (6.82) has the following first integral:

$$\frac{g'^4}{4} + \omega^2 \frac{g^2}{2} = A_1, \tag{6.84}$$

where A_1 is a constant to be determined. This first integral can further be written down as

$$\frac{\mathrm{d}G}{\left(1 - G^2\right)^{1/4}} = \frac{\sqrt{2}}{A} (A_1)^{1/4} \mathrm{d}X, \tag{6.85}$$

where $G = g/A$ and $A = \sqrt{2A_1/\omega^2}$. Note that this parameter A has a physical interpretation: it represents the value of the spatial displacement for which $g' = 0$. Thus, it represents its extreme value, which will be of interest later on for the relationship with boundary conditions and the mode shape.

In order to integrate Eq. (6.85), one can recognize that

$$\int \frac{\mathrm{d}G}{\left(1 - G^2\right)^{1/4}} = \frac{1}{2} B\left(\frac{1}{2}, \frac{3}{4}\right) \text{BetaRegularized}\left(G^2, \frac{1}{2}, \frac{3}{4}\right), \tag{6.86}$$

where the beta (B) function [30] and Beta Regularized function [31] appear (see Appendix A). Thus, after integrating Eq. (6.85), one can derive:

$$\text{BetaRegularized}\left(G^2, \frac{1}{2}, \frac{3}{4}\right) = \frac{2^{3/2}}{A\, B\left(\frac{1}{2}, \frac{3}{4}\right)} (A_1)^{1/4} X + A_2, \tag{6.87}$$

where A_2 is a new constant to be determined. Expressing G^2 from the previous equation, one can obtain

$$G^2 = I_z\left(\frac{1}{2}, \frac{3}{4}\right), \quad z = \frac{2^{3/2}}{A\, B\left(\frac{1}{2}, \frac{3}{4}\right)} (A_1)^{1/4} X + A_2, \tag{6.88}$$

where I_z stands for the Inverse Beta Regularized function [32]. Returning back to g, one can finally derive

$$g(X) = A\left[I_z\left(\frac{1}{2}, \frac{3}{4}\right)\right]^{1/2}, \quad z = \frac{2^{3/2}}{A\, B\left(\frac{1}{2}, \frac{3}{4}\right)} (A_1)^{1/4} X + A_2. \tag{6.89}$$

Let us show now how the existing unknown constants can be determined in two examples.

6.4.1 Clamped–Clamped Bar

As a first example, a clamped–clamped bar is considered (Fig. 6.13c). The boundary conditions are $U(0, T) = 0$ and $U(1, T) = 0$, which yield $g(0) = 0$ and $g(1) = 0$. The former satisfies Eq. (6.89) if $A_2 = 0$ (note that this is the consequence of the fact that the Inverse Beta Regularized function is equal to zero when its argument z is equal to zero [32]).

As the clamped–clamped bar is symmetric and we have learnt from the analogous chain (Figs. 6.17 and 6.18) that the first mode configuration has a maximum in the middle, we will impose the condition that such configuration is created. This means that $g(1/2) = A$, leading to $g(1/2) = \sqrt{2A_1/\omega^2} = A$, and also imposing

$$1 = \left[I_z\left(\frac{1}{2}, \frac{3}{4}\right)\right]^{1/2}, \quad z = \frac{\sqrt{2}}{A\,B\left(\frac{1}{2}, \frac{3}{4}\right)}(A_1)^{1/4}. \tag{6.90}$$

Knowing that $I_1\left(\frac{1}{2}, \frac{3}{4}\right) = 1$ [32], one can obtain the following value for A_1:

$$A_1 = \frac{A^4\,B^4\left(\frac{1}{2}, \frac{3}{4}\right)}{4}, \tag{6.91}$$

where the expression for ω is:

$$\omega = \frac{A\,B^2\left(\frac{1}{2}, \frac{3}{4}\right)}{\sqrt{2}}. \tag{6.92}$$

The expression (6.89) defining the first mode shape for the left part is now found to be

$$g_L(X) = A\left[I_z\left(\frac{1}{2}, \frac{3}{4}\right)\right]^{1/2}, \quad z = 2X, \quad 0 \le X \le \frac{1}{2}. \tag{6.93}$$

The expression for the right part can be obtained based on the symmetry properties:

$$g_R(X) = A\left[I_z\left(\frac{1}{2}, \frac{3}{4}\right)\right]^{1/2}, \quad z = 2(1 - X), \quad \frac{1}{2} \le X \le 1. \tag{6.94}$$

The expressions (6.93) and (6.94) are plotted in Fig. 6.25 as the red dashed line. The result for the chain from Sect. 6.3.2.1 with different N is also presented as the black solid lines with the discrete masses shown in black. These results are adjusted in a way that $X = 1$ and that the amplitude of the middle mass is also unity. With these adjustments, the results obtained for the chain can be compared with the one for the bar. It is seen that the mode shape for the clamped–clamped bar represents the envelope of those for the chain and that they coincide for a higher number of N. However, this comparison serves as a kind of validation of the results derived for the bar as the governing nonlinear partial differential equation could not be solved neither in Wolfram Mathematica nor in Matlab directly. Figure 6.26 is produced to show how the first mode of the pure cubic bar (red dashed line) differs from the well-known solution of a linear clamped–clamped bar [33] (black dotted line).

The exact solution for $U(T, X)$ corresponding to the case when the initial displacement of the bar is equal to the first mode determined ($A = 1$, $A_t = 1$) is found to be [29]:

$$U(T, X) = \operatorname{cn}\left(\frac{B^2\left(\frac{1}{2}, \frac{3}{4}\right)}{\sqrt{2}} T \,\middle|\, \frac{1}{2} \right) g_{L,R}(X). \tag{6.95}$$

This solution is plotted in Fig. 6.27 as a function of X and T. The mode shape can be recognized when viewed from the direction orthogonal to the $X - U$ plane as well as the oscillatory character of the response when viewed from the direction orthogonal to the $T - U$ plane.

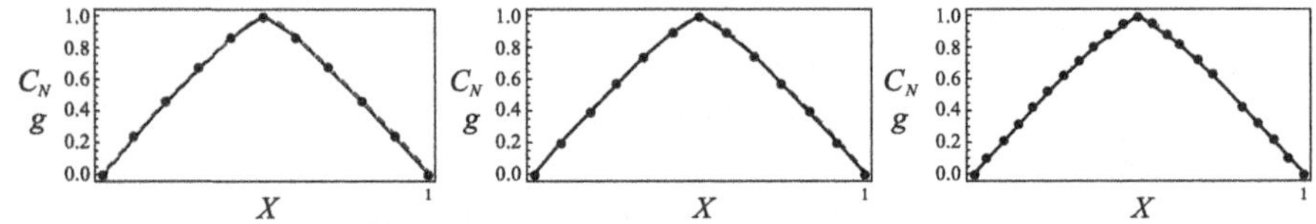

Fig. 6.25 Mode shape for the first mode of a clamped–clamped bar with pure cubic nonlinearity (red dashed line), a chain anchored at both ends with different number of masses N (black solid lines)

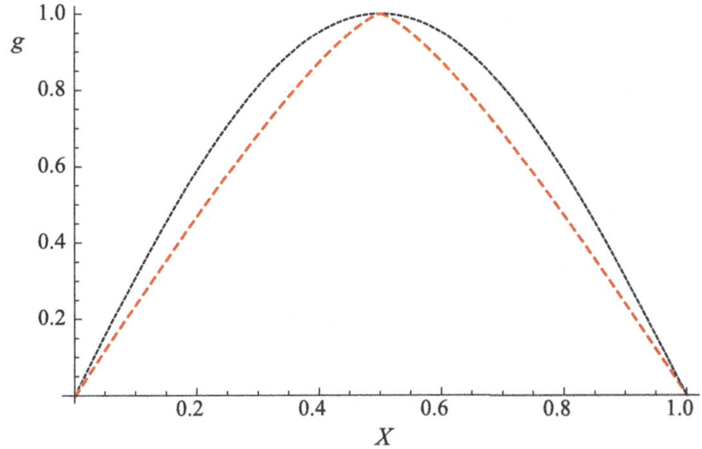

Fig. 6.26 Mode shape for the first mode of a clamped–clamped bar with pure cubic nonlinearity (red dashed line) and a linear clamped-free bar (black dotted line)

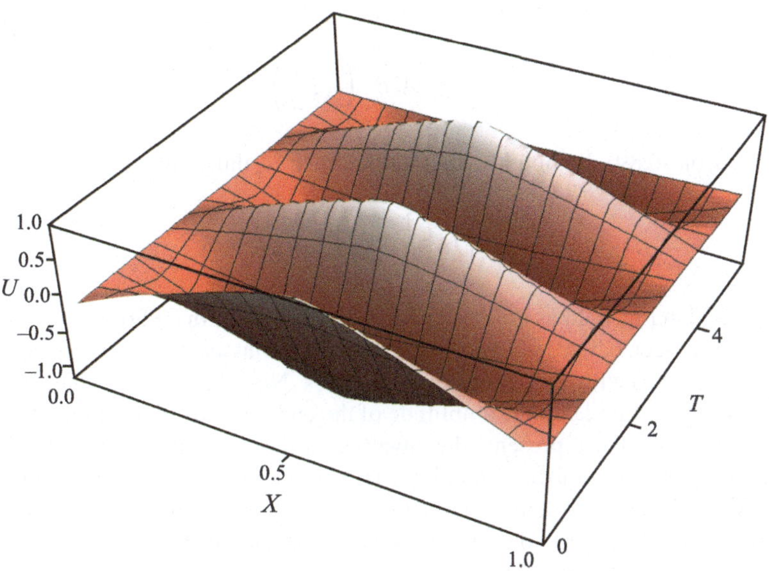

Fig. 6.27 Axial displacement of the clamped–clamped bar, Eq. (6.95)

6.4.2 Clamped-Free Bar

Considering a clamped-free bar (Fig. 6.13d), the boundary conditions are $U(0, T) = 0$ and $U'(1, T) = 0$. They must be satisfied regardless of T and can be transformed into $g(0) = 0$ and $g'(1) = 0$. As previously, the first condition is satisfied if $A_2 = 0$. The second boundary condition can be used in conjunction with the first integral (6.84) to derive $g(1) = \sqrt{2A_1/\omega^2}$. Substituting this expression into Eq. (6.89), one finds that

$$I_z\left(\frac{1}{2}, \frac{3}{4}\right) = 1, \quad z = \frac{2^{3/2}}{A\, B\left(\frac{1}{2}, \frac{3}{4}\right)} (A_1)^{1/4}, \tag{6.96}$$

which is satisfied when $z = 1$ [32]. This gives

$$A_1 = 2^{-6} A^4\, B^4\left(\frac{1}{2}, \frac{3}{4}\right). \tag{6.97}$$

Imposing the requirement that $A = g(1)$, one can obtain ω as a function of A:

$$\omega = 2^{-5/2} A\, B^2\left(\frac{1}{2}, \frac{3}{4}\right). \tag{6.98}$$

The expression (6.89) defining the first mode shape is, thus, given by

$$g(X) = A\left[I_z\left(\frac{1}{2}, \frac{3}{4}\right)\right]^{1/2}, \quad z = X. \tag{6.99}$$

This is plotted in Fig. 6.28 as the red dashed line for $A = 1$. In addition, the result for the chain from Sect. 6.3.2.2 with different number of masses N are also presented as the black solid lines with masses presented in black. Note that they are again adjusted in a way that $X = 1$ and that the amplitude of the ending mass is also unity. It is seen that the one for the bar represents the envelope of those for the chain and that they coincide for a higher number of N. Figure 6.29 contains the first mode shape of the linear [33] and pure cubic clamped-free bar to show how much they differ mutually.

By using Eqs. (6.83) as in the case of the clamped–clamped bar, the exact solution for $U(T, X)$ is found to be [29]:

$$U(T, X) = \mathrm{cn}\left(2^{-5/2}\, B^2\left(\frac{1}{2}, \frac{3}{4}\right) T \,\Big|\, \frac{1}{2}\right)\left[I_X\left(\frac{1}{2}, \frac{3}{4}\right)\right]^{1/2}. \tag{6.100}$$

This solution is plotted in Fig. 6.30 as a function of X and T.

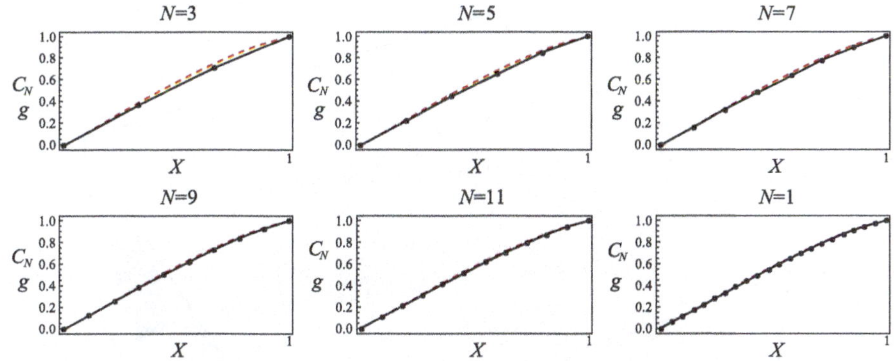

Fig. 6.28 Mode shape for the first mode of a clamped-free bar with pure cubic nonlinearity (red dashed line), a chain anchored at one end with different number of masses N (black solid lines)

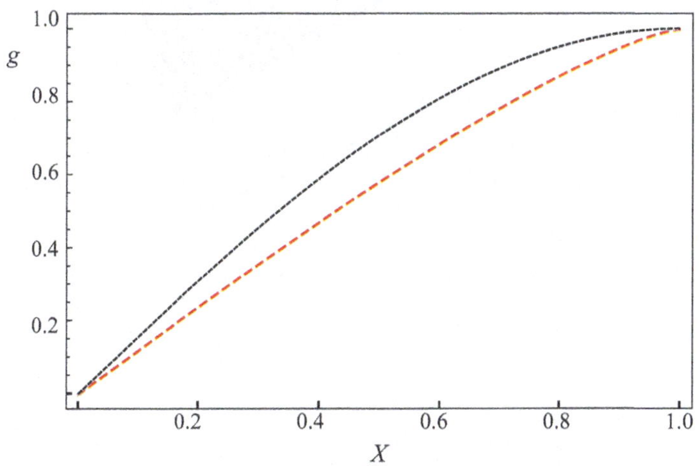

Fig. 6.29 Mode shape for the first mode of a clamped-free bar with pure cubic nonlinearity (red dashed line) and a linear clamped-free bar (black dotted line)

The previous analysis has been concerned with the first mode only, but it can also be extended to obtain analytically other modes as well and compare them with those obtained for the system with discrete masses. Another possible generalization stems from the exact solutions presented for other types of nonlinearity in Chap. 2, so the continuous systems with such type of nonlinearity (for example, quadratic or purely nonlinear) can be treated in this way as well.

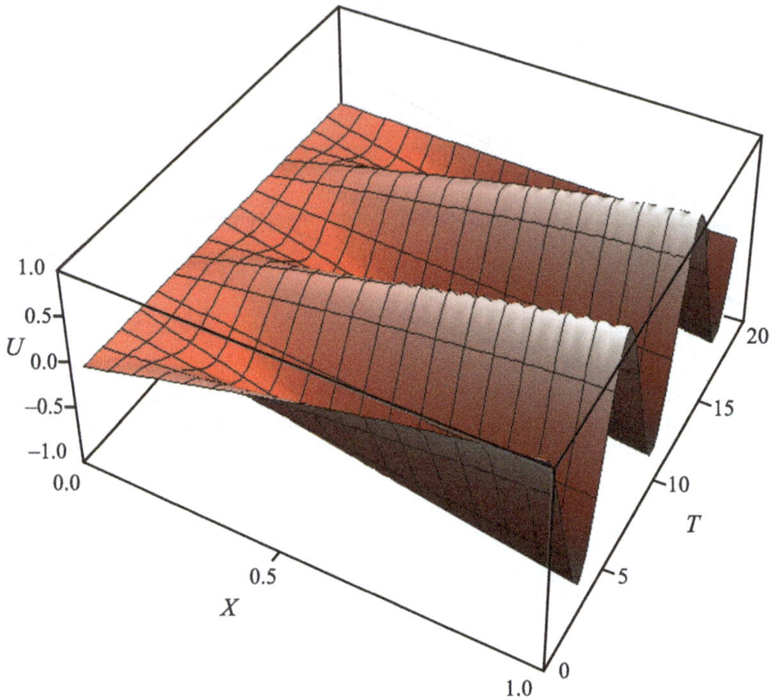

Fig. 6.30 Axial displacement of the clamped-free bar, Eq. (6.100)

References

1. Strogatz SH (2000) From Kuramoto to Crawford: exploring the onset of synchronization in populations of coupled oscillators. Phys D 143:1–20
2. Naik S, Hikihara T, Vu H, Palacios A, In V, Longhini P (2012) Local bifurcations of synchronization in self-excited and forced unidirectionally coupled micromechanical resonators. J Sound Vib 331:1127–1142
3. Sabater AB, Rhoads JF (2012) On the dynamics of two mutually-coupled, electromagnetically-actuated microbeam oscillators. J Comput Nonlinear Dyn 7. Art. No. 031012
4. Jones TB, Nenadic NG (2013) Electromechanics and MEMS. Cambridge University Press, New York
5. Agarwal DK, Woodhouse J, Seshia AA (2014) Synchronization in a coupled architecture of microelectromechanical oscillators. J Appl Phys 115. Art. No. 164904
6. Cortopassi C, Englander O (2010) Nonlinear springs for increasing the maximum stable deflection of MEMS electrostatic gap closing actuators. University of Berkeley, Berkeley
7. de Sudipto K, Aluru NR (2006) Complex nonlinear oscillations in electrostatically actuated microstructures. IEEE J Microelectromechanical Syst 5:355–369
8. de Sudipto K, Aluru NR (2006) U-sequence in electrostatic micromechanical systems (MEMS). Proc R Soc A 462:3435–3464
9. Kurt M, Eriten M, McFarland DM, Bergman LA, Vakakis AF (2014) Frequency-energy plots of steady state solutions for forced and damped systems, and vibration isolation by nonlinear mode localization. Commun Nonlinear Sci Numer Simul 19:2905–2917

10. Rosenberg RM (1966) On non-linear vibration of systems with many degrees of freedom. Adv Appl Mech 9:155–242
11. Rosenberg RM (1966) Steady state forced vibrations. Int J Non-Linear Mech 1:95–108
12. Hsu CS (1960) On the application of elliptic functions in nonlinear forced oscillations. Q Appl Math 17:393–407
13. Rakaric Z, Kovacic I, Cartmell M (2017) On the design of external excitations in order to make nonlinear oscillators respond as free oscillators of the same or different type. Int J Non-Linear Mech 94:323–333
14. Caughey TK, Vakakis AF (1991) A method for examining steady state solutions of forced discrete systems with strong non-linearities. Int J Non-Linear Mech 26:89–103
15. van Groesen EWC (1983) On normal modes in classical Hamiltonian systems. Int J Non-Linear Mech 18:55–70
16. Rand RH, Vito R (1972) Nonlinear vibrations of two degree of freedom systems with repeated linearized natural frequencies. J Appl Mech 39:296–297
17. Month LA, Rand RH (1977) The stability of bifurcating periodic solutions in a two degree of freedom nonlinear system. J Appl Mech 44:782–783
18. Mikhlin YV, Zhupiev AL (1997) An application of the Ince algebraization to the stability of non-linear normal vibration modes. Int J Non-Linear Mech 32:493–509
19. Vakakis AF, Manevitch LI, Mlkhlin YV, Pilipchuk VN, Zevin AA (1996) Normal modes and localization in nonlinear systems. Wiley, New York
20. Kovacic I, Zukovic M (2017) Coupled purely nonlinear oscillators: normal modes and exact solutions for free and forced responses. Nonlinear Dyn 87:713–726
21. Kovacic I, Zukovic M (2017) From a chain of nonlinear oscillators to nonlinear longitudinal vibrations of an elastic bar: the case of pure nonlinearity. Procedia Eng 199:687–692
22. Rosenberg RM (1963) The Ateb(h)-functions and their properties. Q Appl Math 21:37–47
23. Kovacic I, Cveticanin L, Zukovic M, Rakaric Z (2016) Jacobi elliptic functions: a review of nonlinear oscillatory application problems. J Sound Vib 380:1–36
24. Vakakis AF, Caughey TK (1989) Some topics in the free and forced oscillations of a class of nonlinear systems, Dynamics laboratory report DYNL-89-1, California Institute of Technology, Pasadena, California
25. Beléndez A, Francés J, Beléndez T, Bleda S, Pascual C, Arribas E (2015) Nonlinear oscillator with power-form elastic-term: Fourier series expansion of the exact solution. Commun Nonlinear Sci Numer Simul 22:134–148
26. Rivin E (1999) Stiffness and damping in mechanical design. Marcel Dekker Inc., New York
27. Kovacic I, Brennan MJ (eds) (2011) The Duffing equation: nonlinear oscillators and their behaviour. Wiley, Chichester
28. Rakaric Z, Kovacic I (2013) An elliptic averaging method for harmonically excited oscillators with a purely nonlinear non-negative real-power restoring force. Commun Nonlinear Sci Numer Simul 18:1888–1901
29. Kovacic I, Zukovic M (2018) On the response of some discrete and continuous oscillatory systems with pure cubic nonlinearity: exact solutions. Int J Non-Linear Mech 98:13–22
30. http://mathworld.wolfram.com/BetaFunction.html. Accessed 15 Apr 2020
31. http://mathworld.wolfram.com/RegularizedBetaFunction.html. Accessed 15 Apr 2020
32. http://functions.wolfram.com/GammaBetaErf/InverseBetaRegularized. Accessed 15 Apr 2020
33. Kovacic I, Radomirovic D (2017) Mechanical vibrations: fundamentals with solved examples. Wiley, Chichester

Appendix A
On Beta, Gamma and Hypergeometric Functions

The gamma function $\Gamma(z)$ is also known as the complete gamma function or the Euler integral of the second kind. There are several possibilities to present its definition and two of them are given below:

$$\Gamma(z) = \int_0^\infty u^{z-1} e^{-u} \mathrm{d}u, \quad \Re(z) > 0, \tag{A.1}$$

and

$$\Gamma(z) = \int_0^1 \left[\ln\left(\frac{1}{u}\right)\right]^{z-1} \mathrm{d}u, \tag{A.2}$$

If z is an integer $n = 1, 2, 3, \ldots$, then

$$\Gamma(n) = (n-1)\Gamma(n-1) = (n-1)(n-2)\Gamma(n-2) = \cdots = (n-1)!, \tag{A.3}$$

so the gamma function reduces to the factorial for a positive integer argument. Note that (A.3) also implies $\Gamma(n+1) = n\Gamma(n)$, which can be a useful identity during certain transformations.

Some particular values of the gamma function are as follows:

$$\Gamma\left(\frac{1}{3}\right) = 2.67894, \tag{A.4}$$

$$\Gamma\left(\frac{1}{2}\right) = \sqrt{\pi}, \tag{A.5}$$

$$\Gamma(1) = 1, \tag{A.6}$$

$$\Gamma\left(\frac{3}{2}\right) = \frac{\sqrt{\pi}}{2}, \tag{A.7}$$

$$\Gamma(2) = 1. \tag{A.8}$$

I. Kovacic, *Nonlinear Oscillations*,
https://doi.org/10.1007/978-3-030-53172-0

The complete gamma function $\Gamma(a)$ can be generalized to the upper incomplete gamma function $\Gamma[a, x]$ such that $\Gamma(a) = \Gamma[a, 0]$. This 'upper' incomplete gamma function is defined by

$$\Gamma[a, x] = \int_x^\infty u^{a-1} e^{-u} du. \tag{A.9}$$

The hypergeometric function ${}_2F_1$ is a special function represented by the hypergeometric series. It includes many other special functions as specific or limiting cases. Its integral representation is related to the gamma function in the following way:

$$ {}_2F_1(a, b, c, z) = \frac{\Gamma(c)}{\Gamma(b)\Gamma(c-b)} \int_0^1 \frac{u^{b-1}(1-u)^{c-b-1}}{(1-uz)^a} du. \tag{A.10}$$

For the special case when $z = 2$, it can be expressed completely via the gamma function:

$$ {}_2F_1[a, b, c, 1] = \frac{\Gamma(c)\Gamma(c-a-b)}{\Gamma(c-a)\Gamma(c-b)}. \tag{A.11}$$

The beta function, also called the Euler integral of the first kind, is a special function defined by

$$B(x, y) = \int_0^1 u^{x-1}(1-u)^{y-1} du, \quad \Re(x) > 0, \Re(y) > 0. \tag{A.12}$$

The beta function is symmetric, meaning that

$$B(x, y) = B(y, x). \tag{A.13}$$

The beta function can be expressed in terms of the gamma function:

$$B(x, y) = \frac{\Gamma(x)\Gamma(y)}{\Gamma(x+y)}. \tag{A.14}$$

The incomplete beta function $B(z, x, b)$ is defined by

$$B(z, x, y) \equiv B_z(x, y) = \int_0^z u^{x-1}(1-u)^{y-1} du. \tag{A.15}$$

It reduces to the beta function B when $z = 1$:

$$B(1, x, y) \equiv B_1(x, y) \equiv B(x, y). \tag{A.16}$$

The regularized incomplete beta function $I(z, x, y)$ (or regularized beta function for short) is defined as the ratio between the incomplete beta function and the beta function:

$$I(z, x, y) \equiv \frac{B(z, x, y)}{B(x, y)}. \tag{A.17}$$

Appendix B
On Jacobi Elliptic Functions

The Jacobian elliptic functions are usually defined as unique doubly periodic, meromorphic functions satisfying certain properties. Another approach is to present them as the inverses of the incomplete elliptic integral of the first kind u:

$$u = \int_0^{\varphi} \frac{\mathrm{d}\psi}{\sqrt{1 - m\sin^2\psi}}, \tag{B.1}$$

where m is usually called the *elliptic parameter* and φ is called the *Jacobi elliptic amplitude* $\varphi = am\,(u)$. Jacobian elliptic functions are functions of two variables: the first variable is given in terms of the amplitude φ, or more commonly, in terms of u given above; the second variable might be given in terms of the elliptic parameter m, or as the *elliptic modulus* $k^2 = m$. The three basic Jacobian elliptic functions are: cn, sn and dn.

The cn function or the elliptic cosine (Latin: *cosinus amplitudinis*) is

$$\mathrm{cn}\,(u\,|m) = \cos\varphi. \tag{B.2}$$

The sn function or the elliptic sine (Latin: *sinus amplitudinis*) is

$$\mathrm{sn}\,(u\,|m) = \sin\varphi, \tag{B.3}$$

The dn function or the delta amplitude dn (Latin: *delta amplitudinis*) is

$$\mathrm{dn}\,(u\,|m) = \sqrt{1 - m\mathrm{sn}^2\,(u\,|m)}. \tag{B.4}$$

Equations (B.2) and (B.3) give the identity

$$\mathrm{cn}^2\,(u\,|m) + \mathrm{sn}^2\,(u\,|m) = 1, \tag{B.5}$$

I. Kovacic, *Nonlinear Oscillations*,
https://doi.org/10.1007/978-3-030-53172-0

which generalizes $\cos^2 u + \sin^2 u = 1$ when $m = 0$. Equation (B.4) leads to

$$\mathrm{dn}^2\left(u\,|m\right) + m\,\mathrm{sn}^2\left(u\,|m\right) = 1. \tag{B.6}$$

The first derivative of the Jacobi cn elliptic function with respect to the argument u is

$$\frac{\mathrm{d}}{\mathrm{d}u}\mathrm{sn}\left(u\,|m\right) = \frac{\mathrm{d}\varphi}{\mathrm{d}u}\frac{\mathrm{d}\left(\sin\varphi\right)}{\mathrm{d}\varphi} = \sqrt{1 - m\sin^2\varphi}\cos\varphi = -\mathrm{cn}\left(u\,|m\right)\mathrm{dn}\left(u\,|m\right). \tag{B.7}$$

Similarly, one can derive:

$$\frac{\mathrm{d}}{\mathrm{d}u}\mathrm{cn}\left(u\,|m\right) = -\mathrm{sn}\left(u\,|m\right)\mathrm{dn}\left(u\,|m\right), \tag{B.8}$$

and

$$\frac{\mathrm{d}}{\mathrm{d}u}\mathrm{dn}\left(u\,|m\right) = -m\;\mathrm{sn}\left(u\,|m\right)\mathrm{cn}\left(u\,|m\right). \tag{B.9}$$

The (real) period of the Jacobi elliptic functions is related to the complete elliptic integral of the first kind:

$$K\left(m\right) = \int_0^{\pi/2}\frac{\mathrm{d}\psi}{\sqrt{1 - m\sin^2\psi}}. \tag{B.10}$$

The period for the cn and sn function is $4K\left(m\right)$, and for the dn function it is equal to $2K\left(m\right)$. Note that the Jacobi elliptic functions are doubly periodic in the complex plane.

The following expansion for $K\left(m\right)$ gives the insight into its dependence on m:

$$K\left(m\right) \approx \frac{\pi}{2}\left[1 + \frac{1}{4}m + \frac{9}{64}m^2 + \frac{25}{256}m^3 + O\left(m^4\right)\right]. \tag{B.11}$$

It is obvious that the first two functions (the cn and sn functions) generalize the sine and cosine, while the third one (the dn function) ‘generalizes’ unity:

$$\mathrm{cn}\left(u\,|m\right) = \cos u, \tag{B.12}$$

$$\mathrm{sn}\left(u\,|m\right) = \sin u, \tag{B.13}$$

$$\mathrm{dn}\left(u\,|m\right) = 1. \tag{B.14}$$

These three Jacobi elliptic functions are bounded as follows:

$$-1 \le \mathrm{cn}\left(u\,|m\right) \le 1, \tag{B.15}$$

$$-1 \le \mathrm{sn}\left(u\,|m\right) \le 1, \tag{B.16}$$

$$\sqrt{1 - m} \le \mathrm{dn}\left(u\,|m\right) \le 1. \tag{B.17}$$

Special values include

$$\operatorname{cn}(0\,|m) = 1, \quad \operatorname{sn}(0\,|m) = 0, \quad \operatorname{dn}(0\,|m) = 1, \tag{B.18}$$

$$\operatorname{cn}(K\,|m) = 0, \quad \operatorname{sn}(K\,|m) = 1, \quad \operatorname{dn}(K\,|m) = \sqrt{1-m}. \tag{B.19}$$

The cn function solves the differential equation

$$\frac{\mathrm{d}^2 x}{\mathrm{d}u^2} + (1-2m)\,x + 2mx^3 = 0. \tag{B.20}$$

The sn function solves the differential equation

$$\frac{\mathrm{d}^2 x}{\mathrm{d}u^2} + (1+m)\,x - 2mx^3 = 0. \tag{B.21}$$

The dn function solves the differential equation

$$\frac{\mathrm{d}^2 x}{\mathrm{d}u^2} - (2-m)\,x + 2x^3 = 0. \tag{B.22}$$

Derivatives of the sn, dn and cn elliptic functions with respect to the parameter m are given by

$$\frac{\partial \operatorname{cn}(u\,|m)}{\partial m} = \frac{\operatorname{sn}(u\,|m)\operatorname{dn}(u\,|m)\,\{(m-1)u + E(\varphi,m) - m\operatorname{sn}(u\,|m)\operatorname{cd}(u\,|m)\}}{2m(1-m)}, \tag{B.23}$$

$$\frac{\partial \operatorname{sn}(u\,|m)}{\partial m} = \frac{\operatorname{cn}(u\,|m)\operatorname{dn}(u\,|m)\,\{(1-m)u - E(\varphi,m) + m\operatorname{sn}(u\,|m)\operatorname{cd}(u\,|m)\}}{2m(1-m)}, \tag{B.24}$$

$$\frac{\partial \operatorname{dn}(u\,|m)}{\partial m} = \frac{\operatorname{sn}(u\,|m)\operatorname{cn}(u\,|m)\,\{(m-1)\psi + E(\varphi,m) - \operatorname{dn}(u\,|m)\operatorname{sc}(u\,|m)\}}{2(1-m)}, \tag{B.25}$$

where $\operatorname{cd}(\psi\,|m) = \operatorname{cn}(\psi\,|m)/\operatorname{dn}(\psi\,|m)$, $\operatorname{sc}(\psi\,|m) = \operatorname{sn}(\psi\,|m)/\operatorname{cn}(\psi\,|m)$ and $E(\varphi, m)$ is the elliptic integral of the second kind.

Appendix C
Fourier Series: Definition and Examples

A Fourier series is an expansion of a periodic function $f(t)$ in terms of an infinite sum of sines and cosines. For a function $f(t)$ that is periodic on an interval $[-Z, Z]$, the Fourier series has the form:

$$f(t) = \frac{a_0}{2} + \sum_{n=1}^{\infty} a_n \cos\left(\frac{n\pi t}{Z}\right) + \sum_{n=1}^{\infty} b_n \sin\left(\frac{n\pi t}{Z}\right), \tag{C.1}$$

where the Fourier coefficients are

$$a_0 = \frac{1}{Z}\int_{-Z}^{Z} f(t)\mathrm{d}t, \tag{C.2}$$

$$a_n = \frac{1}{Z}\int_{-Z}^{Z} f(t)\cos\left(\frac{n\pi t}{Z}\right)\mathrm{d}t, \tag{C.3}$$

$$b_n = \frac{1}{Z}\int_{-Z}^{Z} f(t)\sin\left(\frac{n\pi t}{Z}\right)\mathrm{d}t. \tag{C.4}$$

C.1 Fourier Series for the Jacobi Elliptic Functions

A Fourier series for the Jacobi cn elliptic function contains cosines with odd multiplications of $\pi t/(2K)$

$$\mathrm{cn}\,(t\,|m) = \sum_{N=1}^{\infty} C_N \cos\left((2N-1)\frac{\pi}{2K}t\right), \tag{C.5}$$

where the Fourier coefficients C_N are

I. Kovacic, *Nonlinear Oscillations*,
https://doi.org/10.1007/978-3-030-53172-0

$$C_N = \frac{2\pi}{K\sqrt{m}} \frac{q^{N-1/2}}{1+q^{2N-1}}, \tag{C.6}$$

where q is the Nome—a special function defined by

$$q = \exp\left(-\frac{\pi K'}{K}\right), \tag{C.7}$$

with K being the complete elliptic integral of the first kind, and K' being its associated complete elliptic integral of the first kind, $K' = K(1-m)$. By the definition, K depends on the elliptic parameter as follows:

$$K \equiv K(m) = \int_0^{\pi/2} \frac{\mathrm{d}\psi}{\sqrt{1 - m\sin^2\psi}}. \tag{C.8}$$

The function $\mathrm{cn}\left(t\left|\frac{1}{2}\right.\right)$ can be represented as the following Fourier series (note that here the second argument is the elliptic modulus $m = 1/2$)

$$\mathrm{cn}\left(t\left|\frac{1}{2}\right.\right) = \sum_{N=1}^{\infty} c_{2N-1}\cos\left((2N-1)\frac{\pi}{2K}t\right), \tag{C.9}$$

where the Fourier coefficients are defined by

$$c_{2N-1} = \frac{2\sqrt{2}\pi}{K} \frac{q^{N-1/2}}{1+q^{2N-1}}, \tag{C.10}$$

with the Nome q being

$$q\left(m = \frac{1}{2}\right) = \exp(-\pi). \tag{C.11}$$

In this case, one has $K\left(m = \frac{1}{2}\right) = 1.85407$. This implies that the corresponding Fourier series expansion has the form

$$\begin{aligned}\mathrm{cn}\left(t\left|\frac{1}{2}\right.\right) &\approx 0.955006\cos(0.847213t) + 0.0430495\cos(2.54164t) \\ &+ 0.00186049\cos(4.23607t) + \\ &0.0000803976\cos(5.93049t) + \cdots\end{aligned} \tag{C.12}$$

The Jacobi elliptic sn function $\mathrm{sn}(t\,|m)$ can be represented as the following Fourier series containing sines with odd multiplications of $\pi t/(2K)$

$$\mathrm{sn}(t\,|m) = \sum_{N=1}^{\infty} S_N \sin\left((2N-1)\frac{\pi}{2K}t\right), \tag{C.13}$$

where the Fourier coefficients are defined by

$$S_N = \frac{2\pi}{K\sqrt{m}} \frac{q^{N-1/2}}{1-q^{2N-1}}. \tag{C.14}$$

The Jacobi elliptic dn function dn($t\,|m$) can be represented as the following Fourier series containing an offset and cosines with odd and even multiplications of $\pi t/K$

$$\mathrm{dn}\,(t\,|m\,) = D_0 + \sum_{N=1}^{\infty} D_N \cos\left(N\frac{\pi}{K}t\right), \tag{C.15}$$

where the Fourier coefficients are defined by

$$D_0 = \frac{\pi}{2K}, \quad D_N = \frac{2\pi}{K}\frac{q^N}{1+q^{2N}}. \tag{C.16}$$

The Fourier series expansion of the function $\mathrm{cn}^3\,(t\,|m\,)$ is

$$\mathrm{cn}^3\,(t\,|m\,) = \sum_{N=1}^{\infty} b_{2N-1} \cos\left((2N-1)\frac{\pi}{2K}t\right), \tag{C.17}$$

where the Fourier coefficients b_{2N-1} are

$$b_{2N-1} = \frac{1}{2m}\left\{2m-1+(2N-1)^2\left(\frac{\pi}{2K}\right)^2\right\}\frac{2\pi}{K\sqrt{m}}\frac{q^{N-1/2}}{1+q^{2N-1}}. \tag{C.18}$$

C.2 Fourier Series for the ca Ateb Function

A Fourier series of the ca Ateb function comprises odd cosine harmonics only, and it can be written down as

$$\mathrm{ca}\,(\alpha, 1, t) = \sum_{N=1}^{\infty} \bar{C}_{2N-1}\,(\alpha) \cos\left[(2N-1)\,\frac{2\pi}{T}t\right], \tag{C.19}$$

where the Fourier coefficients $\bar{C}_{2N-1}$ depend on the parameter α, and are defined by

$$\bar{C}_{2N-1}\,(\alpha) = \frac{4}{T}\int_0^{T/2} \mathrm{ca}\,(\alpha, 1, t) \cos\left[(2N-1)\,\frac{2\pi}{T}t\right] \mathrm{d}t, \tag{C.20}$$

where T is defined by

$$T = \sqrt{\frac{8\pi}{\alpha+1}} \frac{\Gamma\left(\frac{1}{\alpha+1}\right)}{\Gamma\left(\frac{\alpha+3}{2(\alpha+1)}\right)} |A|^{(1-\alpha)/2}, \tag{C.21}$$

and where A is the amplitude of oscillations and Γ is the Euler gamma function (see Appendix A). This is the period of vibration of the oscillators governed by

$$\ddot{x} + \operatorname{sgn}(x)\,|x|^{\alpha} = 0, \tag{C.22}$$

with the power α being any positive real number, which is characterized by the following energy conservation law for the zero-initial velocity and non-zero-initial displacement:

$$\frac{1}{2}\dot{x}^2 + \frac{|x|^{\alpha+1}}{\alpha+1} = \frac{|A|^{\alpha+1}}{\alpha+1}. \tag{C.23}$$

Now, the displacement in Eq. (C.22) is rescaled by the initial amplitude A: $X = x/A$, yielding

$$\bar{C}_{2N-1}(\alpha) = \frac{8}{T}\int_0^{T/4} X(\alpha, t)\cos\left[(2N-1)\frac{2\pi}{T}t\right]\mathrm{d}t. \tag{C.24}$$

To find now the expression for $\mathrm{d}t$, the first integral analogous to Eq. (C.23) is formed, and the following is derived

$$\mathrm{d}t = \sqrt{\frac{\alpha+1}{2}}\,|A|^{(1-\alpha)/2}\frac{\mathrm{d}X}{\sqrt{1-|X|^{\alpha+1}}}. \tag{C.25}$$

This enables one to find how t depends on X (noting that this holds for $X \geq 0$):

$$t(X) = \sqrt{\frac{\alpha+1}{2}}\,|A|^{(1-\alpha)/2}\int_X^1 \frac{\mathrm{d}y}{\sqrt{1-y^{\alpha+1}}}. \tag{C.26}$$

Performing some transformations, one can derive

$$t(X) = \sqrt{\frac{\pi}{2(\alpha+1)}} \frac{\Gamma\left(\frac{1}{\alpha+1}\right)}{\Gamma\left(\frac{\alpha+3}{2(\alpha+1)}\right)} |A|^{(1-\alpha)/2}\, I\left(1-X^{\alpha+1}, \frac{1}{2}, \frac{1}{\alpha+1}\right), \tag{C.27}$$

where I stands for the regularized incomplete beta function (see Appendix A).

Finally, substituting Eq. (C.25) into Eq. (C.24) as well as Eq. (C.27) into the argument of the cosine function in Eq. (C.24), one derives

$$\bar{C}_{2N-1}(\alpha) = \frac{2(\alpha+1)\Gamma\left(\frac{\alpha+3}{2(\alpha+1)}\right)}{\sqrt{\pi}\Gamma\left(\frac{1}{\alpha+1}\right)} \int_0^1 \frac{X}{\sqrt{1-X^{\alpha+1}}} \cos\left[\frac{(2n-1)\pi}{2} I\left(1-X^{\alpha+1}, \frac{1}{2}, \frac{1}{\alpha+1}\right)\right] \mathrm{d}X. \tag{C.28}$$

By using the substitution $z = 1 - X^{\alpha+1}$, the following expression for the Fourier coefficients is obtained:

$$\bar{C}_{2N-1}(\alpha) = \frac{2\Gamma\left(\frac{\alpha+3}{2(\alpha+1)}\right)}{\sqrt{\pi}\Gamma\left(\frac{1}{\alpha+1}\right)} \int_0^1 \frac{(1-z)^{(1-\alpha)/(1+\alpha)}}{\sqrt{z}} \cos\left[\frac{(2N-1)\pi}{2} I\left(z, \frac{1}{2}, \frac{1}{\alpha+1}\right)\right] \mathrm{d}z. \tag{C.29}$$

Their values can be calculated by carrying out numerical integration.

The manufacturer's authorised representative in the EU is Springer Nature Customer Service Centre GmbH, Europaplatz 3, 69115 Heidelberg, Germany. If you have any concerns regarding our products, please contact ProductSafety@springernature.com

Printed and bound by CPI Group (UK) Ltd, Croydon, CR0 4YY
30/07/2026
02181994-0004